VOLUME TWO

DEVELOPMENTS IN NEUROETHICS AND BIOETHICS

Mental Health as Public Health: Interdisciplinary Perspectives on the Ethics of Prevention

Serial Editor

Judy Illes, CM, PHD

Division of Neurology, Department of Medicine
The University of British Columbia
Neuroethics Canada
2211 Wesbrook Mall, Koerner S124
Vancouver BC
Canada
V6T 2B5

Katherine Bassil, Editorial Assistant to Judy Illes

VOLUME TWO

DEVELOPMENTS IN NEUROETHICS AND BIOETHICS

Mental Health as Public Health: Interdisciplinary Perspectives on the Ethics of Prevention

Edited by

KELSO CRATSLEY

Kennedy Institute of Ethics, Georgetown University;
Department of Philosophy & Religion, American University,
Washington, DC, United States

JENNIFER RADDEN

Department of Philosophy,
University of Massachusetts Boston,
Boston, MA, United States

ELSEVIER

ACADEMIC PRESS
An imprint of Elsevier

Academic Press is an imprint of Elsevier
125 London Wall, London EC2Y 5AS, United Kingdom
525 B Street, Suite 1650, San Diego, CA 92101, United States
50 Hampshire Street, 5th Floor, Cambridge, MA 02139, United States
The Boulevard, Langford Lane, Kidlington, Oxford OX5 1GB, United Kingdom

First edition 2019

Notices

Knowledge and best practice in this field are constantly changing. As new research and experience broaden our understanding, changes in research methods, professional practices, or medical treatment may become necessary.

Practitioners and researchers must always rely on their own experience and knowledge in evaluating and using any information, methods, compounds, or experiments described herein. In using such information or methods they should be mindful of their own safety and the safety of others, including parties for whom they have a professional responsibility.

To the fullest extent of the law, neither the Publisher nor the authors, contributors, or editors, assume any liability for any injury and/or damage to persons or property as a matter of products liability, negligence or otherwise, or from any use or operation of any methods, products, instructions, or ideas contained in the material herein.

ISBN: 978-0-12-816756-4
ISSN: 2589-2959

For information on all Academic Press publications visit our website at https://www.elsevier.com/books-and-journals

Publisher: Zoe Kruze
Acquisition Editor: Sam Mahfoudh
Editorial Project Manager: Peter Llewellyn
Production Project Manager: Denny Mansingh
Cover Designer: Alan Studholme

Typeset by TNQ Technologies

CONTENTS

CONTRIBUTORS

Rachel Bingham
Clinical Advisor, Medical Justice, London, United Kingdom; General Practitioner, NHS, Islington, London, United Kingdom

Matthew R. Broome
Institute for Mental Health, University of Birmingham, Birmingham, England, United Kingdom

Christine E. Cooper-Vince
The Chester M. Pierce, MD Division of Global Psychiatry, Massachusetts General Hospital, Boston, MA, United States; Department of Psychiatry, Harvard Medical School, Boston, MA, United States

Suzana Corciova
Higher Trainee in Child and Adolescent Psychiatry, London, United Kingdom

Kelso Cratsley
Kennedy Institute of Ethics, Georgetown University, Washington, DC, United States; Department of Philosophy & Religion, American University, Washington, DC, United States

Phoebe Friesen
Ethox Centre, University of Oxford, Oxford, United Kingdom

Kamal Gautam
Transcultural Psychosocial Organization Nepal, Kathmandu, Nepal

Mona Gupta
Département de Psychiatrie et d'Addictologie, Université de Montréal, Montréal, QC, Canada

Benjamin L. Harris
A.M. Dogliotti College of Medicine, University of Liberia, Monrovia, Liberia

Rebecca S. Hock
The Chester M. Pierce, MD Division of Global Psychiatry, Massachusetts General Hospital, Boston, MA, United States; Department of Psychiatry, Harvard Medical School, Boston, MA, United States

Ginger A. Hoffman
Department of Philosophy, Saint Joseph's University, Philadelphia, PA, United States

Ulrich Koch
Clinical Research and Leadership, School of Medicine and Health Sciences, The George Washington University, Washington DC, United States

Brandon A. Kohrt
Division of Global Mental Health, George Washington University, Washington DC, United States

Dany Lamothe
Université de Sherbrooke, Centre Hospitalier Universitaire de Sherbrooke, Sherbooke, QC, Canada

Elizabeth J. Levey
The Chester M. Pierce, MD Division of Global Psychiatry, Massachusetts General Hospital, Boston, MA, United States; Department of Psychiatry, Harvard Medical School, Boston, MA, United States; Department of Psychiatry, University of Illinois College of Medicine, Chicago, IL, United States

Arianna Manzini
Department of Psychiatry and Wellcome Centre for Ethics & Humanities, University of Oxford, Oxford, United Kingdom

Charlotte A.L. Marriott
Worcestershire Health and Care NHS Trust, Worcester, England, United Kingdom

Alex McKeown
Department of Psychiatry and Wellcome Centre for Ethics & Humanities, University of Oxford, Oxford, United Kingdom

Rose Mortimer
Department of Psychiatry and Wellcome Centre for Ethics & Humanities, University of Oxford, Oxford, United Kingdom

Dasha Nicholls
Imperial College, London, United Kingdom

Katie O'Leary
The Information School, University of Washington, Seattle WA, United States

Gloria A. Pedersen
Division of Global Mental Health, George Washington University, Washington DC, United States

Jennifer Radden
Department of Philosophy, University of Massachusetts Boston, Boston, MA, United States

Mohammed Abouelleil Rashed
Wellcome Trust ISSF Research Fellow, Department of Philosophy, Birkbeck College, University of London, United Kingdom; Visiting Lecturer, Department of Philosophy, King's College London, London, United Kingdom

Ilina Singh
Department of Psychiatry and Wellcome Centre for Ethics & Humanities, University of Oxford, Oxford, United Kingdom

Jacinta O.A. Tan
Specialist Child and Adolescent Mental Health Service, Aneurin Bevan University Health Board, Gwent, Wales, United Kingdom

Lauren A. Taylor
Harvard Business School, Boston, MA, United States

Syed Shabab Wahid
Division of Global Mental Health, George Washington University, Washington DC, United States

FOREWORD

There is a growing appetite for ethics inquiry and discourse in health research today, clinical medicine, and even related product development and commercialization. Ethics curricula have expanded in both breadth and depth – the International Brain Research Organization (IBRO) for example, now requires the integration of a minimum number of neuroethics hours into all its schools – the number of trainees engaged in ethics scholarship in biomedicine is growing alongside expanding numbers of programs worldwide, the importance of the voice and engagement of patients in health decision-making has been recognized as a core element of the translational process from discovery science to clinical trials, and the OECD and many other government and non-government groups have successfully brought together representatives from industry with academics and policy-makers into fruitful dialogue about responsible innovation.

This book series dedicated to *Developments in Neuroethics and Bioethics* serves to address what must be perceived, therefore, as a growing need and interest in ethical content for biomedicine, and this volume on ethical issues in public health for mental health could not be more timely. With high prevalence rates, disturbing rates of morbidity and mortality, and vast costs to economies and societies globally, volume editors Professors Cratsley, Radden and their colleagues present and elaborate on the urgent public health challenges the world is facing today. Indeed, as they describe it, it may well be a full-blown crisis. Depression is the leading cause of global disability; mental, neurological, and substance use disorders account for a vast proportion of disability. Prevention and treatment gaps, scarcity of funding in many regions of the world, and stigma are pervasive.

Solutions to these challenges, or at least course corrections in a positive direction, however, can be facilitated through principled, ethically grounded approaches to recognizing and addressing them. Volume 1 of the series took this challenge on for pain (Buchman and Davis, 2019). This second volume does so expertly for global mental health. I expect subsequent volumes in the series on do-it-yourself neurotechnology (Volume 3; Bard and Hildt, volume editors) and neurolaw (Volume 4; Hevia, volume editor) will similarly tackle their subjects successfully.

There are rarely hard lines to draw in the sands of ethical controversies. Values, perceptions, and priorities appropriately differ with people and culture, problematics and desires. This volume, like the others, embraces diversity of thought, style, and perspective. I hope readers similarly embrace this valuable contribution to our burgeoning field.

Dr. Judy Illes, CM
Canada Research Chair in Neuroethics
Professor of Neurology, Department of Medicine
Director, Neuroethics Canada
University of British Columbia

CHAPTER ONE

Introduction

Jennifer Radden[a] and Kelso Cratsley[b,c,*]
[a]Department of Philosophy, University of Massachusetts Boston, Boston, MA, United States
[b]Kennedy Institute of Ethics, Georgetown University, Washington, DC, United States
[c]Department of Philosophy & Religion, American University, Washington, DC, United States
*Corresponding author: E-mail: kelso.cratsley@gmail.com

Contents

The global burden of mental disorder currently represents one of the more urgent public health challenges. Recent data on prevalence rates, associated morbidity and mortality, and economic costs are substantial (Bloom et al., 2011; Walker, McGee, & Druss, 2015; Whiteford et al., 2013), suggesting something close to a full-blown crisis. Depression is the leading cause of global disability, and it is estimated that mental, neurological, and substance use disorders are responsible for over a third of all years lived with disability from non-fatal diseases (Mnookin, 2016; WHO, 2017).[1] There is also a large treatment gap, in rich and poor countries alike, with most patients failing to receive treatment due to service scarcity, barriers to access, and lack of funding relative to other parts of the health sector (Saxena, Thornicroft, Knapp, & Whiteford, 2007; Wang et al., 2007; WHO, 2018). In addition, the ongoing stigmatisation, discrimination, and other human rights violations experienced by those with mental health conditions represent what has been rightly described as a 'failure of humanity' (Kleinman, 2009, p. 603; also see Drew et al., 2011).

These issues are not neatly circumscribed given the cross-cutting nature of mental health. It has become increasingly clear that mental health conditions have significant social determinants such as stress, migration, and economic disadvantage (Lund et al., 2018). Indeed, explanatory models of many forms of disorder increasingly incorporate social and environmental factors (eg. Harkness, Hayden, & Lopez-Duran, 2015; van Os, Kenis, &

[1] The global burden may be underestimated (Vigo, Thornifcroft, & Atun, 2016).

Developments in Neuroethics and Bioethics, Volume 2
ISSN 2589-2959
https://doi.org/10.1016/bs.dnb.2019.07.001

Rutten, 2010; van den Bosch & Meyer-Lindenberg, 2019). This is part of a broader bidirectional relationship, with mental health ultimately impacting many other domains. These include physical health, given the association between mental health and a range of health-related outcomes (Prince et al., 2007), as well as educational and employment opportunities and a host of other socioeconomic indicators (Lund et al., 2011). Recognition of this interconnectedness places mental health squarely at the centre of public health, global health, and international development more generally. Given the realities of social inequality, it also indicates that a commitment to social justice and political engagement must be part of public mental health initiatives (Patel, 2015; Priebe, 2015).

Recent efforts to respond to these problems have been led by multilateral agencies, national health ministries, non-governmental organisations, and civil society groups. Many countries have developed new policies and programs in public mental health (reviewed in Wahlbeck, 2015).[2] Higher profile examples include the *headspace* youth centres in Australia (www.headspace.org.au), and current efforts in the US to address the opioid crisis (Haffajee & Frank, 2018). The increased incidence of suicide has also led to some noteworthy preventive programming, including Facebook's controversial screening program (Marks, 2019; Singer, 2018). On a global scale, the World Health Organization (WHO) has funded a number of collaborative initiatives such as the Mental Health Gap Action Programme (WHO, 2008), provided influential policy guidance with its Mental Health Action Plan 2013–20 (WHO, 2013), and tracked international progress through its biennial Mental Health Atlas surveys (WHO, 2018). The United Nations' 2030 Agenda for Sustainable Development (2015) now includes one target specific to mental health.[3] The US National Institute of Mental Health, UK Department for International Development, and Grand Challenges Canada have delivered new funding for research in low- and middle-income countries (Collins & Saxena, 2016), and a Lancet Commission has started to promote many of these initiatives (Patel et al., 2018).

The aim of this edited volume is to address the ethical questions raised by many of the aforementioned issues. But if public health ethics is still a

[2] A broad sampling of initiatives can also be found in the following collections: Cottler (2011); Knifton and Quinn (2013); Patel, Minas, Cohen, and Prince (2013); Bährer-Kohler and Carod-Artal (2017); Bhugra, Bhui, Wong, and Gilman (2018); Eaton and Fallin (2019).

[3] For commentary on global mental health policy, see: Tsai and Tomlinson (2015); Thornicroft and Votruba (2016); Cratsley and Mackey (2018).

relatively new field of study, public *mental* health ethics is even more so. Only recently has the full empirical description of the prevailing disease burden become available, and most attempts to redress the lack of appropriate services are fairly new. What this means is that the relevant issues are newly emerging and ethical analysis is in its early stages. There have been several recent contributions (Bonnie & Zelle, 2019; Fernando & Sumathipala, 2015; Silva, Forlini, & Meurk, 2018), but overall the literature in this area remains scant. The present collection is therefore designed as an initial attempt to examine the ethical implications of public health policies and programs targeting mental disorders. It provides analysis of a selection of important international initiatives and is far from exhaustive or definitive, but our hope is that it can help set the agenda for further research and debate. There are at least two important themes running consistently throughout all of the contributed chapters. The first is a focus on prevention as the defining feature of public mental health, and the second is the notion that the field of mental health poses special challenges for public health-style interventions.[4]

Interdisciplinary projects such as this employ a range of methodologies, models and assumptions, and rather than trying for uniformity, one of the secondary aims of our collection is to highlight and demonstrate the need for such diversity of approach. As the complex, person-oriented concerns of mental health are adapted and extended to models of public health; subject to the goals of prevention; and assessed from the perspective of ethics, the efforts of clinicians, social and medical scientists, and more theoretical researchers must all be brought to bear. Added to these kinds of complexity, public health is itself a hybrid discipline. It is situated between a traditional medical orientation and a broader emphasis on the social and economic determinants of health, as well as between an individual-focused clinical approach and societal or even global concerns. Moreover, public health decisions always rest on a mix of ethical and epistemic factors. They entail a balancing act between 'what needs to be known and what ought to be done' (Weed, 2004, p. 314). To these trade-offs must be added those between the desirable and possible, constrained by budgets and political will.

[4] It should be clear at this point that the scope of this volume is global, encompassing work in both public mental health and global mental health. For analysis of the history, priorities, and guiding principles of these closely related fields, see Cohen et al. (2013); Mehta, Croudace, and Davies (2015); Misra, Stevenson, Haroz, de Menil, and Koenen (2019); Stewart-Brown (2018).

Several chapters in what follows focus on outcomes, and particular prevention-based programs are shown to have unintended and even paradoxical consequences that must be carefully weighed against the positive benefits the programs can be expected to bring, however desirable their goals. Other chapters identify and direct attention towards faulty and inappropriate methodology, with potentially misleading, ineffectual or actively harmful effects. The goals of preventive programs are themselves scrutinised and, in some instances, shown to be conceptually unclear, misplaced or in need of revision. As this range illustrates, public mental health will also be a locus for, and guided by, mixed ethical frameworks. The universal human right to healthcare embodied in the grounding document of public health (WHO, 1946) is derived from an assortment of Western religious, natural law, and Enlightenment tenets. Within this tradition lie the widely-employed principles of bioethics (autonomy, benevolence, nonmaleficence, and justice), adherence to which is believed to assist ethical decision making in biomedical contexts (Beauchamp & Childress, 2013). Although these principles form the basis for several of the chapters in this volume, concern has been expressed that they alone are not sufficient for the ethical analysis of contemporary healthcare (eg. Fiester, 2007; Friesen, Kearn, Redman, & Caplan, 2017).

Often added, in the context of public health ethics, has been the principle of solidarity, defined as a sense of representing, advocating for and acting and speaking on behalf of, the Other (Dawson & Jennings, 2012; ter Meulen, 2017).[5] Solidarity is a foundational concept: made up of both normative and descriptive components, it has been compared to the concepts of community and trust, "essential requirements for the very possibility of ethical decision-making" (Dawson & Jennings, 2012, p. 74). Solidarity 'allows us to see that your condition is actually inextricably related to my condition…The culture and society within which we live influences, shapes and controls the determinants of health … it makes no sense to being an analysis of health with individuals, with "you" and "me". *We should start with us*' (Dawson & Jennings, 2012, p. 77; emphasis added). In this way solidarity focuses away from individually-based towards more collectivist models of disorder, where emphasis is on shared causal determinants. Also essential to solidarity, it has been explained, is a willingness to support

[5] Ethical theorising in public health is increasingly diverse (Ortmann et al., 2016), including appeals to virtue ethics (Rozier, 2016).

vulnerable individuals. Solidarity dictates that health care policies 'go beyond merely meeting needs and rights by exploring how people's personal dignity and sense of belonging can be sustained within relations of recognition, reciprocity and support' (ter Meulen, 2017, p. 141).

By contrast with adherence to principle-based ethical guidelines, evidence-based policies in public health practice tend to be justified by strictly consequentialist, risk-benefit calculations. At a level of theory, philosophical ethics usually emphasises the incompatibility between these contrasting frameworks. But a healthcare ethics of prevention, our authors recognise, calls for equal scrutiny from both perspectives. And in pointing to the ethical underpinnings and goals of preventive initiatives directed towards public mental health, as well as their lacunae, limitations, and unforeseen consequences, the chapters in this volume concern themselves as much with effectiveness as with ethical justification.

The first two chapters address fundamental issues raised by prevention-based programs, including questions concerning the ethics of influence and the significance individual values and culture. In Chapter 2, we provide a broad overview of key ethical considerations, including relevant debates in public health ethics and the concept of prevention, as well as the heightened importance of specific issues in mental health, such as coercion, stigma and issues of personal identity. Any coercive interventions with young children who may exhibit early signs of attention deficit or disobedience, McKeown, Mortimer, Manzini, and Singh illustrate in Chapter 3, require a considerable measure of caution: there must be recognition that such behaviour may not augur later mental health conditions, and awareness of the individual circumstances surrounding each case.

Additional ethical issues involve cross-cultural work in mental health. To illustrate potential dangers as Western medical ideas are exported, in Chapter 4 Rashed and Bingham focus on the culture of the Dakhla Oasis of Egypt, where symptoms resembling Western depression are shown to be differently understood in relation to basic descriptions of causes and risk factors, as well as to appropriate interventions. Issues of translation, data protection, and work with young research subjects arise when, as a preliminary to developing preventive initiatives, research is conducted on both social factors and biological markers. Ethical guidelines mindful of those dangers as they apply to low- and middle-income countries are discussed by Gautum, Pedersen, Wahid, and Kohrt in Chapter 5, with particular attention to biological psychiatric research in Nepal. Preventive policies that import best parenting practices from one culture to another with the aim of improving adult mental health through early intervention are the focus of Hock, Levey, Cooper-Vince, and Harris in Chapter 6. They point to risks in such cross-cultural efforts, and emphasise respectful partnerships and collaborative engagement with the communities in which these initiatives are undertaken.

Other ethical challenges identify and direct attention towards faulty and inappropriate methodology, particularly in regard to screening and early detection of mental disorders, with potentially misleading, ineffectual or harmful effects. To reduce the societal and personal burden of depression, screening is an important preventive tool, used as a preliminary to early diagnosis and treatment. Yet, as Lamothe and Gupta explain in Chapter 7, such procedures are based on inexact methodology, since they as yet provide no way to identify cases of diagnostic error. By pointing to ethical complexities associated with technologies designed to predict and intervene on suicide, O'Leary and Friesen emphasise the differing impacts of context and expectations in Chapter 8. These are illustrated through a contrast drawn between two programs, the first relying on data from longitudinal electronic health records undertaken in a clinical setting, and the other involving content posted within an online community. With some evidence of effectiveness, projects involving early intervention in psychosis have received recent support, Marriott and Broome explain in Chapter 9. Ethical considerations that arise with these initiatives are examined, and the need for further research on their impact is stressed in relation to relapse rates, functional outcomes, socioeconomic costs and the usefulness of follow-up treatments they can be expected to make possible.

The goals of specific preventive programs are the focus of the last four chapters, where they are presented as conceptually unclear, misplaced or in need of revision. The aim of reducing or preventing post traumatic suffering through enhanced resilience and growth, Koch explains in Chapter 10, rests on a complex of questionable assumptions from positive psychology that are neither empirically nor ethically well-supported. In Chapter 11, Tan, Corciova, and Nicholls draw attention to dangers inherent in insufficiently well thought out public health messages involving weight and eating, showing how these strategies must avoid misleading ideas about personal responsibility, and messages whose effects may be dangerous and discriminatory. In Chapter 12, Taylor employs a case study method to recount the dilemma faced by the government of Ghana as it was forced to decide whether to build its inadequately dispersed mental health care programs by cooperating with regionally distributed, but notorious, prayer camps. Some frames of reference directly challenge tenets of the widely-accepted focus on public mental health that lies at the center of this volume. These are introduced in Chapter 13, where Hoffman points to unwarranted assumptions underlying the goals sought by preventive programs in public mental health, with particular attention to the widely announced equation of health and well-being in the founding documents of the WHO.

As editors, our hope is that this volume will serve to introduce the boundless possibilities that arise when mental health is sought using the means, methods, and preventive emphases of public health. At the same time, we aim to introduce and draw attention to the ethical range, complexity and risks incumbent in adopting such measures. There are any number of important issues that are not directly addressed here, however. Perhaps the most topical include: the mental health implications of migration and humanitarian crises (Charlson et al., 2019); substance use and the opioid epidemic (Babu, Brent, & Juurlink, 2019); environmental considerations relevant to mental health, particularly pollution (Roberts et al., 2019); the intersection of mental health and discrimination on the basis of gender, race, and ethnicity (Hansen, Riano, Meadows, & Mangurian, 2018); the use of surveillance in public health more generally (Fairchild et al., 2017); the contested relation between mental health and gun violence (Hirschtritt & Binder, 2018); and mental health within the criminal justice system (Fazel, Ramesh, & Hawton, 2017). That said, this collection is meant to serve as a starting point for further research and expanded interdisciplinary collaboration as the enormous challenge of developing public mental health is increasingly recognised and embraced.

Acknowledgments

We would like to thank Judy Illes for her support, and for including this volume in the series, as well as Peter Llewellyn for his help managing the project. Neither of the editors declares any conflicts of interest.

References

Babu, K. M., Brent, J., & Juurlink, D. N. (2019). Prevention of opioid overdose. *New England Journal of Medicine, 380*(23), 2246–2255.

Bährer-Kohler, S., & Carod-Artal, F. J. (2017). *Global mental health: Prevention and promotion*. Springer.

Beauchamp, T. L., & Childress, J. F. (2013). *Principles of biomedical ethics* (7th ed.). Oxford: Oxford University Press.

Bhugra, D., Bhui, K., Wong, S. Y. S., & Gilman, S. E. (Eds.). (2018). *Oxford textbook of public mental health*. Oxford: Oxford University Press.

Bloom, D. E., Cafiero, E. T., Jané-Llopis, E., Abrahams-Gessel, S., Bloom, L. R., Fathima, S., … Weinstein, C. (2011). *The global economic burden of noncommunicable diseases*. Geneva, Switzerland: World Economic Forum.

Bonnie, R. J., & Zelle, H. (2019). Ethics in mental health care: A public health perspective. In A. C. Mastroanni, J. P. Kahn, & N. E. Kass (Eds.), *The Oxford handbook of public health ethics*. Oxford: Oxford University Press.

Charlson, F., van Ommeren, M., Flaxman, A., Cornett, J., Whiteford, H., & Saxena, S. (2019). New WHO prevalence estimates of mental disorders in conflict settings: A systematic review and meta-analysis. *Lancet*. Advance online publication. https://doi.org/10.1016/S0140-6736(19)30934-1.

Cohen, A., Patel, V., & Minas, H. (2013). A brief history of global mental health. In V. Patel, H. Minas, A. Cohen, & M. J. Prince (Eds.), *Global mental health: Principles and practice*. Oxford: Oxford University Press.

Collins, P. Y., & Saxena, S. (2016). Action on mental health needs global cooperation. *Nature, 532*, 25–27.

Cottler, L. B. (Ed.). (2011). *Mental health in public health: The next 100 years*. American Psychopathological Association.

Cratsley, K., & Mackey, T. K. (2018). Global mental health and the United Nations' sustainable development goals. *Families, Systems & Health, 36*(2), 225–229.

Dawson, A., & Jennings, B. (2012). The place of solidarity in public health ethics. *Public Health Reviews, 34*, 65–79.

Drew, N., Funk, M., Tang, S., Lamichhane, J., Chavez, E., Katonoka, ... Pathare, S. (2011). Human rights violations of people with mental and psychosocial disabilities: An unresolved global crisis. *Lancet, 378*, 1664–1675.

Eaton, W. W., & Fallin, M. D. (Eds.). (2019). *Public mental health* (2nd ed.). Oxford: Oxford University Press.

Fairchild, A. L., Haghdoost, A. A., Bayer, R., Selgelid, M. J., Dawson, A., Saxena, A., & Reis, A. (2017). Ethics of public health surveillance: New guidelines. *Lancet Public Health, 2*(8), e348–e349.

Fazel, S., Ramesh, T., & Hawton, K. (2017). Suicide in prisons: An international study of prevalence and contributory factors. *Lancet Psychiatry, 4*, 946–952.

Fernando, B. L., & Sumathipala, A. (2015). Ethics of public mental health in developing societies. In J. Sadler, W. Van Staden, & K. W. M. Fulford (Eds.), *Oxford handbook of psychiatric ethics*. Oxford: Oxford University Press.

Fiester, A. (2007). Why the clinical ethics we teach fails patients. *Academic Medicine, 82*(7), 684–689.

Friesen, P., Kearn, L., Redman, B., & Caplan, A. L. (2017). Rethinking the Belmont report? *American Journal of Bioethics, 17*(7), 15–21.

Haffajee, R. L., & Frank, R. G. (2018). Making the opioid public health emergency effective. *JAMA Psychiatry, 75*(8), 767–768.

Hansen, H., Riano, N. S., Meadows, T., & Mangurian, C. (2018). Alleviating the mental health burden of structural discrimination and hate crimes: The role of psychiatrists. *American Journal of Psychiatry, 175*(10), 929–933.

Harkness, K.,L., Hayden, E. P., & Lopez-Duran, N. L. (2015). Special section: Stress sensitivity in psychopathology: Mechanisms and consequences. *Journal of Abnormal Psychology, 124*(1), 1–231.

Hirschtritt, M. E., & Binder, R. L. (2018). A reassessment of blaming mass shootings on mental illness. *JAMA Psychiatry, 75*(4), 311–312.

Kleinman, A. (2009). Global mental health: A failure of humanity. *Lancet, 374*, 603–604.

Knifton, L., & Quinn, N. (Eds.). (2013). *Public mental health: Global perspectives*. Open University Press.

Lund, C., Brooke-Sumner, C., Baingana, F., Baron, E. C., Breuer, E., Chandra, P., ... Saxena, S. (2018). Social determinants of mental disorders and the sustainable development goals: A systematic review of reviews. *Lancet Psychiatry, 5*, 357–369.

Lund, C., DeSilva, M., Plagerson, S., Cooper, S., Chisholm, D., Das, J., ... Patel, V. (2011). Poverty and mental disorders: Breaking the cycle in low-income and middle-income countries. *Lancet, 378*, 1502–1514.

Marks, M. (2019). Facebook is predicting if you'll kill yourself. That's wrong. *The Guardian*. January 30.

Mehta, N., Croudace, T., & Davies, S. C. (2015). Public mental health: Evidence-based priorities. *Lancet, 385*, 1472–1475.

Misra, S., Stevenson, A., Haroz, E. E., de Menil, V., & Koenen, K. C. (2019). 'Global mental health': Systematic review of the terms and its implicit priorities. *British Journal of Psychiatry, 5*(e47), 1–8.

Mnookin, S. (2016). *Out of the shadows: Making mental health a global development priority*. Washington DC: World Bank Group.

Ortmann, L. W., Barrett, D. H., Saenz, C., Bernheim, R. G., Dawson, A., Valentine, J. A., & Reis, A. (2016). Public health ethics: Global cases, practice, and context. In D. H. Barrett, L. W. Ortmann, A. Dawson, C. Saenz, A. Reis, & G. Bolan (Eds.), *Public health ethics: Cases spanning the globe*. Springer.

Patel, V. (2015). Addressing social injustice: A key public mental health strategy. *World Psychiatry, 14*(1), 43–44.

Patel, V., Minas, H., Cohen, A., & Prince, M. J. (Eds.). (2013). *Global mental health: Principles and practice*. Oxford: Oxford University Press.

Patel, V., Saxena, S., Lund, C., Thornicroft, G., Baingana, F., Bolton, P., ... Unutzer, J. (2018). The Lancet Commission on global mental health and sustainable development. *Lancet, 392*, 1553–1598.

Priebe, S. (2015). The political mission of psychiatry. *World Psychiatry, 14*(1), 1–2.

Prince, M., Patel, V., Saxena, S., Maj, M., Maselko, J., Phillips, M. R., & Rahman, A. (2007). No health without mental health. *Lancet, 370*(9590), 859–877.

Roberts, S., Arseneault, L., Barratt, B., Beevers, S., Danese, A., Odgers, C. L., ... Fisher, H. L. (2019). Exploration of NO_2 and $PM_{2.5}$ air pollution and mental health problems using high-resolution data in London-based children from a UK longitudinal cohort study. *Psychiatry Research, 272*, 8–17.

Rozier, M. D. (2016). Structure of virtues as a framework for public health ethics. *Public Health Ethics, 9*(1), 37–45.

Saxena, S., Thornicroft, G., Knapp, M., & Whiteford, H. (2007). Resources for mental health: Scarcity, inequity, and inefficiency. *Lancet, 370*(9590), 878–889.

Silva, D., Forlini, C., & Meurk, C. (2018). Public mental health ethics: Helping improve health for individuals and communities. *Public Health Ethics, 11*, 121–125.

Singer, N. (2018). In screening for suicide risk, Facebook takes on tricky public health role. *The New York Times*. December 31, 2018.

Stewart-Brown, S. (2018). Principles of public health: Application to public mental health. In D. Bhugra, K. Bhui, S. Y. S. Wong, & S. E. Gilman (Eds.), *Oxford textbook of public mental health*. Oxford: Oxford University Press.

ter Meulen, R. (2017). *Solidarity and justice in health and social care*. Cambridge: Cambridge University Press.

Thornicroft, G., & Votruba, N. (2016). Does the United Nations care about mental health? *Lancet Psychiatry, 3*, 599–600.

Tsai, A. C., & Tomlinson, M. (2015). Inequitable and ineffective: Exclusion of mental health from the post-2015 development agenda. *PLoS Medicine, 12*, e1001846.

United Nations. (2015). *2030 Agenda for sustainable development*.

van den Bosch, M., & Meyer-Lindenberg, A. (2019). Environmental exposures and depression: Biological mechanisms and epidemiological evidence. *Annual Review of Public Health, 40*, 18.1–18.21.

van Os, J., Kenis, G., & Rutten, B. P. F. (2010). The environment and schizophrenia. *Nature, 468*, 203–212.

Vigo, D., Thornicroft, G., & Atun, R. (2016). Estimating the true global burden of mental illness. *Lancet Psychiatry, 3*, 171–178.

Wahlbeck, K. (2015). Public mental health: The time is ripe for translation of evidence into practice. *World Psychiatry, 14*(1), 36–42.

Walker, E., McGee, R. E., & Druss, B. G. (2015). Mortality in mental disorders and global disease burden implications: A systematic review and meta-analysis. *JAMA Psychiatry, 72*, 334–341.

Wang, P. S., Aguilar-Gaxiola, S., Alonso, J., Angermeyer, M. C., Borges, G., Bromet, E. J., … Wells, J. E. (2007). Use of mental health services for anxiety, mood, and substance disorders in 17 countries in the WHO World Mental Health Surveys. *Lancet, 370*, 841–850.

Weed, D. (2004). Precaution, prevention, and public health ethics. *Journal of Medicine and Philosophy, 29*, 313–332.

Whiteford, H. A., Degenhardt, L., Rehm, J., Baxter, A. J., Ferrari, A. J., Erskine, … Vos, T. (2013). Global burden of disease attributable to mental and substance use disorders: Findings from the Global Burden of Disease Study 2010. *Lancet, 382*, 1575–1586.

World Health Organization. (1946). *Constitution of the world health organization.*

World Health Organization. (2008). *Mental health gap action programme: Scaling up care for mental, neurological, and substance use disorders.*

World Health Organization. (2013). *Comprehensive mental health action plan 2013-2020.*

World Health Organization. (2017). *Depression and other common mental disorders: Global health estimates.*

World Health Organization. (2018). *Mental health atlas 2017.*

CHAPTER TWO

Public mental health ethics: An overview

Kelso Cratsley[a,c,*] and Jennifer Radden[b]
[a]Kennedy Institute of Ethics, Georgetown University, Washington, DC, United States
[b]Department of Philosophy, University of Massachusetts Boston, Boston, MA, United States
[c]Department of Philosophy & Religion, American University, Washington, DC, United States
*Corresponding author: E-mail: kelso.cratsley@gmail.com

Contents

1. Introduction

In this chapter we outline ethical issues raised by the application of public health approaches to the field of mental health. Though not a comprehensive review, it surveys a wide range of topics in order to provide theoretical background for this newly emerging area of research. We maintain a relatively abstract level of analysis, as other chapters in the volume detail specific policies and programs. First, we set out some of the basics of public health ethics that are particularly relevant to mental health (Section 2), with

Developments in Neuroethics and Bioethics, Volume 2
ISSN 2589-2959
https://doi.org/10.1016/bs.dnb.2019.07.002

special attention to the ongoing debate over the traditional presumption of non-infringement, increased recognition of the social determinants of health, and the concept of prevention. Then we turn directly to the moral particularities of mental health (Section 3), focussing on questions concerning coercion and treatment pressure, personal identity and sociocultural factors (including stigma), and unresolved conceptual and methodological issues in psychiatry that complicate its extension into the domain of public health.

2. From public health to mental health

The population-wide scope of public health distinguishes it from other areas of medicine and either renders public health ethics distinct from bioethics or, at the very least, significantly expands the latter's remit. Several ethical considerations are fundamental, including questions regarding the nature and reach of governmental regulation, the just allocation of resources, and the negotiation between collective demands and individual burdens (Childress et al., 2002; Faden & Shebaya, 2015; Kass, 2001). All of these concerns apply to the field of mental health, in which public health models are increasingly applied. Of course, there is a sense in which psychiatry has always served a public health function in its role as social and legal arbiter, particularly with its use of involuntary detainment and treatment. One need not endorse the entirety of Foucault's corpus, for instance, to recognise the plausibility of the charge that psychiatry has historically concerned itself with 'public hygiene' (Foucault, 1980). Such worries are still imperative, but the contemporary landscape of mental health also poses new questions, as interventions move out of the clinic and into communities, across borders, and even onto virtual platforms.

This is part of a broader shift in public health. Often described as a move towards 'new' public health, concerns about protections against infectious disease have been gradually overtaken by an interest in identifying and intervening on risk factors for chronic conditions (Petersen & Lupton, 1996; Tulchinsky & Varavikova, 2010). This is not to deny the continued importance of interventions that address communicable pathogens, including the use of quarantines, active monitoring, and contact tracing. But it does signal a general trend towards developing responses to the growing evidence-base on the behavioural, social, and environmental variables involved in many disease processes. On a national level, there have been efforts to influence aspects of individual lifestyles; perhaps the most well-publicised have been attempts to regulate nutrition and physical activity in New York City. These initiatives have been subject to a great deal of debate due to concerns about governmental overreach, though many have defended the general motivations for changing social norms and the 'culture of health' (Gostin, 2013; Jacobson, 2014). Within global health, the shift is reflected in the increased push to address non-communicable diseases, and mental disorders, reflected in recent international policy instruments (UN, 2015; WHO, 2018).

The ethical implications of this broader turn are rather obvious, largely concerning the challenge of justifying restrictions on everyday behaviour deemed beneficial to the health of the individual and/or society. Indeed, much of the debate in public health ethics centres around traditional Millian worries, particularly the presumption, expressed in the harm principle, that challenges to self-regarding interests demand extraordinary grounds (in contrast to other-regarding interests). Some versions of this worry may be legitimate, but the details matter greatly. Public health programs vary significantly in form and implementation, and thus also in degree of interference (as well as target population). Public mental health is no different in this regard. The type of initiative, and each specific proposal, will ultimately need to be evaluated on its own merits.

2.1 The debate over non-infringement

Justifications for public health programs come in many forms, and often involve a mix of closely related claims, at times adjusted to particular problems and populations (Faden & Shebaya, 2015). They usually start with an appeal to expected benefits at the population-level, and can include reference to the necessity of collective participation in order to accomplish health-related objectives. But they also increasingly include direct challenges to the Millian presumption of non-infringement (absent harm to others).[1] Some early treatments remain influential, particularly the idea that in some cases the least restrictive form of paternalistic restriction can allow for the preservation of an individual's future freedoms (Dworkin, 1972). The relevance of this argument for health is plain, and may be more so

[1] A common response challenges the very notion of strictly self-regarding behaviour, given the likelihood of indirect impact on others (eg. second-hand smoke; increased insurance and medical costs). Mill anticipated this criticism, allowing that few actions are 'entirely isolated', emphasized while emphasizing that for those actions that are only self-injurious, or that simply offend rather than significantly harm others, '... the inconvenience is one which society can afford to bear, for the sake of the greater human freedom' (Mill, 1859/1977, p. 282).

for mental health, where the severe nature of many symptoms can compromise an individual's decision-making, even if only temporarily or intermittently. Hence the longstanding view that coercive paternalism in psychiatry is justifiable as an intervention of last resort to restore capacity and well-being (we return to this in Section 3.1).

More recently, the debate has taken new and interesting turns, not unrelated to broader criticism of the ascendancy of the principle of autonomy in bioethics.[2] An increasingly common move in the literature is to justify the expansion of infringements in public health by questioning the quality of freedoms under consideration. Roughly put, the expressions of liberty that Mill was originally concerned with protecting are likely quite distinct from the types of behaviours subject to the interventions of contemporary public health (Powers, Faden, & Saghai, 2012). Regulations governing everything from seatbelt use to smoking in public places — or proposed restrictions on serving size — arguably represent at most annoyances, or inconveniences, rather than substantial infringements on liberty. Thus there may be a wide-ranging set of interventions that represent negligible infringements offset by their potential benefits.

This is entirely consistent with earlier treatments of paternalism and seems quite sensible, as far as it goes. But of course it still leaves plenty of work to establish which interventions cross the line from mere annoyance to ethically problematic interference. In fact, this may be particularly challenging in the domain of health and mental health. At the heart of these matters is the challenge of assigning appropriate moral weight to an individual's preferences for ordering their lives. Mill's primary concern about interfering with self-regarding interests, often cited in these debates, was that it would inevitably *get things wrong*. That is, regulators and policymakers will incorrectly rank the preferences of the individuals involved. Along these lines, the traditional concern within clinical bioethics has been that individuals may rank-order their values in dramatically different ways than would healthcare practitioners (Goldman, 1980).

Similar concerns apply in mental health, including public mental health. Within clinical mental health, especially psychosocial treatment, not only is highly confidential information involved, often the therapeutic target is an individual's mental life, her beliefs, desires, and emotions. Here you might

[2] This is probably best represented by the O'Neill's (2002) influential critique of autonomy as formalised in informed consent procedures.

think that Mill's original worry gains ground. On the one hand, skilled practitioners will gradually acquire an understanding of an individual's inner life and core values, which may allow for more accurate ascription of relevant preferences (and subjective conception of best interests). On the other hand, the practitioner's professional obligation also calls for a degree of interference with — or reshaping of — a client's innermost thoughts and feelings, which raises the ethical stakes. This may be less of an issue in public mental health programs, given the broad scope of most interventions. But here too the target is the content of mental states, although without the personal familiarity and individual specification allowed for by the clinical encounter. There remains a real possibility, then, that an individual's values and personal projects will be misinterpreted or overlooked entirely.

Now, it is true that mental health is in a distinct position with regard to these debates because of pronounced concerns about competence and capacity. This relates to another feature of the Millian tradition that has been subject to prolonged disagreement: the question of exactly what constitutes 'mature faculties'.[3] The sciences of the mind have made significant contributions to progress on this question, but without reaching anything close to a consensus. And it is important to note that mental health conditions do not have a monopoly on impairments to reasoning and decisional capacity. Indeed, recognition of the limitations of *typical* reasoners has been one of the primary sources of calls for increased paternalistic interventions in healthcare, based on a deflationary view of individual autonomy (Conly, 2012). Much of this literature draws on the heuristic and biases tradition in order to call attention to flaws in cognitive performance. The rough idea, contra Mill, is that people actually don't have a good grasp on what they really want or need, or what choices will be in their long-term best interest.[4]

What should we make of this type of empirically-informed line of argument? Does Mill's commitment to personal, proprietary knowledge fail as a bulwark against infringement, easing the way for restrictive interventions? To be sure, empirical confirmation of the bounds of rationality has been

[3] For Mill, the harm principle applies to each individual '... unless he is a child, or delirious, or in some state of excitement or absorption incompatible with the full use of the reflecting faculty ...' (Mill, 1859/1977, p. 294).

[4] There is interesting empirical work on related issues: eg. the inclination to view human behaviour as genetically and environmentally constrained (ie. determinism) correlates with reduced opposition to paternalistic policies (Hannaikainen, Cabral, Machery, & Struchiner, 2017).

one of the most important contributions of recent social science. It is now a truism that human cognition is limited; not to mention the trivial observation that an individual's inner life tends towards something less than full transparency, and that personal projects can be difficult to ascertain and articulate, even to the individual themself. But ultimately such observations do not speak directly to the normative aspects of the questions under consideration. That is, the data is not morally decisive.

Regardless of the systematic biases that render human reasoning imperfect, we might hold that individuals be accorded a significant degree of self-determination. This is because a presumption of non-infringement on self-regarding behaviour has significance as a moral ideal. As some have argued in response to the calls for increased paternalism in health policy, residing alongside Mill's focus on outcomes is a commitment to the notion that '... liberty itself is a constitutive part of the good for persons' (Dworkin, 2013). Thus, to Dworkin, we should not only be concerned with *promoting* healthy living, perhaps by way of restrictions that help preserve future freedom, but also with *honouring* an individual's freedom to choose. By these terms, liberty is not just another value. An exclusive focus on helping individuals achieve positive health-related outcomes, such as increased longevity, runs the risk of neglecting the more fundamental goal of protecting liberty. This suggests that there is also a tension in some of the arguments for paternalism made in Mill's name, in that aspiring to promote liberty by at times restricting liberty, the goal is often to achieve some *other* value. Therefore, even given significant limitations to self-knowledge, we might still be committed to the idea that individuals should be able to choose how to live, including the possibility of imprudent behaviour (if others are not affected).

Not all are convinced, though, and an increasingly influential move has been to argue for health as an uncontroversial, legitimate value to promote (Wilson, 2011), or for the opportunity to pursue health as the 'primary freedom' (Gostin, 2013). This approach directly challenges the notion of a right for individuals to favour other values over health. Closely related, some have also drawn on the idea of a 'right to health' to make the claim that infringements on individual preferences are justified if they serve the interests of public health (Wilson, 2016). The rough idea is that failure to enact interventions that reduce the risk of ill-health represents a violation of the right to health. Such a claim requires a good deal of additional specification of both the relevant risks and the degree of justifiable infringement, but it does offer plausible grounds for broad and substantial public health intervention.[5]

Those still inclined towards a Millian conception of fundamental rights can respond by emphasising the numerous ways in which most societies permit risky and reckless behaviour, even when health and well-being are at stake. Examples include everything from participation in extreme sports to the refusal of blood transfusions or other medical procedures. It does indeed seem like values other than health at times take normative precedence for some individuals in certain situations (as well as for those who do not support regulation of such activities). Thus it's possible to hold steadfast to the right of non-infringement in order to satisfy a higher commitment to self-determination. This need not require an absolute prohibition on paternalistic intervention. But again, it may very well imply a fundamental right to err. In fact, in a more extreme sense, it may also imply a right to ill-health, or even a right to be mentally unwell.

The debate does not end there, of course. Advocates of paternalistic interventions, including within mental health, can maintain that health must not be sacrificed for an absolutist, misguided commitment to rights. This is compelling if the interventions under consideration are not excessively restrictive or intrusive. And from the other direction there is the additional Millian argument for self-determination based on the risks of governmental abuse. In mental health there are good reasons for taking such a worry seriously, given the historical misuse of psychiatry for political purposes (van Voren & Keukens, 2015). As in public health more generally, an emphasis on the right to pursue ones self-regarding interests without interference can serve as an important form of protection against the possibility of various forms of abuse (Faden & Shebaya, 2015).

2.2 Social determinants of health

Rights-based approaches to public health and global health, particularly those based on a 'right to health,' have been influenced by findings confirming that the social determinants of health are unequally distributed (WHO, 2008; Yamin & Constantin, 2018). Most strikingly, there appears to be a 'social gradient' to health, whereby all levels of socioeconomic stratification are correlated with divergent health-related outcomes, with research continuing

[5] This is closely related to the suggestion that the right to health is fundamental given that it is necessary for the pursuit and attainment of other rights and equal opportunity more generally (Daniels, 2008).

to build on the influential findings of the Whitehall studies (Marmot, 2010). Mental health is no exception. There is a growing body of literature that points to a role for social and environmental factors in mental health conditions (Bell & Marmot, 2018; WHO, 2014a). Everything from socioeconomic disadvantage, urbanicity, pollution, migration, and childhood trauma have all been implicated in various forms of psychopathology (Lund et al., 2018). Many explanatory models now incorporate social and environmental factors (eg. van den Bosch & Meyer-Lindenberg, 2019).[6]

The social turn in psychiatry has been slow to take hold, giving rise to the criticism that models exclusively focused on biological variables are incomplete and continue to impede the development of appropriate treatments, services, policies, and research (Kinderman, Schwannauer, Pontin, & Tai, 2013; Sampson & Galea, 2018; Wade & Halligan, 2004). The empirical challenges are substantial, both in untangling the relevant causal relations and devising innovative interventions on social variables. And the ethical questions immediately present themselves, concerning our broader obligation to redress the social inequalities associated with certain health outcomes. On some of these issues it may at first appear as if normative consternation is not all that warranted. While social interventions can certainly target specific individuals, they are more likely to address large groups of people, operating at the population level. Concerns about infringement may be less salient, then, depending on the program in question. But questions about justice are difficult to avoid, given that certain groups are disproportionately affected by the social disadvantage and environmental degradation that contribute to the development of mental health conditions.

The recent convention in public health ethics has been to frame these issues as concerning inequity, as opposed to inequality (Marmot, 2007; Wilson, 2009). This distinction is intended to capture the idea that some inequalities are the product of unavoidable processes, such as genetic disease, whereas negative health outcomes that result from unjust social arrangements are avoidable and thus represent inequities that demand redress. Some commentators have challenged this move, arguing that the avoidability of any health outcome does not by itself establish that it is unjust, while suggesting that claims of social justice are sufficient to motivate action on

[6] Recent research has incorporated social factors in novel ways, such as the concept of 'syndemics', based on the interplay of poverty, depression, and diabetes (Mendenhall, Kohrt, Norris, Ndetei, & Prabhakaran, 2017), and frameworks that characterise mental disorders as communicable through ecological, familial, and sociocultural pathways (Wainberg et al., 2018).

social inequalities, independent of concerns about health (see Preda & Voigt, 2015 and commentaries). This is a constructive clarification, but there are also good reasons to think that certain political arrangements build in an obligation to treat all forms of ill-health, regardless of whether they are the result of unavoidable or unfair conditions (eg. Daniels, 2008).[7]

Such considerations are just as relevant to mental health as any other part of the health sector, whether one adopts an overarching social justice view or one founded on the right to health. The descriptive picture remains incomplete, but we are gradually getting a better understanding of just how deeply connected mental health is to socioeconomic disparities. A recent systematic review, for example, found that increased income inequality is associated with higher rates of depression (Patel et al., 2018). The pragmatic question is what to do about it. There is clearly a role for political engagement, but much more also needs to be known before research can be translated into effective programming. For example, there is evidence that economic interventions — such as cash transfers contingent on treatment adherence — can improve mental health outcomes, but overall the findings are equivocal (Lund et al., 2011). There has also been work on housing-based interventions, including study of the impact of family housing vouchers on depression, with mixed results (Kessler et al., 2014). (For reviews of interventions on upstream social determinants, and other preventive strategies, see Petersen, Barry, Lund, & Bhana, 2013; Wahlbeck, 2015; Mendelson et al., 2019).

2.3 Prevention

Preventive interventions take many different forms; their commonality lies in their shared goal of enhancing health and reducing or averting ill-health. All public health measures involve prevention, broadly understood this way, and prevention is rightly regarded as a grounding principle of public health. The same emphasis on prevention guides initiatives involving public mental health. Indeed, due to the stigma and discrimination associated with mental health issues, explicitly preventive measures around the increased awareness and so-called mental health literacy are one component of most major public mental health programs. Examples include Australia's Beyond Blue initiative (www.beyondblue.org.au) and the Mental Health First Aid project employed in the US (Kitchener & Jorm, 2008); each emphasises awareness

[7] This approach draws on a Rawlsian conception of justice; for discussion, see Wester (2018).

of mental illness and of the availability of effective treatments. While ubiquitous in policy initiatives and research literature directed towards public mental health, however, prevention is an elastic term, used in different ways.[8]

2.3.1 Stages of disorder and intervention

As it is generally employed, prevention refers to reducing the likelihood of developing initial symptoms, but also to delaying the severity of disease symptoms and of slowing the return, or reducing the severity, of symptoms in subsequent episodes (it is used this way in the clinical expression 'preventing relapse'). In public mental health, as elsewhere, the language of 'primary', 'secondary', and 'tertiary' prevention has been introduced to indicate this contrast (Mendelson et al., 2019; Mrazek & Haggerty, 1994; Saxena, Jane-Llopis, & Hosman, 2006). Primary prevention aims to avert or avoid the incidence of new cases; secondary prevention occurs once diagnosable disease is present. Prevention is also sometimes said to be tertiary, to address outcomes after illness onset, such as rehabilitation, or prevention of associated downstream complications of the disease. Defined this way, prevention maps onto a range of clinical situations, and some secondary prevention includes identification of the disease before it is symptomatic, or before help is sought (although few if any mental disorders can be asymptomatic this way, it should be added). To date, attempts to regiment usage around primary, secondary and tertiary prevention have not proven entirely effective. Current research in psychiatry and public health reveals a fairly equal balance between the use of the term prevention to indicate interventions aimed to avert or avoid the initial occurrence of disease (primary) and those covering subsequent (secondary and tertiary) efforts.

The divisions between forms of prevention rest on a standard model of disease that guides much of medicine, whereby each discrete disorder is associated with underlying entities or processes (defined in terms of biological dysfunction or genetic risk) that persist despite their sometimes fluctuating symptom manifestations, and arise within the individual affected. The model may better fit disorders with a more discretely bounded phenotypic profile; its application to the common, often milder, and self-limiting symptoms of anxiety and depression has been challenged (Borsboom, 2017; Kendler, Zachar, & Craver, 2011). However, along with other features of the

[8] For further analysis of the conceptual features of prevention, see Radden (2016, 2018).

parallelism equating mental and physical diseases, the standard disease model from psychiatry is presupposed by the classification into primary and secondary prevention. Typically episodic rather than continuous in their course, most mental disorders are understood to be chronic diseases, so that even when an intervention averts or delays re-occurrence in a later episode long separated from the previous one, subsequent interventions will be secondary, according to that classification.

With mental disorders, the onset of disease is rarely easily or conclusively determined, instead emerging only gradually from normal states, and often anticipated (and afterwards trailed) by symptom-like phenomena. The earliest symptoms that auger the full syndrome of a diagnosable condition thus form the halo effect known as the prodrome. This is also true of many organic conditions (eg. diabetes, and hypertension), but not all. Any preventive action directed towards the prodrome will be primary. So-called 'indicated' prevention recognises this blurred area of the prodrome lying between the absence and presence of disorder that cannot yet be subject to the secondary prevention occurring once diagnosed or diagnosable disease is present (Mrazek & Haggerty, 1994; Saxena et al., 2006). Prevention that is indicated is directed towards high risk individuals with minimal but detectable signs or symptoms foreshadowing disorder, or, in other medical disorders found to possess biological markers strongly indicating predisposition for it, while not — or not yet — meeting the criteria for diagnosis. In contrast to more 'universal' approaches, that target whole populations, what is known as 'selective' prevention also intervenes with populations possessing particular risk factors (National Research Council & Institute of Medicine, 2009).[9]

2.3.2 Treatment and health promotion

The contrast between prevention and treatment has not always been respected either by universal rights rhetoric or in the research and clinical literature on mental disorder, where prevention and treatment are often conflated. When prevention is placed in contrast to treatment, the case has been made that global policies should prioritise prevention, since it is

[9] Few universal strategies have yet been tested (apart from school-based interventions). But there is no dearth of interesting data that factor into recent proposals, such as the association between nutrition and exercise and reduced rates of depression and anxiety (Jacka, Mykletun, & Berk, 2012), or the correlation between higher levels of lithium in the water supply and reduced suicide rates (Ng, Sjostrand, & Eyal, 2019).

practical and cost-effective, compatible with the scaled-up responses necessary to meet the challenge (and possible due to the use of governmental resources), and consistent with the ethical priorities around universal human rights (Tol, 2015). Once preventive principles are translated into particular policies, prevention and treatment are often conjoined.

The European Alliance Against Depression is typical of such interventions: it seeks better care and treatment in combination with public health education to improve community attitudes about depression and suicide (Hegerl & Wittenburg, 2009). Moreover, according to the strict separation identified between primary and other forms of prevention, anything except primary prevention must be said to be forms of treatment. Following this usage (secondary and tertiary efforts are forms of treatment), indicated prevention — although it is usually regarded as secondary — remains anomalous. When the binary between the presence/absence of disease determines our usage, indicated disorder is not fully in either camp. From the perspective of their practical effects, prevention and treatment may be one: preventive measures serving to treat, and treatments to prevent. Conceptually, however, preventing by averting or avoiding disorder is different from — and will, if successful, obviate — treating it. Primary prevention addresses events that have not yet occurred or states that are presently non-existent; treatment addresses those that have (or are), when these events or states correspond to the presence or occurrence of diagnosable disorder.

The language of modifiable risk is increasingly employed to refer to the targets of primary prevention. Such risks are conceptually distinguishable from the presence of disease; at most, and usually in conjunction with a range of factors (including social and economic conditions), they offer estimates of the probability that the condition is present or will develop. Focus on this approach has been dated to new research methods that have become available since the second half of the twentieth century (Schwartz, 2008). Because terminological confusions arise from proposing the imposition of treatments on non-diseases, it will be helpful to maintain the separation employed here, where (primary) prevention applies to risks, and diseases alone can be said to be treated.[10] The blurring of prevention and treatment has raised other, more general ethical concerns: it has been portrayed as a conceptual sleight of hand that may jeopardise individual autonomy and

[10] For alternative language, and the proposal that modifiable risks are appropriately subject to procedures acknowledged to be treatments, see Schwartz (2008).

self-identity. By inviting false positives, it is seen to open the way for unwarranted medicalisation, defined as a process by which human differences are transformed into pathologies (Rose, 1998; Schwartz, 2008; Conrad & Slodden, 2013; McLachlan, 2014).[11] If persuasive, analyses such as these carry far-reaching ethical implications for debates over the proper role of science and government and the place of experts in dictating social norms of health and ill-health.

A further contrast is between disease prevention and health promotion, emphasising that health sometimes connotes more than the absence of disease (Saxena et al., 2006; Vilhelmsson, Svensson, & Meeuwisse, 2011; Wahlbeck, 2015). Mental health may include the presence of positive well-being, for example. The WHO Constitution (1946) states that health is 'not merely the absence of disease or infirmity;' and the enjoyment of the 'highest attainable standard of health' is described as 'one of the fundamental rights of every human being without distinction of race, religion, political belief, economic or social condition'. In the WHO Mental Action Plan (2013), mental health is further characterised as 'a state of well-being in which an individual realises his or her own abilities, can cope with the normal stresses of life, can work productively and is able to make a contribution to his or her community' (the contrast between positive health and absence of illness is discussed in Section 3.2.2).

In relation to prevention, the positive goals of promotion will serve the ends of prevention here also, and co-occur at the practical level within the same programs and activities. Due to the complementary effects of these two, the contrast between illness prevention and health promotion may not be as ethically freighted as that between prevention and treatment, with its obviously competing demands on resources and attention. But the eudaemonic and functional goals of mental health positively construed reflect values and priorities (Vilhelmsson et al., 2011). As such, and because while widespread they may not be universally shared, they require ethical analysis and justification.

11 Programs such as campaigns to promote depression literacy or resilience-building efforts, have been described as 'problematizing the future' in particular ways by 'transforming potential threats into present concerns and action' (McLachlan, 2014, p. 816). These trends by which the risk of disease has replaced disease as the target of therapeutic intervention are identified as examples of other 'bio-keeping technologies' (Foucault, 1988) that reflect the overreach of medical science's authority and power (Naidoo & Wills, 2000) and jeopardise personal identity by producing hapless 'new selves' (McLachlan, 2014, p. 831).

The prevention-treatment division is employed throughout health policy in relation to more standard medical conditions; it is a separation of both practical and ethical importance. In the practical arena of mental healthcare, however, diagnostic vagaries complicate when (primary) prevention ends and treatment begins, as we've seen. Thus far, in psychiatry, the close clinical observation of symptomatic behaviour provides the sole means of determining the presence of diagnosable disorder. In practice, without an independent method of verification, and as long as the binary separation between prevention and treatment is upheld, the question of whether interventions have the status of (primary) prevention or treatment must often remain unresolved.

2.3.3 Focus on depression and suicide

In today's prevention-focused public mental health, depression and suicide have thus far received the lion's share of attention. Depression is often said to be the common cold of mental ill-health; its prevalence imposes an extreme and growing global economic burden as measured by morbidity and disability, and a detrimental personal impact on the quality of life of patients and their families (WHO, 2017). Clinical services for depression, however, have only been estimated to reduce the disease burden by one-third (Andrews, Sanderson, Corry, & Lapsley, 2000). The rationale for undertaking an alternative strategy in the form of prevention thus recommends itself, and has given rise to a large range of psychological interventions targeting depression and related disorders. These vary in their scope (some universal, others more selective and indicated), their approach (including whether they employ pedagogical or other methods), in the age range of the recipients of the interventions, and the methodology employed.

Evaluations of the effectiveness of this range of initiatives have yet to provide clear guidance. (For a meta-analytic review of randomised controlled trials comparing the effects of preventive psychological interventions, see van Zoonen et al., 2014; for a review of school-based programs, see Werner-Seidler, Perry, Calear, Newby, & Christensen, 2017.) Although incomplete, such findings have been taken to indicate that isolated preventive initiatives to reduce the incidence of depression can be expected to have only modest success, which instead recommends large-scale, long-term structural and social embedding of universal and remedial prevention programs beginning early in life, within a suitably-funded, society-wide response (Ormel, Cuijpers, Jorm, & Schoevers, 2019). This includes broad educational initiatives aimed at both parents and children, as well as more selective or indicated interventions. Part of this may also involve peer and lay counselling, especially in low- and middle-income countries (LMICs) where research has started to show the preventive efficacy of such task-shifting strategies (Dias et al., 2019).

Increased rates of suicide have prompted a range of preventive programming, most with comparably limited success.[12] Going forward, research supports several different forms of prevention, including means restriction, school-based awareness programs, and pharmacological interventions (Zalsman et al., 2016). But no one prevailing approach recommends itself, and so more research is required to evaluate integrated programs.[13] In addition, the shortcomings of current programs have been attributed to strategies based on individual-level theorising that serve to make individuals and communities solely responsible, and to direct attention away from suicide risk factors involving general, systemic inequities requiring more comprehensive action (Fitzpatrick, 2018). One model for this more comprehensive approach involves multiple levels of government, where specific provisions are woven into legislative and regulatory requirements in relation to all sectors (housing, transportation, etc.) with impact on population health (Fitzpatrick, 2018).

3. Issues specific to mental health

Among the factors contributing to the distinctiveness of mental healthcare, many of which have significant ethical implications, the widespread use of coercive measures stands out. Related factors also arise out of historical, social, and cultural understanding of the mind, self-identity and personhood. And additional factors include the contested definitions of health and un-health, and the still incomplete explanations and methodology employed by psychiatry, clinical psychology, and allied disciplines in mental health. With the exception of issues around involuntary treatment, the distinctiveness of mental health may rest with no single one of these features, but rather with their combination, in conjunction with the effects

[12] The WHO has started to focus on suicide prevention (WHO, 2014b), and the UN has prioritised suicide by making it the indicator for the one target specific to mental health within current development goals (UN, 2015). Suicide prevention is the theme for World Mental Health Day 2019, an initiative of the World Federation for Mental Health.

[13] Worries have also been raised about recent calls for universal screening, particularly when focused on detecting suicidal ideation, which is not predictive of suicide mortality (Nestadt, Triplett, Mojtabai, & Berman, 2018).

of stigmatising social attitudes (Radden, 2015). In part because of this conjunction, however, physical health and mental health have often been relegated to separate domains (a fact that likely explains the dearth, until recently, of initiatives to ensure public mental health). With belated recognition of the importance of mental healthcare as one aspect of public health, and as a preliminary to advancing the agenda of public mental health, distinguishing features of mental health such as these call for special attention and scrutiny.

3.1 Coercion and treatment pressure

The concept of coercion has broad application, subject to longstanding debate in political philosophy and legal theory (Anderson, 2011). Within bioethics, its ethical implications have received the most attention in relation to involuntary psychiatric treatment. Indeed, the use of coercive measures – including legally mandated hospitalisation, forced medication, and physical restraint – continues to be the defining issue of mental health ethics. This holds to such a degree that it marks out psychiatry from other areas of medicine, seen mostly clearly in the fact that individuals with mental health conditions are subject to legal sanctions that those with physical conditions are not. Some have argued that this is discriminatory, compounded by the fact that in most jurisdictions the legal criteria appeal to empirically dubious notions of risk or dangerousness. Barring further substantiation with more relevant criteria, such as assessments of decisional capacity, this arguably makes involuntary psychiatric treatment little more than an unjustified form of preventive detention (Szmukler, 2015).[14]

Some of these concerns are less relevant to public mental health ethics. Most initiatives fall far short of legally mandated treatment, or even the types of regulation subject to recent debate in public health. There have been no proposals to tax foodstuffs or products associated with psychopathology (in fact, one of the substances most worrying in this regard – cannabis – is gradually being deregulated). Many of the new initiatives simply involve health

[14] One widely-discussed solution is a 'fusion law', which applies standards of capacity and best interests to both mental and physical disorders (Dawson & Szmukler, 2006). These issues have only become more complicated due to the United Nations' Convention on the Rights of Persons with Disabilities (UN, 2006), which dictates that individuals always retain legal capacity regardless of disability or mental disorder (for commentary, see Scholten & Gather, 2018).

promotion strategies, such as informational campaigns and programs that encourage voluntary participation (such as the headspace drop-in centres in Australia: https://headspace.org.au/). That said, some of the traditional worries about pressure and influence still apply. The most immediate involve the extension of coercion into the community in the form of mandated community treatment; such methods are used as a form of secondary prevention, despite suffering from lack of empirical support (Molodynski, , Rugkasa, & Burns, 2016). Other worries are more speculative, probably best understood in relation to more subtle types of influence traditionally categorised under the headings of manipulation and persuasion.[15] Here we briefly mention several potential concerns, drawing from reviews of the literature on coercion in psychiatry (Cratsley, 2019; Kallert, Mezzich, & Monahan, 2011; Newton-Howes & Mullen, 2011).

For a start, it is widely recognised that the mere possibility of coercive psychiatric treatment can represent an implicit threat (or 'coercive shadow'; Szmukler, 2015). This can colour an individual's experience of clinical encounters with mental health services and very well may extend beyond this. For example, those who attend voluntary drop-in centres may be subject to this effect; even health promotion materials might carry this negative connotation. In addition, studies have shown that both voluntary and involuntary psychiatric inpatients report a lack of transparent and fair processes regarding the terms of their hospitalisation. This has been characterised as an issue of 'procedural justice', which highlights the importance of first contacts with mental health services and the need for clear and accessible information (Cascardi, Poythress, & Hall, 2000; O'Donoghue et al., 2014). These lessons are also likely to extend beyond the clinic. Lastly, there is a growing body of qualitative data on the experiences of mental health patients. Such findings indicate that many patients do not agree with the use of incentives or offers from mental health professionals, including in community treatment; relationships with healthcare workers and broader societal norms are perceived as key sources of unwelcome pressure; and freedom of choice and respect from staff are deemed the most important values during treatment (for review of 'informal coercion': Hotzy & Jaeger, 2016).

[15] While the traditional approach in bioethics is to distinguish between coercion, manipulation, and persuasion (Faden & Beauchamp, 1986), there are other typologies of influence. For example, in public health, see the Nuffield Council's Intervention Ladder (2007), in mental health, see Szmukler and Appelbaum (2008).

All of these considerations have potential relevance for public mental health. The outsized role of coercion in mental health can have a chilling effect on help-seeking behaviour and may influence the way that individuals respond to interventions that attempt to influence behaviour, including seemingly benign informational campaigns. In order to design effective preventive programming, efforts should be made to address this issue, through the provision of clear and accessible information and by simplifying and improving pathways to care (this is particularly important given the often-tenuous boundary between prevention and treatment; there are many cases when prevention quickly becomes a need for social or clinical services). These are not novel recommendations, but they do emphasise the variety, and complexity, of individual's responses to mental health services. This is helpfully recognised by at least one recently proposed ethical framework for clinical psychiatry, which recommends that an appeal to personal values should only be used as a source of influence when there is sufficient knowledge of the individual patient (Blumenthal-Barby, McCullough, Krieger, & Coverdale, 2013). But note that this will be missing entirely in public mental health initiatives that operate at the population level.

A relatively new variant of these issues is the ongoing debate over the use of nudges in public health. The ethical implications have been discussed at length elsewhere, but application to mental health has received relatively little attention. Much of the broader debate has centred on the permissibility of interventions that rely upon an opacity condition; that is, what makes many nudges effective is that they bypass conscious awareness. If one is committed to a requirement of informed deliberation, and a right of refusal or withdrawal from health services, then this may seem problematic. Nudgers typically respond that freedom of choice is preserved because the manipulation of choice architecture is resistible (and/or that the nature of interference is negligible). But this takes on added resonance in mental health, where individuals are more vulnerable to cognitive irregularities and thus may have more difficulty attending to and resisting nudges. In addition, notice that nudges often merely harness or make use of systematic reasoning biases rather than attempting to correct or amend them. Therefore, the professional obligation to treat cognitive impairments associated with mental disorder may be difficult to reconcile with the use of nudges, and psychiatric patients may have reasonable expectations that features of their disorder will not be used as a means of influence.[16]

3.2 Individual, social, and cultural variables

To an extent matched in few medical specialities, mental health care and psychiatry are socially and culturally embedded practices. As well as inseparable from a range of deeply held and often contested values, they are person-centred, self-focused, and holistically understood. These features include, but are also compounded by, the stigma attached to mental disorder, and its consequences in discriminatory attitudes, practices and systems. Along with its focus on risk factors for chronic disease, the 'new' public health rightly seeks to distance itself from many of the moralistic judgements of earlier eras (Petersen & Lupton, 1996). Still, as products of history and culture, conceptions of mental disorder have been formed around perceived individual deficiencies of self-identity, personhood and personal agency. And the ideals of human flourishing and reduced psychological suffering that guide therapeutic endeavour are similarly conceptualised with reference to such ideas, and integrally linked to the widely-cherished moral categories of personhood, responsibility and agency that have shaped, and continue to frame, civic life in Western societies.

3.2.1 Personal identity

The subjective experience of mental disorder consistently affects the individual's most personal, private, and self-identifying traits. This feature explains the elevated confidentiality and data protection required in mental health care and treatment, which exceeds the ethical demands imposed in the rest of medicine (American Psychiatric Association, 2013) Because of the centrality of gender to self-identity, it also explains emphasis on gender bias in influential norms governing mentally healthy responses, social roles, and appropriate behaviour (Bluhm, Jacobson & Maibom, 2012; Clark & Horton, 2019). The treatment of those with mental health problems is

[16] Note that much of the early work on heuristics and biases was driven by an interest in *improving* decision-making in medical contexts (eg. Kahneman, 2011; for commentary on related issues, see; Quigley, 2014; Cratsley, 2015, 2019). And while there is evidence that the general population tends to approve of nudging, particularly if it is less intrusive and originates from trustworthy sources, such as experts rather than policymakers (Evers, Marchiori, Junghans, Cremers, & De Ridder, 2018), there is also evidence of significant discrepancies between doctor and patient perspectives on the acceptability of nudges in clinical settings (Avitzour, Barnea, Avitzour, Cohen, & Nissan-Rozen, 2019).

comparably holistic. Person-centred care is always a desirable element in healing, if rarely entirely necessary. By contrast, except when imposed forcibly, psychiatric care, engaging the patient's affective states, personality, personal history, motivation, and self-framed identity, demands at least minimal personal (in the sense of holistic) cooperation from the patient.

The ethical significance of this emphasis on self-identity is evident when, in an approach favoured due to the additive health advantages it affords throughout a lifespan (Tol, 2015), preventive public health initiatives often target juveniles with prodromal or indicated – rather than full-blown – disorder. Issues of consent and assent arise whenever juveniles are subjects of intervention. Other issues centre around self-identity. The self-identity of early adolescents is incomplete, and some interventions will affect its formation. Importantly, this will be true, for example, both in youth who subsequently prove to warrant such early attention and in those (false positives) who do not. Yet identities in each group, during this formative stage, will likely be negatively affected by stigmatising attitudes. The gains of imposing preventive intervention for some in this group, the true positives, have thus been recognised to need balancing against the costs for others, the false positives, of negatively affecting identity formation (Broome, Fusar-Poli, & Wuyts, 2013; Radden, 2018).

The precision medicine framework of psychiatric genomics, which utilises big data to develop preventive and targeted treatments, is focused on even younger children at risk for mental and neurodevelopmental disorders, when identity formation is at rudimentary stages. Ethical issues also arise here: over genetic identity, data sharing and privacy factors, with particular concern over the effects of 'geneticizing' behavioural phenotypes in light of stigma around both mental disorder and intellectual disability (Kong & Singh, 2018). Whether directed towards reducing personal, psychological suffering, or aspiring to positive mental health more broadly understood, mental healthcare engages the affective experience and social functioning of the whole person. Hedonic wellbeing and traits such as the capacity to withstand adversity are similarly attributed to persons holistically understood, and to the affective experience and social functioning of individuals perceived in these person-centred terms (Glas, 2019; Tekin, 2019; Waring, 2016).

3.2.2 Values & culture

The goals sought by all mental healthcare and prevention are explicitly value-laden and sometimes contentious. The inescapably evaluative and cultural elements attaching to mental disorder form part of an influential definition by which disorder is ascribed only when some failure of psychological function finds expression in symptoms also judged to be harmful according to societal value norms (ie. the 'harmful dysfunction' analysis: Wakefield, 1997). There is nearly universal agreement over the desirability, if not the means to achieve, bodily health. By contrast, values in mental healthcare are often challenged and sometimes quite local, involving divergent priorities, interpretations, and ideals (Fulford, 2011; Kingma, 2014). Researchers and experts on public mental health speak of wellbeing as the goal of healthcare, defining that category as feeling good, 'hedonic' wellbeing, including but not limited to reduced suffering, in addition to functioning well, 'eudaemonic' wellbeing or flourishing (Wahlbeck, 2015). Even when interpreted minimally as reduced personal suffering (Vilhelmsson et al., 2011), hedonic wellbeing is also a culturally-influenced, normative category. When and how we suffer and whether justifiably or not, are determined by cultural values and local meanings.

Both the goods of hedonic and eudemonic wellbeing, and the reduction of personal suffering, reflect Western cultural traditions from the last several hundred years. Excluded from these present-day characterisations are many goods and virtues — including some of those by which Aristotle understood *eudaimonia*. When history and circumstances have shaped the traditions out of which such values grew, they cannot be expected to always correspond to the linked values, religious and metaphysical lore, and healing systems found in other cultures. And they do not, as medical anthropology has shown (Kirmayer, Lemelson, & Cummings, 2016; Kirmayer & Pedersen, 2014; Kleinman & Benson, 2006).

Sorting out whether and how Western public healthcare goals fit the rest of the world demands ethnography as much as medicine, and a fine alertness to differences not readily translated into Western science, values or metaphysics. For instance, some of the resistance met by global mental health (GMH) initiatives depicts them as unwarrantedly imposing Western conceptions of health, illness, healing and personhood, and suppressing contrary views (Cooper, 2016; Fernando, 2014; Mills, 2014; Murray, Kass, Mendelson, & Bass, 2016). Only an intercultural encounter based on 'cultural competency' may allow global mental health to answer accusations such as these (Kleinman & Benson, 2006). Although the ethical issues accompanying public health approaches to mental health are raised with particular force in the context of GMH, public mental health also raises ethical challenges nearer to home, some magnified by the GMH agenda, others unchanged by it, and still others distinctive to the Western cultural, social and medical context.

3.2.3 Stigma

Public mental health programs regularly comprise two elements with differing goals: those designed to reduce the incidence, severity, and disability-related effects of mental disorders, and those designed to reduce stigma about such disorders. Typical of the latter are the public education campaigns and school curricula emphasising mental health literacy. Encouraged by a report from the National Academy of Sciences (2016), and by the limited success of what has been called the 'educational fix' (Corrigan, 2018; Corrigan, 2016a,b; Corrigan, Rusch, & Scior, 2018), efforts to achieve this second goal have also included alternative, non-pedagogical strategies. Canada's contact-based Opening Minds project is one of these. It aims to reduce stigma through personal exposure to those with mental disorder, using small, contact-based programs tailored to particular settings (with youth, health care providers, citizens in the workplace, and the media; Stuart, Chen, & Christie, 2014.) Rather than a technical expert teaching facts of illness that reduce stigma, such non-pedagogical approaches proceed by providing contact with people sharing stories of personal experience with illness and recovery (Corrigan, 2018).

Like preventive measures that target mental disorder directly, programs to reduce stigma remain in relatively early stages of development. Few such programs date to before the first years of the 21st century (for reviews, see Thornicroft et al., 2016; Heim, Kohrt, Koschorke, Milenova, & Thornicroft, 2018).[17] Their assessment of penetration and effectiveness is ongoing, and thus far, it has served to expose some paradoxical and potentially unethical sequelae, as well as inherent attitudinal and methodological complexities that will affect the design and approach of future anti-stigma efforts. After preliminary analysis of the kinds of stigma that these projects aim to reduce, we note some of the impediments to achieving these goals.

Stigma involves an out-group being distinguished from and discredited by the majority on the basis of undesired differences (Goffman, 1963). Its mechanism is described as engendering prejudicial beliefs which in turn

[17] Social contact appears to be the most effective ant-stigma intervention, although its long-term benefits remain unclear (Thornicroft et al., 2016). In LMICs, there is need for additional research, given that extant data requires careful interpretation due to lack of culturally appropriate adaptations of anti-stigma programs (Heim et al., 2018).

promote discriminatory behaviour, a characteristic attributed to individuals conveying 'a social identity that is devalued, or a mark of disgrace associated with a particular circumstance, quality, or person' (Thornicroft, 2006, p. 9). Social psychology has developed several explanatory models for the phenomenon as it arises across disrespected groups identified by race, sex and sexual orientation (Link & Phelan, 2001; Pescosolido, Martin, Lang, & Olafsdottir, 2008). For example, stereotypes are defined as knowledge structures often framed in seemingly fact-based beliefs with negative evaluative components, that become prejudice when they are generalised to the respective group involved – from which they are often realised in discriminatory action (Corrigan, 2016a). A contrast is drawn between mental illness related stigma that is public (when the general population endorses the stereotype, often resulting in avoidant behaviour) and the self-stigma involved when individuals internalise prejudicial attitudes and respond towards themselves in negative ways; a third form of stigma, 'label avoidance', is defined as 'stigma that undermines service use for individuals in need of psychiatric care' (Corrigan, 2016b, p. 68).[18]

Stigma is an inherently unjust attribute; the presence of prejudicial beliefs and discriminatory behaviour thus provide strong ethical justification and urgency for efforts taken towards its reduction. Stigma is also a multi-faceted phenomenon, however, and uneven in its effects. Magnifying the challenges involved in identifying and engaging with mental illness stigma, in relation to the substance of their message and also to the methods employed, is recent acknowledgement of its often tacit and unconscious nature. Experiments have triggered biases over even arbitrary and randomly assigned group identities with no meaningful content or history, suggesting that bias of this kind is very deeply rooted in human psychology (Dawes, van de Kragt, & Orbell, 1990). Moreover, those possessing more than one stigmatised trait are subject to intersectionality effects, which cannot be explained by a merely additive model (Cole, 2009).

There is a tension within many anti-stigma initiatives that confuses, and seems likely to compromise, the message conveyed in mental health literacy campaigns – at least those premised on traditional medical models of disease or disorder. By emphasising similarities between mental illness and more ordinary experience, messages that normalise mental disorder counteract fear-

[18] The negative impact of stigma on help-seeking and access to services is difficult to overstate (Schnyder, Panczak, Groth, & Schultze-Lutter, 2017). There is also an association between stigma and symptoms such as suicidal ideation (Oexle et al., 2017).

related prejudices over the imagined Otherness of persons with mental illness. But at the same time they risk the conclusion that mental illness indicates moral weakness, and reduce any benefits conferred by a medical model. Of this tension, likely to detract from the message of anti-stigma campaigns, it has been noted that both likeness-based and unlikeness-based conceptions of psychiatric stigma 'can lead to prejudices stemming from paradoxically opposing assumptions about mental illness', creating a troubling impasse (Gergel, 2014, p. 148).[19]

In a related discussion of the apparently contradictory messages that might form the basis of anti-stigma campaigns, there are parallels with the social groups aiming to revise social attitudes towards once-despised qualities such as skin colour, and sexual orientation, and provided models for social movements around Mad Pride. Rather than normalising — or medicalising — the symptoms of mental illness, these suggest, anti-stigma efforts might be used to support the goal of respecting the differences arising as the result of disorder. While the association with undesirability is eliminated, the differences themselves can be acknowledged, and the identities which they have formed, honoured. Summing up the attitudinal changes required for these messages, it has been proposed that instead of emphasising likeness-based similarities casting those with mental disorder as just like everyone else, these messages might emphasise the value of solidarity. In recognition of solidarity as a key value for public health, the general public is encouraged to stand in solidarity with those who publicly expose — or 'out' — their own stigmatised identity (Corrigan, 2016a,b).

The different settings within which these messages may be more and less effective; their unintended as well as intended consequences; and effects accruing for the individual, the community, and social justice in more general terms, each require and have begun to receive further study, with findings integrated into programs such as the Mental Health Unity campaign (www.activeminds.org/cbs-cares-mental-health-information/1033; Kosyluk, Schmidt, Abelson, Malmon, & Corrigan, 2015). Solidarity-based messages acquire support from analogies with other pride-related projects, and from research findings about the effectiveness of celebrating difference, where, for example, coming out has been shown to correlate with improved mental and physical health (Corrigan & Wassel, 2008). Compared with normalising messages, solidarity-based goals offer some recognition of the

[19] Also see the 'mixed blessings model' (Haslam & Kvaale, 2015).

neuro-diversity position that shifts attention from the above binaries (of health and unhealth, normal and abnormal) towards emphasis on the plurality and variety of human responses, and recognition of the wide range of cognitive abilities (Ortega, 2009).

Given the complexities attaching to stigma, it is not surprising that public mental health campaigns to reduce or eliminate it have an uneven record of success. Some of this may be due to the substance of the messaging, but some will also involve the methods employed. For example, approaches that have been called educational positivism (enhanced public education campaigns, expanded efforts to raise awareness in primary care physicians, and stepped up pedagogies within the school systems) have thus far shown limited success, with unforeseen consequences including too many channels of information that leave the audience overwhelmed (Corrigan, 2018). In this respect, contact-based strategies such as the Opening Minds project deserve further trial.

3.3 Conceptual and methodological challenges

Conceptual and methodological uncertainties further contribute to the distinctness of mental health. Medical psychiatry is a relatively young science, arguably dating to no earlier than the nineteenth century. Its defenders point to this in explaining the lacunae and unresolved disputes surrounding its diagnostic classifications and explanatory models, its seemingly contradictory research findings, problems of definition, lack of verifiability, comorbidity and disorder status (Fernando, 2014; Summerfield, 2012). These unresolved issues have implications for the nature, and extent, of the challenge faced by public health as it turns its attention to mental health.

The two-part harmful dysfunction definition of mental disorder, mentioned above, accommodates both evaluative and factual elements. But whether values are unavoidable in such definitions is a matter of continuing debate (Bolton, 2008; Boorse, 2011; Varga, 2015); similarly disputed are general conceptions of health and ill-health. These definitional controversies have implications for epidemiology, method and practice. For example, the epidemiology and projected healthcare costs for depression, widely cited as having reached epidemic proportions, will be inaccurate if real disorder requiring medical attention has been conflated with the everyday distresses of normal sadness (Horwitz & Wakefield, 2007); or if screening methods are flawed. Unresolved issues of this kind heighten the tensions arising from public health's status as a hybrid discipline combining medicine with social sciences and policy (Beaglehole & Bonita, 2008), demanding its own ground-up, distinctive ethical analysis.

Biomarkers yielded by techniques such as neuroimaging and genetic screening apparently hold future promise as potentially modifiable risk factors for subsequent mental disorder; although methodological and ethical challenges also emerge with the use of this methodology (Singh & Rose, 2009). Meanwhile, without independent biological markers for identifying disorders, psychiatric disease classifications into phenotypic symptom clusters, employed for purposes of diagnostic identification, rely on clinical observation of patients (including first-person reports). This nosology has been the subject of unremitting challenge for more than forty years (recent critiques include: Kendler & Zachar, 2008; Hyman, 2010; Cuthbert & Insel, 2013). Despite reliability among clinical observers, and the partial – although modest and debated – success of these classifications as predictive tools, the validity of symptom-defined psychiatric conditions remains unresolved, bringing a range of concerns over the science underlying psychiatric classifications.

Without known, anchoring aetiology, observed phenotypic trait clusters may result from diverse causes, for example, or the same cause may be responsible for ostensibly unrelated symptoms (Cuthbert & Insel, 2013; Hyman, 2010). Differentiating disorder entities from more normal states risks arbitrariness and false positives, bringing the charge of unwarranted medicalisation (Horwitz & Wakefield, 2007, 2012; Summerfield, 2012). Traits deemed pathological, it has been pointed out, often embody racial and gender norms incompatible with both scientific objectivity and principles of social equality (Bluhm, Jacobson, & Maibom, 2012; Clark & Horton, 2019). The assumption that observed disorder clusters comply with the natural kinds of disorders found in some other parts of medicine leaves unexplained the widespread comorbidities between the disorder categories currently employed (Whittchen & Jacobi, 2005). Whether categorical or dimensional models better fit psychiatric disorders remains similarly unresolved, with implications for research, clinical treatment, and public mental health (Cratsley, 2019; Zachar, 2014).

The incomplete medical science associated with mental disorder has been introduced here to illustrate special challenges arising when mental health is approached as public health. The ongoing debates over classificatory schemes are particularly important because to allow for the effective timing and targeting of interventions, preventive programs need workable definitions of disorder that incorporate aetiology (Nestadt, Seymour, Potash,

& McHugh, 2019). When dealing with adolescents and young adults to whose developmental trajectories such programs must be tailored, this demand is magnified.[20] However, the incomplete empirical picture may not be a decisive ground for limiting current preventive initiatives, we must add. Of the precautionary approach that underlies (primary) preventive measures, it has been observed: 'Where there are significant risks of damage to the public health, we should be prepared to take action to diminish those risks, *even when the scientific knowledge is not conclusive*, if the balance of costs and benefits justifies it' (Weed, 2004, p. 316; emphasis added). Depending on the risk, weakening the rules of evidence may be ethically required.[21]

4. Conclusion

The issues surveyed here are far from exhaustive, but they do suggest that the relatively new field of public mental health has a wide-range of ethical challenges to consider. Our intent here was not to attempt to resolve the many potential problems, or to advance specific proposals. Rather, we wanted to map the terrain in order to establish the theoretical background to research and debate that is rapidly developing in this part of the health sector. That said, several lessons are fairly clear at this point. As in public health more generally, the deliberative processes behind the design and implementation of mental health policy and programming must directly engage the question of which health-related behaviours are ethically appropriate targets of intervention. This includes consideration of the inevitable trade-offs between social utility and individual self-determination. Increased recognition of the social determinants of mental disorder only compounds this particular challenge.

An important part of any work in this area concerns the role of individual, social, and cultural values. The person-centred nature of mental health, along with the fact that it is arguably more social embedded than other aspects of health, means that the impact of coercion and stigma will be highly variable. Different forms of pressure, including the manipulation of

[20] As part of the broader problem of service scarcity, there is a lack of mental health services for adolescents and young adults, particularly for depression (Davey & McGorry, 2019).

[21] Causal inference represents a special challenge for public mental health. For example, the randomised controlled trial (RTC) methodology cannot be an appropriate measure for preventive interventions using multi-component programs in dynamic community settings (Saxena et al., 2006).

information, are likely to be experienced in different ways by different people in different contexts. Likewise, the study of stigma has proven it to be a highly complex phenomenon, and so devising effective responses demands innovative interventions. Further research is necessary. Not only is there a dearth of applicable data in most areas of public mental health, but, as we've seen, the field of mental health is still beset by a host of conceptual and methodological problems. More needs to be done to provide accounts of both the relevant causal mechanisms and the effectiveness of various preventive interventions.

Acknowledgements

We would like to thank Judy Illes and Ulrich Koch for comments on earlier versions of this chapter, as well as Matt Kessler-Cleary for general research support. Neither author declares any conflicts of interest.

References

American Psychiatric Association. (2013). *The principles of medical ethics with annotations especially applicable to psychiatry*.

Anderson, S. (2011). Coercion. In E. Zalta (Ed.), *The Stanford encyclopedia of philosophy*. http://plato.stanford.edu/archives/win2011/entries/coercion/.

Andrews, A., Sanderson, K., Corry, J., & Lapsley, H. M. (2000). Using epidemiological data to model efficiency in reducing the burden of depression. *Journal of Mental Health Policy and Economics, 3*, 175–186.

Avitzour, D., Barnea, R., Avitzour, E., Cohen, H., & Nissan-Rozen, I. (2019). Nudging in the clinic: The ethical implications of differences in doctors' and patients' point of view. *Journal of Medical Ethics, 45*, 183–189.

Beaglehole, R., & Bonita, R. (2008). Global mental health: A scorecard. *The Lancet, 372*, 1988–1996.

Bell, R., & Marmot, M. (2018). Social inequalities and mental health. In D. Bhugra, K. Bhui, S. Y. S. Wong, & S. E. Gilman (Eds.), *Oxford textbook of public mental health*. Oxford: Oxford University Press.

Bluhm, R., Jacobson, A. J., & Maibom, H. (Eds.). (2012). *Neurofeminism: Issues at the intersection of feminist theory and cognitive science*. London: Palgrave Macmillan.

Blumenthal-Barby, J. S., McCullough, L. B., Krieger, H., & Coverdale, J. (2013). Methods of influencing the decisions of psychiatric patients: An ethical analysis. *Harvard Review of Psychiatry, 21*, 275–279.

Bolton, D. (2008). *What is mental disorder? An essay in philosophy, science and values*. Oxford: Oxford University Press.

Boorse, C. (2011). Concepts of health and disease. *Philosophy of Medicine, 16*, 13–64.

Borsboom, D. (2017). A network theory of mental disorders. *World Psychiatry, 16*, 5–13.

van den Bosch, M., & Meyer-Lindenberg, A. (2019). Environmental exposures and depression: Biological mechanisms and epidemiological evidence. *Annual Review of Public Health, 40*, 18.1–18.21.

Broome, M., Fusar-Poli, P., & Wuyts, P. (2013). Conceptual and ethical issues in the prodromal phase of psychosis. In K. W. M. Fulford, M. Davies, G. Graham, J. Sadler, G. Stanghellini, & T. Thornton (Eds.), *Oxford handbook of philosophy and psychiatry*. Oxford: Oxford University Press.

Cascardi, M., Poythress, N. G., & Hall, A. (2000). Procedural justice in the context of civil commitment: An analogue study. *Behavioral Sciences and the Law, 18*, 731–740.

Childress, J. F., Faden, R. R., Gaare, R. D., Gostin, L. O., Kahn, J., Bonnie, R. J., … Kass, N. E. (2002). Public health ethics: Mapping the terrain. *Journal of Law, Medicine & Ethics, 30*, 170–178.

Clark, J., & Horton, R. (2019). What is *the Lancet* doing about gender and diversity? *The Lancet, 393*, 508–510.

Cole, E. R. (2009). Intersectionality and research in psychology. *American Psychologist, 64*, 170–180.

Conly, S. (2012). *Against autonomy: Justifying coercive paternalism*. Cambridge: Cambridge University Press.

Conrad, P., & Slodden, C. (2013). The medicalization of mental disorder. In C. S. Aneshensel, J. C. Phelan, & A. Bierman (Eds.), *Handbook of the sociology of mental health*. New York: Springer.

Cooper, S. (2016). Global mental health and its critics: Moving beyond the impasse. *Critical Public Health, 26*, 355–358.

Corrigan, P. (2016a). Resolving mental illness stigma: Should we seek normalcy or solidarity? *British Journal of Psychiatry, 208*(4), 314–315.

Corrigan, P. (2016b). Lessons learned from unintended consequences about erasing the stigma of mental illness. *World Psychiatry, 15*, 67–73.

Corrigan, P. (2018). Beware the educational fix: Limitations of efforts to promote mental health literacy. *Psychiatric Services, 69*(4), 469–471.

Corrigan, P., Rusch, N., & Scior, K. (2018). Adapting disclosure to reduce the stigma of mental illness. *Psychiatric Services, 69*, 826–828.

Corrigan, P., & Wassel, A. (2008). Understanding and influencing the stigma of mental illness. *Journal of Psychosocial Nursing and Mental Health Services, 46*, 42–48.

Cratsley, K. (2015). Nudges and coercion: Conceptual, empirical, and normative considerations. *Monash Bioethics Review, 33*, 210–218.

Cratsley, K. (2019). The ethical and empirical status of dimensional diagnosis: Implications for public mental health? *Neuroethics, 12*(2), 183–199.

Cratsley, K. (2019). The ethics of coercion and other forms of influence. In R. Bluhm, & S. Tekin (Eds.), *The bloomsbury companion to philosophy of psychiatry*. London: Bloomsbury Publishing.

Cuthbert, B. N., & Insel, T. (2013). Toward prevision medicine in psychiatry: The NIMH research domain criteria project. In D. Charney, et al. (Eds.), *Neurobiology of mental illness* (4th ed.). Oxford: Oxford University Press.

Daniels, N. (2008). *Just health: Meeting health needs fairly*. New York: Cambridge University Press.

Davey, C. G., & McGorry, P. D. (2019). Early intervention for depression in young people: A blind spot in mental health care. *Lancet Psychiatry, 6*(3), 267–272.

Dawes, R., van de Kragt, A., & Orbell, J. M. (1990). Cooperation for the benefit of us – not me, or my conscience. In J. J. Mansbridge (Ed.), *Beyond self-interest* (pp. 97–110). Chicago: University of Chicago Press.

Dawson, J., & Szmukler, G. (2006). Fusion of mental health and incapacity legislation. *British Journal of Psychiatry, 188*, 504–509.

Dias, A., Azariah, F., Anderson, S. J., Sequeira, M., Cohen, A., Morse, J. Q., … Reynolds, C. F. (2019). Effect of a lay counsellor intervention on prevention of major depression in older adults living in low- and middle-income countries: A randomized clinical trial. *JAMA Psychiatry, 76*(1), 13–20.

Dworkin, G. (1972). Paternalism. *The Monist, 56*, 64–84.

Dworkin, G. (2013). Review of 'Against autonomy: Justifying coercive paternalism', by S. Conly. *Notre Dame Philosophical Reviews*. https://ndpr.nd.edu/news/against-autonomy-justifying coercive-paternalism/.

Evers, C., Marchiori, D. R., Junghans, A. F., Cremers, J., & De Ridder, D. T. D. (2018). Citizen approval of nudging interventions promoting health eating: The role of intrusiveness and trustiness. *BMC Public Health, 18*, 1182.
Faden, R. R., & Beauchamp, T. L. (1986). *A history and theory of informed consent*. Oxford: Oxford University Press.
Faden, R., & Shebaya, S. (2015). Public health ethics. In N. Zalta (Ed.), *The Stanford encyclopedia of philosophy*. https://plato.stanford.edu/entries/publichealth-ethics/.
Fernando, S. (2014). Globalization of psychiatry – a barrier to mental health development. *International Review of Psychiatry, 26*, 551–557.
Fitzpatrick, S. (2018). Reshaping the ethics of suicide prevention: Responsibility, inequality and action on the social determinants of suicide. *Public Health Ethics, 11*(2), 179–190.
Foucault, M. (1980). In C. Gordon (Ed.), *Power/Knowledge: Selected interviews and other writings 1972–1977*. New York: Pantheon Books.
Foucault, M. (1988). Technologies of the self. In M. Luther, H. Gutman, & P. Hutton (Eds.), *Technologies of the self: A seminar with Michael Foucault*. Amherst: University of Massachusetts Press.
Fulford, K. W. M. (2011). The value of evidence and evidence of values: Bringing together values-based and evidence-based practice in policy and service development in mental health. *Journal of Evaluation in Clinical Practice, 17*(2), 341–343.
Gergel, T. (2014). Too similar, too different: The paradoxical dualism of psychiatric stigma. *Psychiatric Bulletin, 38*, 148–151.
Glas, G. (2019). *Person centered care in psychiatry: Self-reflective, contextual and normative perspectives*. London: Routledge.
Goffman, E. (1963). *Stigma: Notes on the management of spoiled identity*. Upper Saddle River, NJ: Prentice Hall.
Goldman, A. H. (1980). *The moral foundations of professional ethics*. Towata: Rowman and Littlefield.
Gostin, L. O. (2013). Bloomberg's health legacy: Urban innovator or meddling nanny? *Hastings Center Report, 43*(5), 19–25.
Hannaikainen, I., Cabral, G., Machery, E., & Struchiner, N. (2017). A deterministic worldview promotes approval of state paternalism. *Journal of Experimental Social Psychology, 70*, 251–259.
Haslam, N., & Kvaale, E. P. (2015). Biogenetic explanations of mental disorder: The mixed-blessings model. *Current Directions in Psychological Science, 24*(5), 399–404.
Hegerl, U., & Wittenburg, L. (2009). Focus on mental health care reforms in Europe: The European alliance against depression: A multilevel approach to the prevention of suicidal behavior. *Psychiatric Services, 60*(5), 596–599.
Heim, E., Kohrt, B. A., Koschorke, M., Milenova, M., & Thornicroft, G. (2018). Reducing mental health-related stigma in primary health care settings in low- and middle-income countries: A systematic review. *Epidemiology and Psychiatric Sciences*. http://dx.doi.org/10.1017/S2045796018000458. Advance online publication.
Horwitz, A., & Wakefield, J. (2007). *The loss of sadness: How psychiatry transformed normal sorrow and depressive disorder*. New York: Oxford University Press.
Horwitz, A., & Wakefield, J. (2012). *All we have to fear: Psychiatry's transformation of natural anxieties into mental disorders*. New York: Oxford University Press.
Hotzy, F., & Jaeger, M. (2016). Clinical relevance of informal coercion in psychiatric treatment – a systematic review. *Frontiers in Psychiatry, 7*, 197. http://dx.doi.org/10.3389/fpsyt.2016.00197.
Hyman, S. E. (2010). The diagnosis of mental disorders: The problem of reification. *Annual Review of Clinical Psychology, 6*, 155–179.
Jacka, F. N., Mykletun, A., & Berk, M. (2012). Moving towards a population health approach to the primary prevention of common mental disorders. *BMC Medicine, 10*, 149.

Jacobson, P. D. (2014). Changing the culture of health: One public health mishap at a time. *Society, 51*, 221–228.

Kahneman, D. (2011). *Thinking fast, thinking slow*. New York: Penguin.

Kallert, T. W., Mezzich, J. E., & Monahan, J. (Eds.). (2011). *Coercive treatment in psychiatry: Clinical, legal and ethical aspects*. West Sussex: Wiley.

Kass, N. E. (2001). An ethics framework for public health. *American Journal of Public Health, 91*(11), 1776–1782.

Kendler, K. S., & Zachar, P. (2008). The incredible insecurity of psychiatric nosology. In K. S. Kendler, & J. Parnas (Eds.), *Philosophical issues in psychiatry: Explanation, phenomenology, and nosology*. Baltimore: Johns Hopkins University Press.

Kendler, K. S., Zachar, P., & Craver, C. (2011). What kinds of things are psychiatric disorders? *Psychological Medicine, 41*, 1143–1150.

Kessler, R. C., Duncan, G. J., Gennetian, L. A., Katz, L. G., Kling, J. R., Sampson, N. A., … Ludwig, J. (2014). Associations of housing mobility interventions for children in high-poverty neighborhoods with subsequent mental disorders during adolescence. *JAMA, 311*(9), 937–948.

Kinderman, P., Schwannauer, M., Pontin, E., & Tai, S. (2013). Psychological processes mediate the impact of familial risk, social circumstances and life events in mental health. *PLoS One, 8*, 10. http://dx.doi.org/10.1371/journal.pone.0076564.

Kingma, E. (2014). Naturalism about health and disease: Adding nuance for progress. *Journal of Medicine and Philosophy, 39*(6l), 590–608.

Kirmayer, L., Lemelson, R., & Cummings, C. (Eds.). (2016). *Re-Visioning psychiatry: Cultural phenomenology, critical neuroscience, and global mental health*. Cambridge: Cambridge University Press.

Kirmayer, L., & Pedersen, D. (2014). Toward a new architecture for global mental health. *Transcultural Psychiatry, 5*, 759–776.

Kitchener, B. A., & Jorm, A. F. (2008). Mental health first aid: An international programme for early intervention. *Early Intervention in Psychiatry, 2*, 55–61.

Kleinman, A., & Benson, P. (2006). Anthropology in the clinic: The problem of cultural competency and how to fix it. *PLoS Med, 3*, e294.

Kong, C., & Singh, I. (2018). The ethics of global psychiatric genomics: Multilayered challenges to integrating genomics in global mental health and disability. *American Journal of Medical Genetics*. http://dx.doi.org/10.1002/ajmg.b.32697. Advance online publication.

Kosyluk, K., Schmidt, A., Abelson, S., Malmon, A., & Corrigan, P. (2015). Campus solidarity campaign: Respect and support for students with mental illness. *Mental Health and Social Inclusion, 19*, 180–190.

Link, B., & Phelan, J. (2001). Conceptualizing stigma. *Annual Review Sociology, 27*, 363–385.

Lund, C., Brooke-Sumner, C., Baingana, F., Baron, E. C., Breuer, E., Chandra, P., … Saxena, S. (2018). Social determinants of mental disorders and the sustainable development goals: A systematic review of reviews. *Lancet Psychiatry, 5*, 357–369.

Lund, C., DeSilva, M., Plagerson, S., Cooper, S., Chisholm, D., Das, J., … Patel, V. (2011). Poverty and mental disorders: Breaking the cycle in low-income and middle-income countries. *Lancet, 378*, 1502–1514.

Marmot, M. (2007). Achieving health equity: From root causes to fair outcomes. *Lancet, 370*, 1153–1163.

Marmot, M. (2010). *Fair society, healthy lives – The Marmot review*. Strategic Review of Health Inequalities in England Post-2010 http://www.instituteofhealthequity.org/resources-reports/fair-society-healthy-lives-the-marmot-review.

McLachlan, A. (2014). Preventative therapeutics: A study of risk and prevention in Australian mental health. *Culture Unbound, 6*, 815–837.

Mendelson, T., Pas, E. T., Lee-Winn, A. E., Bradshaw, C. P., Rebok, G. W., Leis, J. A., & Mandell, W. (2019). The logic and practice of the prevention of mental disorders. In W. W. Eaton, & M. D. Fallin (Eds.), *Public mental health*. New York: Oxford University Press.

Mendenhall, E., Kohrt, B., Norris, S. A., Ndetei, D., & Prabhakaran, D. (2017). Non-communicable diseases syndemics: Poverty, depression, and diabetes among low-income populations. *Lancet, 389*, 951–963.

Mill, J. S. (1859/1977). On liberty. In J.S. Mill. In J. M. Robson (Ed.), *Collected works of John Stuart Mill*. Toronto: University of Toronto Press.

Mills, C. (2014). *Decolonizing global mental health: The psychiatrization of the majority world*. London: Routledge.

Molodynski, A., Rugkasa, J., & Burns, T. (Eds.). (2016). *Coercion in community mental health care: International perspectives*. Oxford: Oxford University Press.

Mrazek, P. J., & Haggerty, R. J. (Eds.). (1994). *Reducing risks for mental disorders: Frontiers of preventive intervention research*. Washington DC: National Academy Press.

Murray, S., Kass, N., Mendelson, T., & Bass, J. (2016). The ethics of mental health survey research in low-and middle-income countries. *Global Mental Health, 3*, e12.

Naidoo, J., & Wills, J. (2000). *Health promotion, foundations for practice*. London: Elsevier.

National Academy of Sciences, Engineering, and Medicine. (2016). *Ending discrimination against people with mental and substance use disorders: The evidence for stigma change*.

National Research Council & Institute of Medicine. (2009). *Preventing mental, emotional, and behavioral disorders among young people*. Washington, DC: National Academies Press.

Nestadt, P. S., Seymour, K. E., Potash, J. B., & McHugh, P. R. (2019). Psychiatric nosology: Approaches to the characterization of mental disorders. In W. E. Eaton, & M. D. Fallin (Eds.), *Public mental health* (2nd ed.). New York: Oxford University Press.

Nestadt, P. S., Triplett, P., Mojtabai, R., & Berman, A. (2018). Universal screening may not prevent suicide. *General Hospital Psychiatry*. http://dx.doi.org/10.1016/j.genhosp-psych.2018.06.006. Advance online publication.

Newton-Howes, G., & Mullen, R. (2011). Coercion in psychiatric care: Systematic review of correlates and themes. *Psychiatric Services, 62*, 465–470.

Ng, J., Sjostrand, M., & Eyal, N. (2019). Adding lithium to drinking water for suicide prevention – the ethics. *Public Health Ethics*. http://dx.doi.org/10.1093/phe/phz002. Advance online publication.

Nuffield Council on Bioethics. (2007). *Public health: Ethical issues*. London: Nuffield Council on Bioethics.

O'Donoghue, B., Roche, E., Shannon, S., Lyne, J., Madigan, K., & Feeney, L. (2014). Perceived coercion in voluntary hospital admission. *Psychiatry Research, 215*, 120–126.

O'Neill, O. (2002). *Autonomy and trust in bioethics*. Cambridge: Cambridge University Press.

Oexle, N., Ajdacic-Gross, V., Kilian, R., Muller, M., Rodgers, S., Xu, Z., ... Rusch, N. (2017). Mental illness stigma, secrecy and suicidal ideation. *Epidemiology and Psychiatric Sciences, 26*, 53–60.

Ormel, J., Cuijpers, P., Jorm, A., & Schoevers, R. (2019). Prevention of depression will only succeed when it is structurally embedded and targets big determinants. *World Psychiatry, 18*, 111–112.

Ortega, F. (2009). The cerebral subject and the challenge of neurodiversity. *Biosocieties, 4*(4), 425–445.

Patel, V., Burns, J. K., Dhingra, M., Tarver, L., Kohrt, B. A., & Lund, C. (2018). Income inequality and depression: A systematic review and meta-analysis of the association and a scoping review of mechanisms. *World Psychiatry, 17*(1), 76–89.

Pescosolido, B. A., Martin, J. K., Lang, A., & Olafsdottir, S. M. J. (2008). Rethinking theoretical approaches to stigma: A framework integrating normative influences on stigma (FINIS). *Social Sciences & Medicine, 67*(3), 431–440.

Petersen, I., Barry, M., Lund, C., & Bhana, A. (2013). Mental health promotion and the prevention of mental disorders. In V. Patel, H. Mina, A. Cohen, & M. Prince (Eds.), *Global mental health: Principles and practice*. Oxford: Oxford University Press.
Petersen, A., & Lupton, D. (1996). *The new public health: Health and self in the age of risk*. Thousand Oaks, CA: Sage Publications.
Powers, M., Faden, R., & Saghai, Y. (2012). Liberty, Mill and the framework of public health ethics. *Public Health Ethics, 5*(1), 6–15.
Preda, A., & Voigt, K. (2015). The social determinants of health: Why should we care? *American Journal of Bioethics, 15*(3), 25–36.
Quigley, M. (2014). Are health nudges coercive? *Monash Bioethics Review, 32*, 141–158.
Radden, J. (2015). Unique ethical challenges for psychiatric practice. In J. Sadler, W. Van Staden, & K. W. M. Fulford (Eds.), *Oxford handbook of psychiatric ethics*. Oxford: Oxford University Press.
Radden, J. (2016). Mental health, public health and depression, a bioethical perspective. *Ethics, Medicine and Public Health, 2*, 197–204.
Radden, J. (2018). Public mental health and prevention. *Public Health Ethics, 11*(2), 126–138.
Rose, N. (1998). Governing risky individuals: The role of psychiatry in new regimes of control. *Psychiatry, Philosophy and Law, 5*, 177–195.
Sampson, L., & Galea, S. (2018). An argument for the foundations of population mental health. *Frontiers in Psychiatry, 9*, 600.
Saxena, S., Jane-Llopis, E., & Hosman, C. (2006). Prevention of mental and behavioural disorders: Implications for policy and practice. *World Psychiatry, 5*(1), 5–14.
Schnyder, N., Panczak, R., Groth, N., & Schultze-Lutter, F. (2017). Association between mental health-related stigma and active help-seeking: Systematic review and meta-analysis. *British Journal of Psychiatry, 210*, 261–268.
Scholten, M., & Gather, J. (2018). Adverse consequences of Article 12 of the UN Convention on the Rights of Persons with Disabilities for persons with mental disabilities and an alternative way forward. *Journal of Medical Ethics, 44*, 226–233.
Schwartz, P. (2008). Risk and disease. *Perspectives in Biology and Medicine, 51*, 320–334.
Singh, I., & Rose, N. (2009). Biomarkers in psychiatry. *Nature, 460*, 202–207.
Stuart, H., Chen, S.-P., & Christie, R. (2014). Opening minds in Canada: Targeting change. *Canadian Journal of Psychiatry, 59*, S13–S18.
Summerfield, D. (2012). Against "global mental health". *Transcultural Psychiatry, 49*, 519–530.
Szmukler, G. (2015). Compulsion and "coercion" in mental health care. *World Psychiatry, 14*, 259–261.
Szmukler, G., & Appelbaum, P. (2008). Treatment pressures, leverage, coercion and compulsion in mental health care. *Journal of Mental Health, 17*, 233–244.
Tekin, S. (2019). *The missing self in scientific psychiatry* (pp. 2197–2215). *Synthese*.
Thornicroft, G. (2006). *Shunned: Discrimination against people with mental illness*. Oxford: Oxford University Press.
Thornicroft, G., Mehta, N., Clement, S., Evans-Lacko, S., Doherty, M., Rose, … Henderson, C. (2016). Evidence for effective interventions to reduce mental-health-related stigma and discrimination. *Lancet, 387*, 1123–1132.
Tol, W. A. (2015). Stemming the tide: Promoting mental health and preventing mental disorders in low- and middle-income countries. *Global Mental Health, 2*, 1–10.
Tulchinsky, T. H., & Varavikova, E. A. (2010). What is the "new public health"? *Public Health Reviews, 32*(1), 25–53.
United Nations. (2006). *Convention on the rights of persons with Disabilities*.
United Nations. (2015). *Transforming our world: The 2030 agenda for sustainable development*.
Varga, S. (2015). *Naturalism, interpretation, and mental disorder*. Oxford: Oxford University Press.

Vilhelmsson, A., Svensson, T., & Meeuwisse, A. (2011). Mental ill health, public health and medicalization. *Public Health Ethics, 4*, 207–217.
van Voren, R., & Keukens, R. (2015). Political abuse in psychiatry. In J. Sadler, W. Van Staden, & K. W. M. Fulford (Eds.), *Oxford handbook of psychiatric ethics*. Oxford: Oxford University Press.
Wade, D. T., & Halligan, P. (2004). Do biomedical models of illness make for good healthcare systems? *BMJ, 329*, 1398–1401.
Wahlbeck, K. (2015). Public mental health: The time is ripe for translation of evidence into practice. *World Psychiatry, 14*(1), 36–42.
Wainberg, M. L., Helpman, L., Duarte, C. S., Vermund, S. H., Mootz, J. J., Gouveia, L., ... Cournos, F. (2018). Curtailing the communicability of psychiatric disorders. *Lancet Psychiatry, 5*, 940–944.
Wakefield, J. (1997). Diagnosing DSM-IV – Part I and the concept of disorder. *Behavioral Research and Therapy, 35*, 633–649.
Waring, D. (2016). *The healing virtues: Character ethics in psychotherapy*. Oxford: Oxford University Press.
Weed, D. (2004). Precaution, prevention, and public health ethics. *Journal of Medicine and Philosophy, 29*, 313–332.
Werner-Seidler, A., Perry, Y., Calear, A. L., Newby, J. M., & Christensen, H. (2017). School-based depression and anxiety prevention programs for young people: A systematic review and meta-analysis. *Clinical Psychology Review, 51*, 30–47.
Wester, G. (2018). When are health inequalities unfair? *Public Health Ethics, 11*(3), 346–355.
Whittchen, H.-U., & Jacobi, F. (2005). Size and burden of mental disorders in Europe: A review and appraisal of 27 studies. *European Psychopharmacology, 15*, 357–376.
Wilson, J. (2009). Justice and the social determinants of health: An overview. *Public Health Ethics, 2*(3), 210–213.
Wilson, J. (2011). Why it's time to stop worrying about paternalism in health policy. *Public Health Ethics, 4*(3), 269–279.
Wilson, J. (2016). The right to public health. *Journal of Medical Ethics, 42*, 367–375.
World Health Organization. (1946). *Constitution of the World Health Organization.*
World Health Organization. (2008). *Closing the gap in a generation: Health equity through action on the social determinants of health*. Final Report of the Commission on Social Determinants of Health.
World Health Organization. (2013). *Comprehensive Mental Health Action Plan 2013–2020.*
World Health Organization. (2014b). *Preventing suicide: A global imperative.*
World Health Organization. (2017). *Depression and other common mental disorders: Global health estimates.*
World Health Organization. (2018). *Time to deliver: Report of the WHO Independent High-Level Commission on Noncommunicable diseases.*
World Health Organization and Calouste Gulbenkian Foundation. (2014a). *Social determinants of mental health.*
Yamin, A. E., & Constantin, A. (2018). The evolution of applying human rights frameworks to health. In B. M. Meier, & L. O. Gostin (Eds.), *Human rights in global health: Rights-based governance for a globalizing world*. Oxford: Oxford University Press.
Zachar, P. (2014). *A metaphysics of psychopathology*. Cambridge, MA: MIT Press.
Zalsman, G., Hawton, K., Wasserman, D., van Heeringen, K., Arensman, E., Sarchiapone, M., ... Zohar, J. (2016). Suicide prevention strategies revisited: 10-year systematic review. *Lancet Psychiatry, 3*, 646–659.
van Zoonen, K., Buntrock, C., Ebert, D. D., Smit, F., Reynolds, C. F., Beekman, A. T. F., & Cuijpers, P. (2014). Preventing the onset of major depressive disorder: A meta-analytic review of psychological interventions. *International Journal of Epidemiology, 43*, 318–329.

CHAPTER THREE

Is coercion ever beneficent? Public health ethics in early intervention and prevention for mental health

Alex McKeown*, Rose Mortimer, Arianna Manzini and Ilina Singh
Department of Psychiatry and Wellcome Centre for Ethics & Humanities, University of Oxford, Oxford, United Kingdom
*Corresponding author: E-mail: alexander.mckeown@psych.ox.ac.uk

Contents

1. Introduction

Half of all people who develop mental disorders experience their first symptoms by the age of 14%, and 75% have had their first symptom by their mid-20s (Department of Health, 2017; Khan, Parsonage, & Stubbs, 2015). According to the latest published national study on young people's mental health in UK, 1 out of 10 children aged 5–16 has a diagnosable problem (Mental Health Taskforce, 2016). Yet most of them, 60%–70%, do not receive appropriate and timely interventions (Department of Health, 2015). It follows that society should take child mental health very seriously

Developments in Neuroethics and Bioethics, Volume 2
ISSN 2589-2959
https://doi.org/10.1016/bs.dnb.2019.04.004

and should consider strategies for prevention and early intervention. In this context, prevention involves identifying individuals at risk of developing a disorder and taking action before the condition becomes manifest. Early intervention includes both early detection of disorders, and the use of psychosocial and pharmacological interventions for the phase-specific treatment of the earlier stages of illness (McGorry & Pelosi, 2008).

We will begin by showing that within UK government policy, the case for prevention and early intervention in mental health relies on the existence of a virtuous circle between societal and individual interests, wherein the good of the individual and the good of society are perfectly aligned. We call for critical scrutiny of this alleged virtuous circle, by raising the concern that young people may be coerced to undergo preventative interventions for the good of society, and in the process their individual goals and values may be undermined, with their capacity for authentic agency unjustly constrained.

We will focus on two disorders commonly diagnosed in childhood, attention deficit hyperactivity disorder (ADHD) and conduct disorder (CD), to illustrate how issues of coercion may arise in proposals for managing challenging behaviour towards the realisation of the apparently virtuous circle of individual and collective goods. To do this we will use a theoretical examination of the concept of coercion and apply it to policy proposals for management of these conditions. We will conclude by drawing on insights from critiques of the public health strategy of nudging to show how the moral status of proposals for managing the conditions may differ depending on the socio-economic positioning of those to whom the proposals are made.

2. Early intervention: a moral and an economic case

Early intervention and prevention are increasingly promoted by governments both in the UK and abroad (Mortimer, McKeown, & Singh, 2018). In advocating for early intervention, typically two arguments are proposed - we call these 'the moral case' and 'the economic case'.

The moral case turns on the duty of the state to promote and safeguard the wellbeing of minors. The claim is that poor mental health adversely affects young people's emotional and social development as well as their academic and job performance, both now and in later life. It is argued that the state should intervene to prevent this, not least because young people are vulnerable and cannot always protect themselves. Monitoring young people's mental health is needed not just by reference to individual well-being but as a requirement of justice, since *'children with a persistent mental health problems face unequal chances in life'* (Department of Health, 2017, p. 6), or fewer opportunities than their peers. Thus, it appears that UK policy provides a twofold justification to make the moral case, focussing both on the best interests of individual children, and on social justice.

The moral case is supported by an economic argument that takes into account the broader societal costs of mental health issues starting in early life. This is generally discussed in terms of cost to the taxpayer, and policy documents often focus specifically on the overall lifetime economic costs of behavioural problems, which, if sufficiently severe and persistent, lead to the diagnosis of mental health conditions such as ADHD or CD (Department of Health, 2015). It is estimated that ADHD costs society £100,000 per case; two-thirds of the cost takes the form of additional public expenditure on education and health care, with the remainder being reflected in reduced earnings (Khong, 2014). Costs may also be incurred by the criminal justice system, since individuals with conditions like CD often become involved in crime and antisocial behaviour. This means that young people who develop CD at a young age cost the public sector ten times more by the age of 28 than other children (Scott, Knapp, Henderson, & Maughan, 2001).

The putative virtuous circle asserted between the moral case and the economic case suggests that individuals' interests are necessarily aligned with broader public health and economic interests (Department of Health, 2017). If individual mental health and societal interests *are* seen to coincide, the moral case and the economic case can be used to emphasize the importance of early intervention and prevention of mental issues in minors:

> *Early intervention avoids young people falling into crisis and avoids expensive and longer term interventions in adulthood. There is a compelling moral, social and economic case for change [...] Prevention and early intervention are not only desirable but cost-effective*
>
> **(Department of Health, 2015, p. 23).**

> *The growing interest in early intervention as a policy issue reflects the widespread recognition that it is better to identify problems early and intervene effectively to prevent their escalation than to respond only when the difficulty has become so acute as to demand action. It is better for the individuals concerned, their families and society more broadly; it avoids a lot of personal suffering, reduces social problems and generally, it costs less than remedial action*
>
> **(Davies & Bromley-Derry, 2010, p. 4).**

However, a persistent feature of public health discourse is the recognition that individual and societal interests often conflict. Since public health seeks to achieve gains at the population level, in making utility calculations about what will produce the greatest net or aggregate benefit it is sometimes necessary to sacrifice individuals' interests. In other words, given that wishes, plans, desires, goals, and values are plural, some will inevitably conflict and so strategies deemed beneficial at the societal level will not always align with what particular persons consider good for them *as individuals*. Public health strategies are primarily justified by their utility, so the interests of a minority can be subordinated if doing so yields aggregate benefits at the population level. Therefore, despite the apparent confluence often posited between individual and societal interests with respect to behaviour deemed challenging or in need of intervention, we argue that we should not unreflectively assume that such interests will be mutually consistent.

In light of these considerations, we contend that there are two main reasons to be concerned about policy guidelines that use the kind of language and argument described and demonstrated above. First, they appear unarguable because few would deny that the state has a responsibility to promote child welfare and protect minors from harm where possible. The mental health of minors is a rightly emotive and important issue for society. Nevertheless, arguments for early intervention often involve the translation of complex scientific data and ethical arguments into policy rhetoric that eschews ambiguity (Mortimer et al., 2018) and we should thus be prepared to critically assess policy rhetoric that appears unarguable.

Second, the attempt to demonstrate individual and societal interests as mutually complementary raises the question of in whose interests interventions are designed. We are concerned that the rhetoric of support for people in need may mask the promotion of practices or interventions that seek to shape individuals to display certain traits and behavioural characteristics deemed convenient for society as a whole but according to an excessively idealistic view of the good citizen. In this way there is a risk that interventions will take the form of surveillance, targeting individuals deemed risky to treat behaviour that the state considers deviant. This is concerning not only if the traits and behaviours of the good citizen are imposed via interventions that are themselves ethically controversial - for example, if treatment is likely to stigmatise a given individual (Moldavsky & Sayal, 2013).

Further, interventions are also ethically concerning if the traits and behaviours that the intervention seeks to promote do not align with the kind of character that some young people themselves endeavour to cultivate.

Or conversely, if the defiant traits or those considered deviant that intervention seeks to change are in fact valued by the individual as part of her identity and useful in her life (Potter, 2016). Of course, in saying this we do not claim that young people should be allowed to pursue *any* life they choose; it is right, for example, that the state should intervene to prevent individuals from cultivating cruelty and aggression in order to pursue a life of crime. Rather, we argue that the vision of the good citizen inherent in state early intervention should not be so narrow as to limit the ambitions of individuals who might pursue visions of the good life that are not defined by economic prosperity, academic success, or raising a family, for example.

It is important to note that we do not raise these concerns to advance a straightforward libertarian case for the state to stay out of individuals' private lives and choices; indeed, we accept and endorse the need to promote, support, and improve the mental health of minors. However, it is important to scrutinize the arguments upon which preventative measures are based, since such arguments carry the risk of endorsing policies that are unjustly coercive, and which may ultimately harm those minors they purport to protect.

Before exploring the nature of coercion in more detail, we will outline the two conditions we are considering - ADHD and CD. We introduce each condition before considering some of the forms that early intervention and prevention can take in the context of these two disorders.

3. The two conditions

ADHD and CD are among *'the most common mental health conditions affecting children and adolescents'* (Khan et al., 2015, p. 4). Studies suggest that the worldwide prevalence of childhood ADHD ranges between 3% and 7%; ADHD is more commonly diagnosed in boys than in girls, with a male-to-female ratio of 3:1 in population-based studies and between 5:1 to 9:1 in clinical samples (Skogli, Teicher, Andersen, Hovik, & Øie, 2013). The proportion of children with CD increases with age (Khan et al., 2015), and — like ADHD — they are more common in boys than girls. It is estimated that 7% of boys and 3% of girls aged 5—10 years are diagnosed CD; amongst children aged 11—16 years, the proportion rises to 8% of boys and 5% of girls (NHS, 2013).

ADHD has an estimated comorbidity rate of approximately 16%—20% with CD, although some consider ADHD to be a developmental precursor to CD as well as to Oppositional Defiant Disorder (Beauchaine, Hinshaw, &

Pang, 2010; Biederman et al., 2008; Villodas, Pfiffner, & McBurnett, 2012); in this sense, prevention of one disorder may also prevent the other.

Given the apparent prevalence of ADHD and CD in the UK and the serious implications that these conditions have for individuals' wellbeing and life opportunities (Khan et al., 2015; Khong, 2014), many have advocated early identification of at-risk children and families as a method of prevention, as well as timely support for individuals exhibiting symptoms in the early stages of illness. This advocacy spans policy makers, academic researchers, and voices from the third sector. At the same time, the status of ADHD in particular as a genuine psychiatric disorder is widely contested (Rafalovich, 2004). Both ADHD and CD exist on a spectrum from mild to severe behaviours. In the context of ADHD, critics argue that too many children are diagnosed and treated for behaviours at the relatively milder end, due to contextual factors such as inadequacy of educational resources; biases against children who are poor and/or ethnic minorities; performance pressures on children in high-performing schools, and so forth; however these same factors can also contribute to under-identification and under-treatment in various contexts (Merten, Cwik, Margraf, & Schneider, 2017; Singh, Filipe, Bard, Bergey, & Baker, 2013). Due to these and other complexities related to identification of ADHD and CD, epidemiological statistics we and others cite should be viewed with a critical eye (Singh, 2018).

3.1 Attention Deficit Hyperactivity Disorder

ADHD is diagnosed when an individual displays both inattentive and hyperactive behaviour, and when this interferes with normal functioning in multiple domains of life (i.e. at home, at school, during social activities). DSM-V states that several symptoms must have been present before the child reached age 12 (APA, 2013). Examples of hyperactive behaviour include: fidgeting, failure to remain seated when sitting down is required, and an inability to wait one's turn. Examples of inattentive behaviour include: inability to focus on a task, often failing to check one's work for mistakes, being easily distracted and regularly forgetting or losing items necessary for daily activities.

3.2 Conduct Disorder

CD is diagnosed when an individual displays a persistent and repetitive pattern of behaviour that *'violates the rights of others'* and which *'breaches age-appropriate social norms'* (American Psychiatric Association [APA], 2013,

p. 1). In practice, this often leads to breaking of school or home rules, or law breaking and criminal activity. CD involves a range of anti-social behaviours varying in seriousness, from school truancy and defiant behaviour to arson and sexual assault.[1] When diagnosing CD, DSM-V states that one must specify whether the individual also displays *'limited prosocial emotions'*; in order to fulfil this criteria, the individual must have displayed two of the following over the past 12 months across multiple environments: lack of remorse or guilt; lack of empathy; lack of concern about one's own performance; shallow or deficient effect.

4. What forms do early intervention and prevention take?

Similar preventative interventions are recommended in the case of ADHD and CD. Some focus on changing the environment around the child, by targeting the parents of children who are at risk of developing the conditions. A notable example is the Family Nurse Partnership in England, which provides first-time young parents with regular visits from early pregnancy to when the child is two. The aim is to identify mothers experiencing social disadvantage, to support them to have a healthy pregnancy and to promote their children's development. The virtuous circle discussed above can be seen in the vision of Family Nurse Partnership, as stated on its website, that *'every baby, child and young parent can thrive, fulfil their aspirations and contribute to society* (Family Nurse Partnership, 2018), clearly implying a correlation between good health and good citizenship. Here, 'health' refers both to physical and mental health, and the Family Nurse Partnership explicitly seeks to reduce crime and antisocial behaviour by minimising risk factors that may contribute to the development of mental health conditions like ADHD and CD in young people (Olds, 2007).

Some intervention programmes focus on parents whose child has already been diagnosed with ADHD or CD, or who is beginning to display difficult

[1] Here we will focus on the lower end of the spectrum, because these milder behaviours are more comparable to normal - albeit 'naughty' - childhood behaviour. In the second part of this chapter we consider the possibility that preventative action may coerce young people into accepting the values and ideals of society rather than allowing them to develop freely. While such an argument may be persuasive in the context of mild behaviours like inattentiveness or school truancy, it does not apply to cases of extremely violent or aggressive behaviour. It is not appropriate to argue that we should protect the rights of youth to commit violent crimes, since such behaviours clearly violate the rights of others in ways that are severely harmful and dangerous.

behaviour which might suggest he is at risk of developing these conditions.[2] In this case, parents may be taught discipline strategies, learning how to appropriately praise and reward good behaviour, as well as practicing techniques like distraction or time-out for bad behaviour (Jones, Daley, Hutchings, Bywater, & Eames, 2008; Sonuga-Barke, Daley, Thompson, Laver-Bradbury, & Weeks, 2001). Similar interventions may take place in schools, supporting teachers to adapt the classroom environment or learning outcomes so as to meet the specific needs of children with ADHD or CD (DuPaul & Stoner, 2014).

Finally, there are some pharmacological and psychosocial interventions that target children or young people themselves. Drug treatments for ADHD are common, and the most widely used is Methylphenidate, a stimulant drug that works by increasing activity in the brain, particularly in areas that play a part in controlling attention and behaviour (NHS, 2018). Indeed, a recent systematic review and meta-analysis of the relevant data has found methylphenidate to be the most effective drug treatment for ADHD in young people and recommends it for severe cases of ADHD in view of the significant therapeutic benefit that it offers to affected individuals (Cortese et al., 2018).

To date there is no licenced drug specifically for the treatment of CD, though some medications have demonstrated effectiveness, mainly belonging to three classes of psychotropic drugs: mood stabilizers, neuroleptics, and stimulant (Gérardin, Cohen, Mazet, & Flament, 2002). Psychosocial interventions that target the individual child include therapies such as Cognitive Behavioural Therapy or general counselling, to promote self-esteem or good peer relationships for children. Other psychosocial interventions seek to target the environment around the child, which may be the home or school environment; such interventions often focus on parent training (Khan et al., 2015; Scott, 2008). While none of these interventions purport to cure the disorder, they may help the individual and those around them to manage symptoms. In neither the case of ADHD nor CD are medications typically used as early interventions for minors deemed at risk who do not as yet display symptoms of either condition. It is arguable, however, that treatment of children with milder bad behaviour is preventative, in the sense that it might help curtail current behavioural difficulties that could

[2] Some of these programmes can rightly be considered 'early intervention', whereas others are more accurately termed preventative. The distinction between intervention and prevention is blurred in the context of conditions like ADHD, as we discuss further below.

result in more severe consequences for the child which could eventually trigger more severe mental health problems.

To conclude this point, it is important to remark that preventative action is particularly difficult in the context of mental health conditions like ADHD and CD because it is hard to determine on whom preventative or early interventions should be performed. This is a consequence of the diagnostic uncertainty that characterises psychiatry in general and conditions such as these in particular (Singh & Wessely, 2015). Unlike fields of medicine in which biological markers are available, psychiatric diagnosis is based on a behavioural assessment or on the observation of behavioural symptoms (Singh & Rose, 2009). The behavioural dimension of mental health conditions is one of the reasons why defining boundaries between pathology and 'normality' as well as between psychiatric disorders is more complicated, which often means that psychiatric diagnoses refer to a wide spectrum of presentation of symptoms (Hyman, 2010). As noted, ADHD and CD share some diagnostic criteria and, since these include behaviour on a continuum with normality, it can be difficult to distinguish, for example, 'normal' inattentiveness and ADHD (Parens & Johnston, 2009). By extension, therefore, it can be argued that medications are sometimes used to treat children who do not in fact have a disorder.[3]

These epidemiological, epistemic and nosological complexities explain why in most cases we currently lack objective biological measures to identify with any degree of certainty which at-risk children will actually develop a specific mental health disorder and to what degree, given that ADHD and CD are spectrum disorders. These pose challenges for clearly identifying those individuals that early interventions should target. Thus, it is clear that while early intervention and prevention can in principle offer an important opportunity to improve the wellbeing of children, including some of the most vulnerable and disadvantaged, the reality of correctly identifying and supporting those children who really do need help in the particular case of ADHD and CD can be complex and challenging. In light of these practical hurdles to effective intervention, the virtuous circle between

[3] Indeed, there are reasons to think that the effects of a drug such as methylphenidate is likely to be appealing to individuals without the condition given that it has been shown to be effective in improving concentration and attention and thus has gained popularity amongst students as a study aid or 'smart drug' for cognitive enhancement (Ragan, Bard, & Singh, 2013). A considerable literature exists on the ethics of stimulant enhancement, which falls beyond the purview of this chapter, and which we therefore do not cover here.

individual and societal interests begins to look more complex. In the next section, we will discuss whether, why, and in which cases this may be ethically problematic by critically analysing the ethics of early intervention through the lens of coercion.

5. Coercion

CD and ADHD are disorders with a strong normative dimension; their diagnostic criteria include behaviour and traits considered unruly at the lower end of the spectrum, and cruel or immoral at the more severe end. The state has an interest in preventing such behaviour, particularly when an individual may commit a criminal offence; this creates a context in which the state may support coercive policies to prevent unruly behaviour in the name of public health and wider societal goods. Irrespective of whether a proposal is or is not coercive, policies that endorse the curtailment of individual liberty for the greater good are powerful, since an argument for almost anything can be made once the basis of justification is utility-driven, as is typically the case in public health.

Given this background, coercion-relevant questions arise in the case of ADHD and mild CD. In this section, we address these concerns. First, whether the putative public health benefit[4] justifies coercive interventions in minors beyond the nominal claim that we ought to support the (future) interest of the developing individual where possible. Second, how the right to refuse intervention can be protected, given that however wise we think it might be for people to act in ways that conduce to each other's well being, we should not enforce it in all cases, as this would be unacceptably intrusive. Relatedly, we consider how to protect young people's capacity to develop their own values and goals, rather than assuming they will adopt the values and goals enumerated by the state and adhere to a predetermined account of good citizenship. Towards the end of the paper we briefly refer to the literature on nudging in public health, as a way of illustrating by analogy some of the ideas that we discuss here.

Before continuing, it is important to note that because early interventions target disorders that typically begin in childhood, our reflection on

[4] What counts as 'public health benefit' is itself contestable because of conflicting views about the boundaries of public health and which approach to justice we should employ in optimising it (Coggon, 2012; Verweij & Dawson, 2007). Space constraints forbid a fuller exposition of this here.

coercion will focus on children and young people. Children and young people may make a particularly challenging case in discussions about coercion. It is often assumed that minors lack decision-making capacities and that they often copy adults' opinions when asked to make their own choices; this is one of the reasons why, for instance, research on minors' views and experiences is often criticised as lacking validity and reliability (Singh, 2016). If this is the case, coercion may be more subtle, and hence more difficult to identify, when exercised on individuals who are not fully self-determinant, and may therefore lack the cognitive and emotional skills and the socio-economic and political means to express dissent. This implies that we may need to exercise more careful scrutiny to unveil hidden forms of childhood coercion.

However, assumptions about children's lack of competence have been challenged in the academic literature for some time, particularly since the publication of the United Nations Convention on the Rights of the Child in 1989, whose Article 12 states children's right to express their views in all matters concerning their life (United Nations, 1989). Minors do not constitute a homogenous group; on the contrary, childhood has a 'developmental nature' which means that while growing up children show evolving capacities (Nuffield Council on Bioethics, 2015). This means that it is hard to identify a cut-off point when autonomous decision-making can and ought to be attributed, and there is increasing recognition that some minors, especially the so-called 'mature minors', are capable of forming their own views, for instance regarding treatment choices (Dickens & Cook, 2005; Manzini & Vears, 2018). Therefore, it is important to question generalising views of all children as vulnerable, particularly since labels of vulnerability can be used to justify interventions that risk crossing the subtle distinction between care and control, thus legitimising paternalistic interventions that may have a disciplinary function (Brown, 2014).

There is insufficient space to explore these issues in more detail here, however, the above considerations should serve as a reminder of the importance of keeping in mind both young people's potential vulnerabilities and also their potential for capacity and autonomy, when considering the possibility of coercion in early intervention.

5.1 Is coercion always wrong?

Although its connotations are typically negative, it is arguable that coercion is not necessarily unethical in all cases. Insofar as coercion consists in the limiting of individual freedoms, there are instances in which coercion is *prima facie* permissible. Examples include: deprivation of individual liberty through quarantine to stop the spread of infectious diseases; or hypothetical future disaster scenarios, for instance uncontrollable climate change and resource depletion, that might justify efforts to protect the good of the population by curtailing personal liberty. Although these are relatively straightforward examples of justified coercion, it is fair to note that even they are more complicated than they appear, as well defended opposing analyses exist regarding coercion's normative status that we do not have space to attempt to settle here. For example, Ryan (1980) holds that since the wrongness of coercion is grounded in whether one has a right to resist whatever proposal is being made, and since rights are irreducibly normative insofar as they are necessarily comprised of talk of duties and obligations, so a non-normative account of coercion must be similarly inconsistent. By contrast, summarising the opposite position that coercion is not irreducibly normative, Anderson (2017) notes that *'if coercion is necessarily immoral action, then it is hard to explain how an act of coercion could count as justified'*. In short, Anderson's claim is that if coercion were *necessarily* immoral it could not be successfully defended.

Moreover, not only is the debate regarding the normative status of coercion persistent, but even relatively unproblematic instances of justified coercion may have stringent conditions of permissibility. For example, as Giublini et al. (2018) demonstrate, just because an instance of coercion such as quarantine may be justifiable, this justifiability turns significantly on the state meeting its reciprocal duties to ensure that people held in quarantine are kept in conditions under which their basic needs are met and in which the burden of quarantine is made as easy as possible to bear. This may sound like a stipulation that can be easily satisfied; however, given that instances of quarantine are often correlated with areas in the developing world where health and welfare resources are especially limited, it may in fact be extremely difficult for states to meet the reciprocal obligations which would ensure that the forced curtailment of liberty is satisfactorily compensated for.

In short, there are considerable theoretical and practical challenges inherent to determining the conditions under which coercion may or may not be justified. Nevertheless, and notwithstanding how formidable these challenges might be, if we conclude that there *are* circumstances in which the involuntarily curtailment of freedom is permissible, such instances suggest that the moral status of coercion is not uniform and thus cannot be assumed to be *a priori* morally objectionable without analysis of the details of a particular case.

With the preceding analysis in mind, an instructive example closely connected to the subject of this paper, namely the welfare of minors, is mandatory education. Parents are compelled by the state under pain of prosecution to have their children educated even if they do not wish to (Bostrom & Sandberg, 2009). The benefits for individual children and society are so great, however, that if a parent did not wish to have their children educated the state would be justified in coercing the parent to do so. To underline the point, although our nominal, pre-theoretical, assumptions about coercion are typically negative, the label may not in fact be sufficient for determining whether an intervention is or is not permissible. We therefore need to disentangle the reasons for this when applying it to our context. Wertheimer (1993, p. 239) distinguishes between two philosophical questions relating to coercion:

> ... *(a) what constitutes or counts as coercion? (b) what justifies the use of coercion? To exemplify, consider two questions about the use of kidneys for transplants. Whether poor people who sell their kidneys are coerced by their circumstances to do so (as many have argued) is a type (a) problem. By contrast, whether (as fewer have argued) the state is justified in coercively removing kidneys from healthy persons in order to save others is a type (b) problem*

In our context, whether children and adolescents are coerced to receive state-mandated support, treatment, or punishment (depending on how one defines the interventions proposed) in view of their behaviour is a type (a) question; and whether it is justifiable for the state to do so is a type (b) question. In answering type (a) questions, determining whether someone *is in fact* coerced may be answered by considering whether empirical, psychological, or moral considerations suggest that the person has been made worse off through violation of their baseline rights. Answering type (b) questions similarly requires an appeal to particular circumstances in judging whether or not the state's *reasons* for proposing the coercive act are acceptable.

On what Wertheimer (1993, p. 239) describes as an *'analytic'* (type a) view, minors (or their parents or guardians) would be coerced to receive interventions if the threat of not complying with them is such that it would leave them with no rational choice (for example in the sense that if someone has a gun to my head I only have a limited choice about whether or not to comply with their demands). On an alternative, (type b) *'moralised'* (Wertheimer, 1993, p. 239) view, however, minors would only be coerced to receive a given intervention if the state is wrong to propose it. On this view, if the state is right to make the proposal, the minors (or their parents or guardians) in question are not coerced *even if* they are left with little realistic choice but to comply.

5.2 What counts as coercion?

Let us consider the type (a) question first, that is, the analytic question of one's realistic freedom to choose between options in the face of a proposal by another. Coercion may take various different forms and have more or less oppressive consequences depending upon the circumstances in which the proposal is made. For example, a minor who displays behaviour deemed concerning is less likely to be at risk of coercive intervention from the state if he is growing up within affluent social circumstances (Viner et al., 2012; Wacquant, 2010). However, although such a young person is less at risk of coercion from the state, he may be at risk of coercion from his parents or peers, who could place pressure on him to access stimulant medication in order to enhance his academic performance, particularly during exam periods (DeSantis, Webb, & Noar, 2008; Hall, Irwin, Bowman, Frankenberger, & Jewitt, 2005). By contrast, a minor is likely to be at greater risk of coming to the attention of state authorities if their circumstances are adverse and they have less socioeconomic capital (Bottrell, 2009; Kemshall, 2008), even if they have not yet displayed what are taken to be symptoms of a disorder. Thus, while both children are at risk of coercion, the consequences of coercion may be more oppressive and severe for the second child, since he lacks the social capital that could give him access to more resources and options (NICE, 2014; Parkin, Long, & Bate, 2018), and because he is more likely to come to the attention of state authorities.

Conceptions of autonomy as relational are relevant here. According to such accounts the nature and extent of autonomy are dependent on the socially-determined choices that are available to an individual. On the relational view, all people are interdependent and exist in webs of relationships, and the individual need not be abstracted from this context before she can be said to be autonomous (Christman, 2014; Mackenzie, 2002, 2008). Relational autonomy thus contrasts with the traditional liberal account in which autonomy is understood in purely theoretical terms as an individual's right to choose abstracted from the contingencies in which a choice must be made. What the two theoretical examples remind us is that one's choices will be constrained or extended by the particular circumstances in which one finds oneself. As such, and unjustly, some will be more fortunate in their range of options than others. We are not saying here, therefore, that it is wrong for the state to intervene in cases where a minor needs support that they do not have. However, this highlights the structural differences which place some children at a disadvantage and at greater risk of state coercion for behaving no worse than other, more fortunate, children whose behaviour poses less of a risk to their liberty in view of their social advantage.

Having said this, one might still think that there are reasons to favour interventions of the kind we are considering. It is, after all, reasonable to want minors to grow up in ways that enable them to flourish and are not held back by their circumstances and it is similarly reasonable to desire a populous comprised of mutually supportive and socially-minded individuals. However, concerns arise when those who advocate a particular intervention *assume* the values, needs, and goals of the minor who is the intended recipient of the intervention. This can be demonstrated via the following quote from the UK Government 'Future in Mind' report:

> *All the professionals you meet should treat you as a whole person, considering your physical and mental health needs together … You are experts in your care and want to be involved in how mental health services are delivered and developed, not just to you and those who support you, but to all the children, young people and families in your area … If things aren't going well, the team providing your care will work with you to make changes to achieve your goals. You have the opportunity to shape the services you receive. That means listening to your experience of your care, how this fits with your life and how you would like services to work with you. It means giving you and those who care for you the opportunity to feedback and make suggestions about the way services are provided*
>
> ***(Department of Health, 2015, p. 11).***

Here the report adopts a language of agency and choice which implies that young people *really do* have the freedom to choose between options and *'shape the services [they] receive'*. However, the report assumes that the minors in question have and are aware of having the kinds of needs and goals that would lead them to want assistance and moreover some well-formed wider view of what is also best for other young people. There is no problem with this to the extent that the minors to whom support is proposed confirm that they have the same kinds of needs and goals envisaged by the architects of the report. However, it is important to consider what would be the case if the minors involved profess to have goals and needs that deviate from that vision and wish to shape their choices towards different ends. Given the breadth of the criteria for what might be considered as having ADHD or

CD, and given that many of the forms of behaviour covered by the diagnostic criteria may, if taken individually, be ultimately benign even if temporarily inconvenient to parents, teachers or other adults in positions of responsibility, the language used in the policy risks shrinking the potential space for young people to exercise their freedom and dissent from intervention.

5.3 Justified coercion?

Having considered whether a specific instance *constitutes* coercion, we will explore the type (b) question by considering whether coercion is ever *justifiable* in the context of early intervention for mental health. One reason to think that coercion is not justifiable is if we were to judge the risk of harm to the individual so great that it is not outweighed by potential benefits to society. This might occur if early intervention or prevention cause stigma by medicalising normal behaviour or exerting excessive control over an individual's autonomous choices such that the infringement of their liberty causes them harm. Indeed, criticism of the type (b) view turns on the claim that in allowing circumstances in which the normative force of coercion is neutralised, the moralised view is (Arnold, 2001, p. 54):

> *... unable to account for the* prima facie *wrongness of coercion. Coercion is* prima facie *harmful because it undermines individual freedom ... coercion is a 'vice word,' one that carries with it a negative normative judgement. This judgment is based on a strong moral presumption against the forced restriction of individual freedom*

As Arnold notes, it conflicts with our intuitions about coercion to consider that there are instances in which it is permissible: coercion does indeed often strike us as *prima facie* wrong. However, that it conflicts with our intuitions does not *prove* that it is wrong. One might respond to Arnold's claim by noting that it implies we should assume that our duties to society are necessarily subordinate to duties to individuals, but this would also be *prima facie* wrong. It is not obvious why we should assume that our duties to one outweigh our duties to the other without reference to the context in which a decision is to be made.

Nevertheless, given the centrality of autonomy to moral thought in general and contemporary moral thought in particular, and with respect to the risk that Arnold underlines, the view that coercion is a *'vice word'* has force. Consistent with concerns about the legitimacy of curtailing individual freedoms in the name of social order, Szasz (1958; 1976) and others have highlighted reasons to be wary of the fundamental validity of applying diagnostic labels to behaviour:

> *… his diagnosis, becomes a 'personal fact' in the child's life which follows him into every nook and cranny of our society and surfaces in countless questionnaires which ask: 'Have you ever seen a psychiatrist?' or 'Have you ever been treated for mental illness?..' (1976, p. 1012)*

Moreover, psychiatric labels can in themselves be coercive, insofar as they can be used to reinforce social norms. The history of psychiatry includes several examples of past injustices in cases such as homosexuality, masturbation, female hysteria, drapetomania, all of which have been labelled as psychiatric illnesses and used perniciously to incarcerate individuals, or to force them into conversion therapies. As Bentall and Pilgrim (1993, p. 71) point out, historical antecedents in which behaviour has been medicalised suggest that it is *'undoubtedly correct to point to the social control function of psychiatry'*.

The history of coercive use of psychiatric labels as well as the restrictive consequences of psychiatric labelling means that what we might call 'justified coercion' is a substantively relevant and important ethical dimension of early intervention in children's problem behaviours. Economic and normative arguments given in favour of intervening to curb certain kinds of behaviour under the label of one of these diagnoses and in the name of public health require scrutiny, as we have said, because of their rhetorical power. Moreover, once we use aggregate benefit as our unit of measurement, as is the case in public health, anything can be justified assuming its beneficiaries are sufficiently numerous.

Of course, we do not want to suggest that *all* behaviour, however unruly or antisocial, is permissible, since this would be counter-productive; nor do we suggest that minors may not benefit from interventions to curtail certain kinds of behaviour for their own good. However, if there are reasons to think that policy is developed with socio-political ends which require scrutiny and reflection, the preceding analysis highlights that *even if* we believe the state may in principle be justified in intervening, there are sufficient grounds to consider it potentially coercive enough in certain cases to require further attention.

Although arguments in favour of curbing certain kinds of behaviour have merit in view of the consequences for individuals and society, policies proposed for this purpose may also risk entrenching the positions of individuals who are already more likely to be in a position of social disadvantage. As we noted, given the looseness of the definitions of ADHD and CD, many children may be labelled with one and/or the other in ways that cause harm. In this respect the validity, reliability and utility of the labels is uncertain, and we should be wary of what they can be exploited to support.

6. Conclusion

We can bring the various strands together here and open our conclusion by drawing briefly on nudge literature. The nudge literature is extensive and constitutes a major body of research in its own right. As such, the analogy we draw here is for the purposes of underlining what is important in our key argument, rather than for placing any particular emphasis on the significance of nudging in the context we are considering.

Nudging exploits cognitive biases to persuade individuals to pursue certain courses of action rather than others, towards an outcome deemed preferable to the alternatives. They are frequently used in public health (as well as in marketing and behavioural economics) as a way to promote healthy lifestyle choices. For example, in shops fruit or healthy snacks rather than sweets may be placed closer to checkouts, on the basis that by the time they reach the checkout people are more likely to choose what is nearby and convenient to reach rather than to resume shopping and seek out less healthy options.

Although nudges may promote healthy choices in ways which yield public health benefits, it has been argued in the literature that the risk is that they do this by infantilising individuals or denying them the opportunity to arrive at fully informed choices by making mistakes and reflecting thereon (Baldwin, 2014; Schmidt, Voigt, & Wikler, 2010; Selinger & Whyte, 2011, 2012; Yeung, 2012).

It is of course fair to note that others have disputed whether nudges infringe autonomy in any meaningful or damaging way. Quigley (2014), for example, holds that nudges such as displaying food options in such a way that healthy ones are easier to select than less healthy cannot count as coercive because, irrespective of any pressure to make the healthier choice, *'The choice of food in the queue is not an "or else" proposition. The new food policy does not say "choose the healthy option or else you do not eat at all"'*. This observation is correct; however, it does not undermine the point that we are making. Of course it is still *possible* to make the unhealthy choice, but this has no bearing on those who *do* make the healthy choice because it is easier. We do not claim that minors should be permitted to do whatever they wish, however personally or interpersonally damaging, because of a belief that this is the only way that they will learn; this would be a caricature of the position. Rather, we contend that the question as to whether a decision can be properly informed and autonomy properly protected if the choice architecture is designed in such a way that certain choices, deemed unacceptable by the architects rather than the consumers, are less likely to be made, and subsequently reflected upon, for reasons of convenience alone.

We do not need to pursue this any further here and space forbids a fuller analysis of the nudge literature. Nevertheless, it is this general line of argument concerning the risks of infantilisation which illustrates what is important in the context that we have been considering in this chapter, since if one is not permitted the latitude to engage in behaviour which it turns out one would not repeat or will perhaps regret, how is one to *know*, in a sense that goes beyond the theoretical, that one should not do it? Indeed, according to some theories of moral development, it is practice and training of moral thinking and action that drive the realisation of the good person, who, by definition, enacts good behaviours (MacIntyre, 1985).

We suggest that the duty to protect the mental wellbeing of young people must therefore also incorporate the duty to ensure that they are able to develop authentic and fully realised agency as adults. In some cases, engaging in certain kinds of behaviour *is* indeed likely to result in adverse outcomes and impede their ability to realise their plans, and this is a risk to be taken seriously. Equally, however, if one goal of public health is to have a citizenry that is, for example, *both* responsible for themselves *and* tolerant of others - who, indeed, may have values and views with which one disagrees or of which one disapproves - then it is also important not to narrow unduly the range of options for minors to acquire these aspects of character while agency and moral capacity are developing.

With this in mind, in some cases the framing of mental health support for minors may risk shrinking their developmental options in ways that are counter-productive. Although certain kinds of behaviour *are* undesirable or inconvenient, it does not follow from this that we should seek their prevention in all cases. For the reasons outlined here, therefore, we contend that the vagueness of the criteria for ADHD and mild CD may be exploited in ways that are malignly, rather than beneficently, coercive. We therefore argue that taking seriously the legitimacy of right to refuse intervention should be protected.

Finally, although the early intervention approach which drives these policies claims to be against the medicalisation of mental health — because social solutions are applied before they become serious enough to warrant clinical intervention – the risks raised by Szasz and others are not neutralised. It is irrelevant whether the proposed care solutions are medical or social in nature (Horstkötter & Berghmans, 2014). With regard to surveillance, once a young person comes to the attention of the state displaying behaviour that it regards as a problem which carries a medical diagnosis, there is a utility-driven justification for mandating interventions denoted as support (Foresight Mental Capital and Wellbeing Project, 2008). We suggest that this justification may cause harm in ways that are hidden by the elegance of the solution in which individual and societal goals are apparently demonstrated as necessarily mutually consistent and harmonious. Rather, dissent should not automatically be assumed to be illegitimate, unreasonable, antithetical to the realisation of outcomes that are all-things-considered desirable, nor necessarily contrary to public mental health.

Given the utility-driven justificatory basis of arguments in public health, for the reasons outlined earlier they clearly offer scope for coercion, since anything is in principle justifiable if it maximises overall good in a statistical sense. Now, as we have argued, coercion is permissible in some circumstances even if we commit to the view that it is *prima facie* wrong. Many things are *prima facie* wrong but permissible depending on the circumstances – deliberate killing in self-defence, for example - and so it is with efforts needed to safeguard people from harm and ensure that they can develop in ways that are good for them and the social context in which they live. If it is the case that compelling people to make certain kinds of decisions does yield these outcomes then the answer to the initial question is that coercion can indeed sometimes be beneficent. However, caution is needed.

The notion of the ideal citizen who always acts in their own interests and those of society, as often advanced by policymakers, is seductive, but precisely because it is an ideal it is important to remember that it cannot be realised. It is of course fair to note here that interventions such as nudges are designed in the way that they are precisely *because* humans are imperfect reasoners and because, therefore, their cognitive biases can be exploited to direct behaviour in certain ways. However, our analysis of the relevant policy material suggests that what is recognised here by behavioural scientists and economists is easily lost in the move from theory to practice. The necessarily political ends of policy making do not always lend themselves to nuance, and consequently it is expedient to present an oversimplified picture in wich it appears obvious that the 'good' citizen is necessarily a person who acts in certain tightly sanctioned ways and, by implication, does not engage in behaviour contrary to it.

We contend that the idealised representations which contribute to the apparent unarguability of certain policy edicts may be counter-productive. All people are fallible and in the interests of protecting each other's well-being it is important that some latitude is given mutually, especially in the case of young people who are developing into adults, rather than automatically considering deviations from the ideal citizen's behaviour as either malign in nature or indicative of the presence of a serious mental health condition requiring correction. In these cases, the circumstantial facts about to whom a proposal to intervene is made and why, and what the likely effects of intervening are likely to be in view of those circumstances, can substantially influence the ethical landscape of individual cases in ways which demand close scrutiny to determine whether any coercion into treatment is or is not permissible.

Acknowledgements

Alex McKeown is supported by The Wellcome Centre for Ethics and Humanities, which is supported by core funding from the Wellcome Trust [203132/Z/16/Z]; and the Medical Research Council Mental Health Data Pathfinder Award [MC_PC_17215]. Rose Mortimer is supported by the Wellcome Trust [104825/Z/14/Z]; and The Wellcome Centre for Ethics and Humanities, which is supported by core funding from the Wellcome Trust [203132/Z/16/Z]. Arianna Manzini is supported by the Wellcome Trust [203329/Z/16/Z], [104825/Z/14/Z], and the Wellcome Centre for Ethics and Humanities, which is supported by core funding from the Wellcome Trust [203132/Z/16/Z]. Ilina Singh is supported by the Wellcome Trust [104825/Z/14/Z], and the Wellcome Centre for Ethics and Humanities, which is supported by core funding from the Wellcome Trust [203132/Z/16/Z]; and the NIHR Oxford Health Biomedical Research Centre [IS-BRC-1215-20005].

References

American Psychiatric Association. (2013). *Conduct disorder*. Retrieved August 15, 2018, from https://www.psychiatry.org/psychiatrists/practice/dsm/educational-resources/dsm-5-fact-sheets.

Anderson, Scott, & Coercion. (2017 Edition). The Stanford Encyclopedia of Philosophy. In Edward N. Zalta (Ed.), *Winter*. https://plato.stanford.edu/archives/win2017/entries/coercion/.

Arnold, D. G. (2001). Coercion and moral responsibility. *American Philosophical Quarterly, 38*(1), 53–67.

Baldwin, R. (2014). From regulation to behaviour change: Giving nudge the third degree. *The Modern Law Review*, 77(6), 831–857.

Beauchaine, T. P., Hinshaw, S. P., & Pang, K. L. (2010). Comorbidity of attention-deficit/hyperactivity disorder and early-onset conduct disorder: Biological, environmental, and developmental mechanisms. *Clinical Psychology: Science and Practice*, 17(4), 327–336.

Bentall, R. P., & Pilgrim, D. (1993). Thomas Szasz, crazy talk and the myth of mental illness. *Journal of Medical Psychology, 66*(1), 69–76.

Biederman, J., Petty, C. R., Dolan, C., Hughes, S., Mick, E., Monuteaux, M. C., & Faraone, S. V. (2008). The long-term longitudinal course of oppositional defiant disorder

and conduct disorder in ADHD boys: Findings from a controlled 10-year prospective longitudinal follow-up study. *Psychological Medicine, 38*(7), 1027–1036.
Bostrom, N., & Sandberg, A. (2009). Cognitive enhancement: Methods, ethics, regulatory challenges. *Science and Engineering Ethics, 15*(3), 311–341.
Bottrell, D. (2009). Dealing with disadvantage: Resilience and the social capital of young people's networks. *Youth and Society, 40*(4), 476–501.
Brown, K. (2014). Questioning the vulnerability zeitgeist: Care and control practices with 'vulnerable' young people. *Social Policy and Society, 13*(3), 371–387.
Christman, J. (2014). Relational autonomy and the social dynamics of paternalism. *Ethical Theory and Moral Practice, 17*(3), 369–382.
Coggon, J. (2012). *What makes health public?A critical evaluation of moral, legal and political claims in public health*. Cambridge: Cambridge University Press.
Cortese, S., Adamo, N., Del Giovane, C., Mohr-Jensen, C., Hayes, A. J., Carucci, S., … Hollis, C. (2018). Comparative efficacy and tolerability of medications for attention-deficit hyperactivity disorder in children, adolescents, and adults: A systematic review and network meta-analysis. *The Lancet Psychiatry, 5*(9), 727–738.
Davies, C., & Bromley-Derry, K. (2010). *Grasping the nettle: Early intervention for children, families and communities*. Centre for Excellence and Outcomes in Children and Young People's Services. Retrieved from https://www.bl.uk/collection-items/grasping-the-nettle-early-intervention-for-children-families-and-communities.
Department of Health. (2015). *Future in mind - promoting, protecting and improving children and young people's mental health and wellbeing*. London. Retrieved from https://assets.publishing.service.gov.uk/government/uploads/system/uploads/attachment_data/file/414024/Childrens_Mental_Health.pdf.
Department of Health. (2017). *A framework for mental health research*. London. Retrieved from https://www.gov.uk/government/publications/a-framework-for-mental-health-research.
DeSantis, A. D., Webb, E. M., & Noar, S. M. (2008). Illicit use of prescription ADHD medications on a college campus: A multi-methodological approach. *Journal of American College Health, 57*(3), 315–324.
Dickens, B. M., & Cook, R. J. (2005). Adolescents and consent to treatment. *International Journal of Gynaecology & Obstetrics, 89*(2), 179–184.
DuPaul, G. J., & Stoner, G. (2014). *ADHD in the schools: Assessment and intervention strategies* (3rd ed.). New York: Guilford Press.
Foresight Mental Capital and Wellbeing Project. (2008). *Final Project report*. London: The Government Office for Science.
Gérardin, P., Cohen, D., Mazet, P., & Flament, M. F. (2002). Drug treatment of conduct disorder in young people. *European Neuropsychopharmacology, 12*(5), 361–370.
Greely, H., Sahakian, B., Harris, J., Kessler, R. C., Gazzaniga, M., Campbell, P., & Farah, M. J. (2008). Towards responsible use of cognitive-enhancing drugs by the healthy. *Nature, 456*(7223), 702.
Hall, K. M., Irwin, M. M., Bowman, K. A., Frankenberger, W., & Jewitt, D. C. (2005). Illicit use of prescribed stimulant medication among college students. *Journal of American College Health, 53*(4), 167–174.
Horstkötter, D., & Berghmans, R. (2014). Early prevention of antisocial behavior (ASB): A comparative ethical analysis of psychosocial and biomedical approaches. *BioSocieties, 9*(1).
Hyman, S. E. (2010). The diagnosis of mental disorders: The problem of reification. *Annual Review of Clinical Psychology, 6*, 155–179.
Jones, K., Daley, D., Hutchings, J., Bywater, T., & Eames, C. (2008). Efficacy of the incredible years programme as an early intervention for children with conduct problems and ADHD: Long-term follow-up. *Child: Care, Health and Development, 34*(3), 380–390.
Kemshall, H. (2008). Risks, rights and justice: Understanding and responding to youth risk. *Youth Justice, 8*(1), 21–37.

Khan, L., Parsonage, M., & Stubbs, J. (2015). *Investing in children's mental health report.* London. Retrieved from https://www.centreformentalhealth.org.uk/investing-in-children-report.

Khong, B. (2014). *The lifetime costs of attention deficit hyperactivity disorder (ADHD).* London: Centre for Mental Health. Retrieved September 6, 2018, from https://www.centreformentalhealth.org.uk/publications/lifetime-costs-of-adhd.

MacIntyre, A. (1985). *After virtue: A study in moral theory.* London: Duckworth.

Mackenzie, C. (2002). Critical reflection, self-knowledge, and the emotions. *Philosophical Explorations, 5*(3), 186–206.

Mackenzie, C. (2008). Relational Autonomy, normative authority, and perfectionism. *Journal of Social Philosophy, 39*(4), 512–533.

Manzini, A., & Vears, D. F. (2018). Predictive psychiatric genetic testing in minors: An exploration of the non-medical benefits. *Journal of Bioethical Inquiry, 15*(1), 111–120.

McGorry, P., & Pelosi, A. (2008). Is early intervention in the major psychiatric disorders justified? *British Medical Journal, 338*, a695.

Mental Health Task Force. (2016). *The five year forward view for mental health.* London. Retrieved September 16, 2018, from https://www.england.nhs.uk/wp-content/uploads/2016/02/Mental-Health-Taskforce-FYFV-final.pdf.

Merten, E. C., Cwik, J. C., Margraf, J., & Schneider, S. (2017). Overdiagnosis of mental disorders in children and adolescents (in developed countries). *Child and Adolescent Psychiatry and Mental Health, 11*, 5.

Moldavsky, M., & Sayal, K. (2013). Knowledge and attitudes about attention-deficit/hyperactivity disorder (ADHD) and its treatment: The views of children, adolescents, parents, teachers and healthcare professionals. *Current Psychiatry Reports, 15*, 337–377.

Mortimer, R., McKeown, A., & Singh, I. (2018). Just policy? An ethical analysis of early intervention policy guidance. *The American Journal of Bioethics, 18*(11), 43–53.

NHS. (2013). *New guidelines on child antisocial behaviour.* Retrieved August 15, 2018, from https://www.nhs.uk/news/pregnancy-and-child/new-guidelines-on-child-antisocial-behaviour/#how-common-are-conduct-disorders.

NHS. (2018). *Attention-deficit hyperactivity disorder (ADHD).* Retrieved September 6, 2018, from https://www.nhs.uk/conditions/attention-deficit-hyperactivity-disorder-adhd/.

NICE. (2014). *NICE support for commissioning for antisocial behaviour and conduct disorders in children and young people.* London.

Nuffield Council on Bioethics. (2015). *Children and clinical research.* London, United Kingdom.

Olds, D. (2007). Preventing crime with prenatal and infancy support of parents: The nurse-family partnership. *Victims & Offenders, 2*(2), 205–225.

Parens, E., & Johnston, J. (2009). Facts, values, and attention-deficit hyperactivity disorder (ADHD): An update on the controversies. *Child and Adolescent Psychiatry and Mental Health, 3*(1).

Parkin, E., Long, R., & Bate, A. (2018). *Children and young people's mental health: CAMHS policy, services and education.* London. Retrieved from https://researchbriefings.parliament.uk/ResearchBriefing/Summary/CBP-7196.

Potter, N. (2016). *The virtue of defiance and psychiatric engagement.* Oxford: Oxford University Press.

Quigley, M. (2014). Are health nudges coercive? *Monash Bioethics Review, 32*(1–2), 141–158.

Rafalovich, A. (2004). *Framing ADHD children: A critical examination of the history, discourse, and everyday experience of attention deficit/hyperactivity disorder.* Lanham: Lexington Books.

Ragan, I., Bard, I., & Singh, I. (2013). What should we do about student use of cognitive enhancers? An analysis of current evidence. *Neuropharmacology, 64*, 588–595.

Ryan, C. C. (1980). The normative concept of coercion. *Mind, 89*(356), 481–498.

Schmidt, H., Voigt, K., & Wikler, D. (2010). Carrots, sticks, and health care reform—problems with wellness incentives. *New England Journal of Medicine, 362*(2), e3.

Scott, S. (2008). An update on interventions for conduct disorder. *Advances in Psychiatric Treatment, 14*(1), 61–70.

Scott, S., Knapp, M., Henderson, J., & Maughan, B. (2001). Financial cost of social exclusion: Follow up study of antisocial children into adulthood. *British Medical Journal, 323*(7306), 191.

Selinger, E., & Whyte, K. (2011). Is there a right way to nudge? The practice and ethics of choice architecture. *Sociology Compass, 5*(10), 923–935.

Selinger, E., & Whyte, K. (2012). Nudging cannot solve complex policy problems. *European Journal of Risk Regulation, 3*(1), 26–31.

Singh, I. (2018). *ADHD in the United Kingdom: Conduct, class, and stigma*. Global perspectives on ADHD: Social dimensions of diagnosis and treatment in sixteen countries: M. R. Bergey, A. M. Filipe, P. Conrad and I. Singh. Johns Hopkins University Press.

Singh, I. (2016). Evidence, epistemology and empirical Bioethics. In J. Ives, M. Dunn, & A. Cribb (Eds.), *Empirical bioethics: Theoretical and practical perspectives* (Vol. 37, pp. 67–83). Cambridge: Cambridge University Press.

Singh, I., Filipe, A. M., Bard, I., Bergey, M., & Baker, L. (2013). Globalization and cognitive enhancement: Emerging social and ethical challenges for ADHD clinicians. *Current Psychiatry Reports, 15*(9), 385.

Singh, I., & Rose, N. (2009). Biomarkers in psychiatry. *Nature, 460*, 202–207.

Singh, I., & Wessely, S. (2015). Childhood: A suitable case for treatment? *The Lancet Psychiatry, 2*(7), 661–666.

Skogli, E. W., Teicher, M. H., Andersen, P. N., Hovik, K. T., & Øie, M. (2013). ADHD in girls and boys - gender differences in co-existing symptoms and executive function measures. *BMC Psychiatry, 13*(298).

Sonuga-Barke, E. J. S., Daley, D., Thompson, M., Laver-Bradbury, C., & Weeks, A. (2001). Parent-based therapies for preschool attention-deficit/hyperactivity disorder: A randomized, controlled trial with a community sample. *Journal of the American Academy of Child & Adolescent Psychiatry, 40*(4), 402–408.

Szasz, T. S. (1958). Psychiatry, ethics, and the criminal law. *Columbia Law Review, 58*(2), 183–198.

Szasz, T. S. (1976). Child as involuntary mental patient: The threat of child therapy to the child's dignity, privacy, and self-esteem. *The San Diego Law Review, 14*, 1005.

The Family Nurse Partnership. (2018). *The Family Nurse Partnership: About us*. Retrieved September 6, 2018, from http://fnp.nhs.uk/about-us/.

United Nations, General Assembly. (1989). Convention on the rights of the child, 20 November 1989. *Annual Review of Population Law, 16*(95), 485–501.

Verweij, M., & Dawson, A. (2007). The meaning of "public" in "public health". In A. Dawson, & M. Verweij (Eds.), *Ethics, prevention and public health* (pp. 13–29). Oxford: Oxford University Press.

Villodas, M. T., Pfiffner, L. J., & McBurnett, K. (2012). Prevention of serious conduct problems in youth with attention deficit/hyperactivity disorder. *Expert Review Neurotherapeutics, 12*(10), 1253–1263.

Viner, R. M., Ozer, E. M., Denny, S., Marmot, M., Resnick, M., Fatusi, A., & Currie, C. (2012). Adolescence and the social determinants of health. *The Lancet, 379*(9826), 1641–1652.

Wacquant, L. (2010). Crafting the neoliberal state: Workfare, prisonfare, and social insecurity. *Sociological Forum, 25*(2), 197–220.

Wertheimer, A. (1993). A philosophical examination of coercion for mental health issues. *Behavioral Sciences & the Law, 11*(3), 239–258.

Yeung, K. (2012). Nudge as fudge. *The Modern Law Review, 75*(1), 122–148.

CHAPTER FOUR

Public mental health across cultures: the ethics of primary prevention of depression, focusing on the Dakhla Oasis of Egypt

Mohammed Abouelleil Rashed[a,b,*] and Rachel Bingham[c,d]

[a]Wellcome Trust ISSF Research Fellow, Department of Philosophy, Birkbeck College, University of London, United Kingdom

[b]Visiting Lecturer, Department of Philosophy, King's College London, London, United Kingdom

[c]Clinical Advisor, Medical Justice, London, United Kingdom

[d]General Practitioner, NHS, Islington, London, United Kingdom

*Corresponding author: E-mail: m.rashed@bbk.ac.uk

Contents

1. Introduction

For over a decade there has been an active and ambitious movement concerned with reducing the 'global burden' of mental disorders in low- and middle-income countries (Horton, 2007; Prince et al., 2007; Saxena, Thornicroft, Knapp, & Whiteford, 2007). Global Mental Health, as its proponents call it, aims to close the 'treatment gap', which is defined as

Developments in Neuroethics and Bioethics, Volume 2
ISSN 2589-2959
https://doi.org/10.1016/bs.dnb.2019.04.005

the percentage of individuals with serious mental disorders who do not receive any mental health care. According to one estimate, this amounts to 75%, rising in sub-Saharan Africa to 90% (Patel & Prince, 2010, p. 1976). In response to this, the movement recommends the 'scaling up' of services in these communities in order to develop effective care and treatment for those who are most in need. This recommendation, the movement states, is founded on two things: (1) a wealth of evidence that medications and psychosocial interventions can reduce the disability associated with mental disorder, and (2) closing the treatment gap restores the human rights of individuals, as described and recommended in the Convention on the Rights of Persons with Disabilities (Patel et al., 2011; Patel & Saxena, 2014).

In addition to its concern with *treatment*, the movement has identified *prevention* among the 'grand challenges' for mental and neurological disorders. It states, among its key goals, the need to identify the 'root causes, risk and protective factors' for mental disorders such as 'modifiable social and biological risk factors across the life course'. Using this knowledge, the goal is to 'advance prevention and implementation of early interventions' by supporting 'community environments that promote physical and mental well-being throughout life' and developing 'an evidence-based set of primary prevention interventions' (Collins et al., 2011, p. 29). Similar objectives have been raised several years before by the World Health Organisation, who identified evidence-based prevention of mental disorders as a 'public health priority' (WHO 2004, p. 15).

Soon after its inception, the movement of Global Mental Health met sustained and substantial critique.[1] Essentially, critics argue that psychiatry has significant problems in the very contexts where it originated and is not a success story that can be enthusiastically transported to the rest of the world (see Fernando, 2011; Mills & Davar, 2016; Summerfield, 2008, 2012, 2013; Whitley, 2015). The conceptual, scientific, and anthropological limitations of psychiatry are well known and critics appeal to them in making their case. Conceptually, psychiatry is unable to define mental disorder, with ongoing debates on the role of values versus facts in distinguishing disorder from its absence (e.g., Bolton, 2008, 2013; Boorse, 2011; Varga, 2015; Kingma, 2013). Scientifically, the lack of discrete biological causes,

[1] Most recently there was vocal opposition to a Global Ministerial Mental Health Summit that was held on the ninth and 10th of October 2018 in London. The National Survivor and User Network (U.K.) sent an open letter to the organisers of the summit, objecting to the premise, approach, and intention of Global Mental Health along the lines outlined in the main text.

or biomarkers, for major psychiatric conditions has resulted in the reliance on phenomenological and symptomatic classifications. This has led to difficulties in defining with precision the boundaries between disorders, and accusations that psychiatric categories lack validity (e.g., Horwitz & Wakefield, 2007; Kinderman et al., 2013; Timimi, 2014).[2] Anthropologically, while the categories themselves are associated with tangible and often severe distress and disability, they remain culturally constructed in that they reflect a 'Western' cultural psychology (including conceptions of the person and overall worldview) (e.g., Fabrega, 1989; Littlewood, 1990; and Rashed, 2013a).[3] Given this, critics see Global Mental Health as a top-down imposition of 'Western' norms of health and ideas of illness on the 'Global South', suppressing long-standing cultural ideas and healing practices that reflect entirely different worldviews. It obscures conditions of extreme poverty that exist throughout many non-Western countries, and which underpin the expressions of distress that Global Mental Health now wants to medicalise. On the whole, Global Mental Health, in the words of the critics, becomes a form of 'medical imperialism' (Summerfield, 2008, p. 992) that 'reproduces (neo)colonial power relationships' (Mills & Davar, 2016, p. 443).

We acknowledge the conceptual, scientific, and anthropological critiques of psychiatry and have written about them elsewhere (e.g., Bingham & Banner, 2014; Rashed, 2013b; Rashed & Bingham, 2014). At the same time we do not wish to speculate about and judge the intention of Global Mental Health, or whether it's a 'neo-colonial' enterprise that serves the interests of pharmaceutical companies. Our concern is to proceed at face-value by examining a particular kind of interaction: on one hand, we have scientifically grounded public mental health prevention campaigns that seek to reduce the incidence of mental disorders in low- and middle-income

[2] Often, the problem is framed by asking whether mental disorders are natural kinds, e.g., Jablensky (2016), Kendell and Jablensky (2003), Zachar (2015), and Simon (2011).

[3] The problems with utilising psychiatric categories across cultures for purposes of diagnosis, research, and treatment have been pointed out and discussed for several decades. Examples include the work of Kleinman (1987), Littlewood (1990), Hopper (1991, 2004), Kirmayer (2006), and Barrett (2004). This research operated at the intersection of Medical Anthropology and Cross-Cultural Psychiatry and had no illusions about the complexities involved. For example, before we can work with psychiatric categories across cultures, there is a need to establish if these categories have validity and that we are not committing a 'category fallacy'. Additionally, questions of translation loom large, for example concerning the meaning of the terms that make up the diagnostic criteria and whether it is possible to render them in different contexts. It is therefore unusual that the movement of Global Mental Health has chosen to bypass this important research and the complexities it definitively raises.

countries; on the other hand, we have the cultural contexts in these countries where there already are entirely different frameworks for categorising, understanding, treating, and preventing various forms of distress and disability. What sort of ethical principles ought to regulate this interaction, where prevention of mental disorders is at stake?

The meaning of prevention with which we are concerned in this chapter is primary, universal prevention, to be distinguished from mental health *promotion*, from *secondary* prevention, and from primary prevention that is of a *selective* or *indicated* nature. Primary prevention 'aims to reduce the onset of mental ill-health, thus reducing the incidence of mental disorders' — it is concerned with tackling the risk factors for mental disorders (Peterson, Barry, Lund, & Bhana, 2014, p. 3; see also; WHO 2004, p. 16). Secondary prevention, on the other hand, 'seeks to lower the rate of established cases of the disorder or illness in the population (prevalence) through early detection and treatment of diagnosable diseases' (Saxena, Jane-Llopis, & Hosman, 2006, p. 5; cf.; Radden, 2018, p. 127). In contrast to prevention, mental health promotion 'employs strategies for strengthening protective factors to enhance the social and emotional well-being and quality of life of the general population' (Peterson et al., 2014, p. 3). It is not directly concerned with risk factors for disorders but with positive mental health. With universal prevention the entire population is within view of the interventions, whereas with selective and indicated prevention, the target groups are, respectively, those 'whose risk for developing the mental health disorder is significantly higher than average' and those who have 'minimal but detectable signs or symptoms' (Evans, Foa, Gur, & Hendin, 2012, p. 5). While there is overlap among these various efforts, we focus on primary, universal prevention. Our decision to do so stems from the fact that such interventions, in being wholly anticipatory and population wide put marked, and perhaps even unique, ethical pressure on the encounter between the cultural context (and existing ideas on risk and prevention of distress and disability) and the biomedical public mental health approach.

It is helpful for ethical analysis to begin with a sufficiently detailed understanding of the contexts and interactions that are the subject of analysis. With these details at hand, what matters in a particular interaction is brought to light and the ethical issues become easier to grasp. Accordingly, we begin in Section 2 with an ethnographic account of the primary prevention of depression in the Dakhla Oasis from the perspective of the community. The Dakhla Oasis is a rural community in the Western desert of Egypt where there is no psychiatric presence, yet like most communities around the world there is no shortage of mental health related distress and disability.[4] It is a paradigmatic example of the kind of community where Global Mental Health would want to implement its campaigns. In Section 3 we move on to the perspective of a Public Health Team concerned with preventing depression in light of scientific and evidence-based risk factors and preventive strategies. Section 4 outlines the conflict between the perspective of the Team and that of the community. Given this conflict, Sections 5 and 6 discuss the ethical issues that arise in the case of two levels of intervention: family and social relationships, and individual interventions.

2. Primary prevention of depression: the perspective from Dakhla

From an emic (or insider) perspective, symptoms and signs that fall under the psychiatric category of depression cut across a range of local categories.[5] The main categories are spirit possession (*labs*) and mental illness (*marad nafsi*).[6] The term depression (*ekte'ab*) is known but is understood as a term-of-art that doctors use. Spirit possession and mental illness, as understood in Dakhla, are serious problems that can manifest in moderate to severe depression-like symptoms. Before describing these categories and their associated risk factors from the local perspective, we distinguish them from everyday expressions of distress.[7]

4 Extending west of the Nile-valley and occupying two-thirds of the land surface of Egypt, the Western Desert — otherwise an arid expanse of over 680,000 square kilometres — is dotted by six depressions: the oases of Siwa, Fayyoum, Bahariyya, Farafra, Dakhla, and Kharga. Beyond the Western political border of Egypt, it becomes the Libyan Desert, and shortly after it merges with the Sahara. With Cairo as a reference point, Fayyoum is the closest, largest and most populated of the six oases. Siwa lies only 50 km east of the Libyan border and, traditionally, has been the most isolated. The other four oases lie on an arc that starts at Cairo, curves West into the desert and 1330 km later returns to the Nile-valley at Luxor. Farafra lies at the Western-most point of that arc, followed by Dakhla.

5 Coined in linguistics to denote the relationship between units of language and linguistic systems, the terms emic and etic were adopted into cultural anthropology by Marvyn Harris in 1964 (Headland, Pike, & Harris, 1990, p. 2). Harris described statements created in a fashion regarded as appropriate by the actors themselves (emic) and statements which depend on distinctions 'judged appropriate by the community of scientific observers' (Harris, 1968, p. 575). The terms gradually took on their more general use in contemporary medical and psychological cross-cultural work to distinguish an insider (emic) from outsider (etic) perspective or understanding (Headland et al., 1990, p. 2).

6 In this section the transliteration of Arabic terms is provided in italics.

7 The ethnography reported on here was carried out by the first author during 2009 and 2010. The full ethnography can be read in Rashed (2012).

Unlike in psychiatry and its cultural milieu, psychological distress in Dakhla is not expressed in terms of a low mood – a 'phenomenological sinking downwards of the self' (Littlewood, 1990, p. 312) – but in terms of a constriction of the space available for the self. In the face of the inevitable troubles and obstacles of day-to-day life such as problems at work, a drying well, impending exams and, in the domestic sphere, the curtailment of freedom and movement that many women experience in the marital home, people may experience a change in mood. This is expressed in several idioms; the most common is *makhnu'* (suffocated) and, less commonly, *deiq* (to constrict or close-in). These idioms refer primarily to space and are accompanied by a desire to 'run away' and 'let in some air'. The advice in such cases is to reclaim one's space, literally, through a drive at dusk to the sand dunes adjacent to town or a trip to the nearest city, Asyut, while married women who are more constrained in what they can do seek respite from home through a visit to their families. All of this is considered a normal reaction to the difficulties of life and does not invite people in Dakhla to think about risk factors and prevention. With spirit possession and mental illness, the stakes are higher.

2.1 Spirit possession (*labs*)

Description: Spirits (known locally as *jinn*) are frequently invoked to account for an individual's mental states and behaviour. For example, mood changes, dysphoric states, and behaviours that are out of character or bizarre, are understood as manifestations of spirit possession. The effects of the spirit can be direct in the sense that the behaviour witnessed is literally the spirit's, or they can be indirect if the spirit influences the person by targeting the centres of volition. Behaviours understood in this way include aimless roaming, frequenting deserted places, isolating oneself, avoiding people, preoccupation with fire, inattention to appearance and grooming, inability to remain in one place for long, irritability, aggressive outbursts, talking to oneself, and praying, washing, and eating either excessively or not at all. In Dakhla, possession is frequently referred to as an illness and the possessed as a patient. In such cases possession is the primary problem and finds expression in a variety of symptoms and signs, some of which recall depression.

Causes: When spirit possession results in depression-like symptoms and signs, the proximal cause is the spirit's activity. But spirits and humans do not ordinarily interact, accordingly a further (distal) cause is required to explain why this person in particular has been targeted by the spirit. In Dakhla, a person can be targeted in several situations, two of which are most common: magic (*se'hr*) and inadvertent touch (*mas'*).

Magic: Suspicions of magic lurk at the background of many attempts to make sense of misfortune. It is thought that a neighbour with a grudge or an envious relative visits a magician who directs a spirit at the victim in order to affect their health and behaviour. The identity of the perpetrator could be revealed during an exorcism episode upon direct questioning by the healer. More commonly, the victim guesses who might have a reason to harm him or her. Usually, that would be a person whom they had slighted in the past and who now has a motive to retaliate. The same process applies to the presumed perpetrators who, should they encounter a problem, might suspect that the original party had blamed them for their misfortune and may harbour an intention to retaliate. This initiates a cycle of mal-intentions that operates implicitly and is fuelled by real or presumed interpersonal conflicts. These conflicts are understood to have initiated a chain of events directly connected to the persons' mood and behaviour.

Inadvertent Touch: Contrasting with the deliberate nature of magic, the inadvertent touch is a pure chance encounter. It involves a human trespassing onto areas where spirits reside and the unintentional infliction of harm upon them, say by falling or stepping on the spirits. Angered by the attack and unwilling to forgive, the spirit responds by possessing the person.

***Risk Factors*:** Whether the cause is magic or an inadvertent touch, weakness of faith renders the individual vulnerable to spirit possession. Faith is characterised as a protective sheath that envelops the individual. The metaphor here is thoroughly mechanical: lapses in religious observance form holes in the sheath thus allowing the spirit to enter. A person must constantly strengthen and repair his sheath through prayer and recitation of the Qur'an. With regards to inadvertent touch specifically, a risk factor is ignorance of the places and times where spirits are common. With regards to magic, a risk factor is social discord; to the extent that I quarrel with others and disrespect or harm them, I provide them with reasons to want to harm me back through magic.

2.2 Mental illness (*marad nafsi*)

***Description*:** The psychological and behavioural problems cited in Section 2.1 are sometimes understood outside the framework of spirit possession. The main category here is *marad nafsi* which translates as mental illness. The distinction between these two categories depends on various factors. For any person *suddenly* presenting with psychological and behavioural changes, the consensus will be that he is possessed. Additionally, possession is a contingent event, and it is expected that a healer or magician would be able to undo it. The longer the problems persist, the presence of a key and intensely difficult event in the person's life, and the consistent ineffectiveness of treatments offered by healers and magicians push the narrative surrounding the person from the domain of possession and magic to that of mental illness.

Causes: *Key shock:* A sudden loss or realisation that a state-of-affairs is contrary to what the person thought it was. For example, a man working abroad has been sending money to his fiancée in Egypt for safe-keeping; he returns home to find that she had married another man and spent all his money. A man regularly stored part of his income under a brick in the basement of his house; a water leak damaged his life-savings. In such cases, it is theorised that the experienced shock results in a rush of blood to the brain. The blood forms a 'drop' that settles on the brain, stimulating and damaging the nerves, and producing the problematic behaviours and experiences.

Spiritual Deviance: An alternative explanation for mental illness ties it to reduction in faith or religious practice. If you do not lead a truly religious life, your values, motives and desires are constituted by the givens and concerns of the material world. You end up fighting with others over the same things and exposing yourself to the petty disappointments and grievances that such fights inevitably lead to. Mental illness is the state of a person who leads this kind of life and who has become so absorbed in his failures to the extent of developing a psychological complex (*'oqda*).

Magic: Magic can be at the root of mental illness as a special category of spirit possession. The community recognises a particular kind of magic known as madness magic.

Risk Factors: With regards to key shock as a cause of mental illness, it is not possible to predict what the risk factors are: anyone can be affected at any time. With regards to madness magic, risk factors are similar to those for spirit possession. With regards to spiritual deviance as a cause of mental illness, the risk factor is living an irreligious life. The account here is grounded in Islamic cultural psychology where the self is understood to have three modes: the inciting self (the source of carnal passions and vices), the peaceful self (the fruit of the individual's moral and spiritual development), and the blaming self (a mediating means of reproach and correction) (see Rashed, 2015). When the person is distracted from worship, begins to have doubts about God, and stops living an ethical life in accordance with the Qur'an and the Sunna (the Traditions attributed to the Prophet

Mohammed), the inciting self takes hold. It is a selfish, egoistic self, absorbed in the world and prone to envy, bitterness, vanity, greed, suspicions, anxieties, and frustrations.

2.3 Primary prevention

Drawing on the aforementioned risk factors, healers in Dakhla suggest a range of preventive measures. These can be regarded as primary universal interventions concerned with reducing the incidence of spirit possession (*labs*) and mental illness (*marad nafsi*) (as they are understood in Dakhla).

Spirit possession: Weakness of faith is a central risk factor for all causes of spirit possession. Prevention consists in strengthening one's faith through prayer, reciting the Qur'an, visiting the mosque, regularly invoking God's name, and avoiding sinful activities. Other protective measures that can reduce the risk for possession and magic include drinking and bathing from holy water, lighting incense at home, and placing amulets with inscribed Qur'anic verses on one's clothing and around the house. A preventive intervention specific to the inadvertent touch is education about the nature of spirts, where they are found, and when to avoid these locations. A preventive intervention specific to magic draws on social discord as a risk factor and recommends measures that promote peaceful social relations thus minimising the occasions where people might want to harm each other.

Mental Illness: With regards to living an irreligious life as a risk factor for mental illness, prevention would consist in reversing this state of affairs. This could be through educational campaigns about, for example, the various Islamic virtues, how to cultivate them in everyday life, and how, through this, one might be able to elevate the self to the mode of the peaceful self.

So far we have described the risk factors for depression and their prevention in the Dakhla Oasis from an emic (or insider) perspective. This is given as an example of a perspective from inside one community that might be the recipient of an external intervention in the event of a roll-out of a public mental health programme in the region. We now turn to the model employed at the level of the Global Mental Health approach, before moving to consider the ethical issues that arise when these two perspectives meet.

3. Primary prevention of depression: the Global Mental Health approach

In their report on the prevention of mental disorders, the World Health Organization (2004) identify two kinds of determinants that have an evidence-based causal influence on the development of disorders: risk factors and protective factors. The former are 'associated with an increased probability of onset, greater severity and longer duration of major health problems', and the latter 'refer to conditions that improve people's resistance to risk factors and disorders' (2004, p. 20). Both kinds of factors, the report continues, 'can be individual, family-related, social, economic and environmental in nature' (World Health Organization (2004)). A related formulation is the 'socio-ecological framework', which identifies 'three broad nested and interactive levels of risk and protective factors'; the innermost level consists in intrapersonal factors (e.g., genetic make-up, cognitions, and skills), the middle level consists in personal factors (e.g., factors related to family, school, and community), and the outermost level consists in distal factors (e.g., economic and environmental policies Peterson et al., 2014, p. 9). Primary prevention of mental disorders must tackle this multitude of factors by countering risk factors and strengthening protective factors 'along the lifespan in order to disrupt those processes that contribute to human mental dysfunction' (Peterson et al., 2014, p. 9).

There are a range of known generic risk and protective factors. With regards to social, environmental and economic risk factors these include displacement, isolation and alienation, lack of housing and education, poverty, and war. Protective factors include social support and community networks, social tolerance, integration, and empowerment. With regards to individual and family-related risk factors these include child abuse, chronic pain and insomnia, substance use, family conflict, bereavement, and sensory disabilities. Protective factors include ability to face adversity, adaptability, exercise, literacy, positive early attachments, self-esteem, problem-solving and conflict managements skills (WHO, 2004, pp. 21, 23).

In relation to depression specifically, preventive interventions need to tackle 'depression-specific risk factors (e.g., parental depression, depressogenic cognitions) and generic risk factors (e.g., inadequate parenting, child abuse and neglect, stressful life events, bullying) and protective factors (e.g., sense of mastery, self-esteem, self-efficacy, stress resistance, social support)' (WHO, 2004, p. 39). Additionally, marital discord is associated with depression among women (ibid.). Given this, preventive interventions can range from seeking to reduce child abuse and extreme poverty to changing thought processes that can lead to depression, and strengthening skills for coping with stress. With regards to the latter two, Evans and colleagues (2012) report on a number of universal prevention programmes involving children, adolescents, and families. While they vary in some of the details, they all consist in brief (10–15 sessions) interventions based on cognitive behavioural therapy and interpersonal therapy principles. The aims of the programmes include reducing 'vulnerability to future depressive episodes', and preventing first episodes of depression through 'cognitive restructuring and problem-solving approaches' (2012, p. 13, 14).[8]

What are the possibilities for interaction between the public health perspective just outlined and the perspective of the community in the Dakhla Oasis?

4. Conflicts in primary prevention

Sections 2 and 3 describe two different visions for the primary prevention of depression. Each vision proceeds, as it must, with a certain view of the condition (for example, depressive disorder or spirit possession), its causes, and the risk factors relevant to these causes. From there, interventions are proposed that can play a role in preventing the condition. In Dakhla these interventions include activities to strengthen one's faith (such as regular prayer, recitation of the Qur'an, and wearing protective amulets), educational efforts concerning spirits and how to avoid them, social efforts to minimise discord (and therefore the risk of harmful magic), and religious efforts to educate people in the Islamic virtues and how to achieve the peaceful self. From a Global Mental Health perspective, interventions to reduce the incidence of depression must tackle the three dimensions identified earlier and which can be characterised as follows: macro interventions (e.g., to address poverty, unemployment, education, and housing), family and social relationships interventions (e.g., to address domestic abuse, restrictions of freedom, and social isolation), and individual interventions (e.g., to improve cognitive skills and abilities to cope with life-events).

Consider a meeting between representatives of the Dakhla community and a Public Health Team with a Global Mental Health perspective. The Team could be international or representing the Ministry of Health in Cairo. Their aim is to launch a project to reduce the incidence of depression and to improve mental health in the community. They have completed

[8] The jury is out on the effectiveness of such psychological interventions. Evans et al. (2012) point towards some evidence that they do reduce the incidence of depression, while a more recent Cochrane review is more circumspect (Hetrick, Cox, Witt, Bir, & Merry, 2016).

their needs assessment, identified a prevalence of depression, identified locally evidenced risk factors together with an ameliorative strategy. They have familiarised themselves with local context and practices as described in Section 2 of this chapter. The Public Health Team come to the table armed with expert knowledge on preventable risk factors and the theory and practice of planning a community level intervention. But as we have seen, there is disagreement between them and the community on all the relevant issues: description of the condition, causes, risk factors, and preventive strategy. How can the Team implement evidence-based interventions in an ethical manner, given the radical difference between their vision and the community's vision with regards to the primary prevention of depression? In what follows we examine this question, dividing our account into the ethical issues raised by two of the three levels of intervention identified earlier: family and social relationships, and individual interventions.[9]

5. Family and social relationships interventions

In many communities, social and family relationships can generate risk factors for mental health problems including depression. In Section 3, we identified family conflict, stressful life events, marital discord, social isolation and alienation as possible risk factors. In Dakhla, the constellation of beliefs, values, and practices surrounding marriage and gender roles generate ample opportunities for these risk factors. Marriage in Dakhla is a non-negotiable milestone that all men and women must achieve by their mid- to late-twenties at the latest. It is regarded as a 'completion of one's religion',

[9] Macro socio-economic and political factors such as poverty, unemployment, lack of access to education, poor quality or overcrowded housing, and conflict or war, are among the most impactful risk factors for depression. Indeed, as noted in the Introduction, the Global Health Movement has been criticised for emphasising the risk of mental health problems at the individual level, thereby facilitating the neglect of the social, economic, and political determinants of wellbeing. Despite their paramount importance, we do not consider macro interventions in this chapter for two reasons: First, due to the nature of such interventions, they lie outside the immediate expertise of a Public Health Team as they concern political and governmental action that has an impact beyond health. What the Team can and should do is advocacy for governmental action at the macro level; indeed, this may be an ethical imperative relating to principles of justice and equality. Second, from an ethical standpoint, interventions to improve access to healthcare and education, create jobs, and increase economic prosperity are ones over which we can expect significant consensus and, therefore, less opportunities for ethical disagreement. With regards to the latter point, however, we must add that there can be controversy over the details, for example in some contexts, education for girls is still not recognised.

'half of one's faith', and the 'natural progression of life". For the men and particularly the women of Dakhla, it is frequently their first and only sexual outlet. To be unmarried in your thirties is considered a disaster (*musiba*) and a blow to a person's worth and social standing. Further, divorce and second marriages are frowned upon and, accordingly, are rare. This lends to the choice of marriage partner the weight of finality, yet it is a choice that cannot be informed in many ways, due to restrictions on interactions between men and women before they get married.

These ideas and practices create significant stress for those who are within the ideal marriage age as well as for those who have passed that age. For the former, the early twenties become a battleground of sorts: on one hand, there is constant pressure from parents on their children to marry; on the other hand, these young men and women must make a more or less final decision under conditions of limited knowledge. The pressure is intensified by the fact that not getting married is a disaster, and that getting married to an unsuitable person can lead to a life of misery that might be hard to get out off. Those who do not marry, risk social isolation and alienation from the community. Given that marriage is a non-negotiable milestone, drifting into your thirties without getting married is not perceived as a possible choice, but as an occasion to inquire into the reasons behind the failure to fulfil this important milestone, much like a child failing to walk at the appropriate age invites speculation as to his constitution and health.[10] Within marriage itself, women face an additional range of stresses owing to rigid gender roles that restrict their life to the confines of the family home with very limited possibilities for interaction with anyone who is not immediately family. For example, married women in Dakhla are only expected to leave the house to visit their parents' home or for a very urgent matter. This loss of freedom creates opportunities for marital discord and plays a direct role in the experience of suffocation and constriction commonly reported by people in the community, and which, as discussed in Section 2, is the main idiom for expressing unhappiness. A public mental health campaign concerned with the prevention of depression may, therefore, concern itself with aspects of the community's ideas and practices surrounding gender roles and marriage that contribute to the risk factors.

[10] The unmarried person risks being seen as socially undesirable. People may reason that he or she was unable to marry due to a tarnished moral reputation or perceived mental health problems.

5.1 Communicating the message: the role of transparency

The first challenge the Public Health Team faces is how to communicate the idea that aspects of gender roles and expectations and practices around marriage are connected to risk factors for a mental health condition known as depression. This is especially complicated given that the community have an entirely different perspective over what the Team calls depression.

In general, information in public health campaigns ought to be communicated *transparently*. Transparency refers to the information provided and to the process of providing information. It is frequently cited in international public health as an important principle. For example, the World Health Organisation define two aspects of transparency in communication in the event of a disease outbreak. The first is the quality of communication, which 'should be factually accurate, timely and easily understood'; this includes information 'about actual or potential health risk, including behaviours [the community] should adopt to avoid disease' (WHO, 2008, p. 12). The second aspect relates to building trust with the public and other stakeholders. This requires transparency about the evidence used to inform policy and programme management decisions, and information about decision-making processes and outcomes.

The ethical principle underlying the importance of transparency is the necessity for communities to be able to give informed consent for the intervention. For example, in a document titled *Principles of the Ethical Practice of Public Health*, the American Public Health Association notes:

> *Public health institutions should provide communities with the information they have that is needed for decisions on policies or programs and should obtain the community's consent for their implementation.*
>
> ***(Public Health Leadership Society, 2002, p. 4).***

Accordingly, in its formulation in the public health ethics literature, transparency has been conceived as an equaliser, regulating the relations between parties and preventing coercion. In an exposition of the principles of public health ethics, Upshur (2002, p. 102) defines the transparency principle as follows:

> *This principle refers to the manner and context in which decisions are made. All legitimate stakeholders should be involved in the decision-making process, have equal input into deliberations, and the manner in which decision-making is made should be as clear and accountable as possible. As much as possible, the decision-making process should be free of political interference and coercion or the domination by specific interests.*

Alongside the ethical justification, there is also a pragmatic rationale for transparency, as the effectiveness of public health programmes and institutions relies on the trust of the public (Oliver & Lewis, 2009, p. 468). Moreover, provision of information is not only a requirement for ethical intervention but an intervention in itself; for some, it is the ideal means to behavioural change. Thus, advocates of a liberal framework for public health ethics argue for public education and information provision as 'non-coercive ways of bringing about improvements in health' (Nuffield Council on Bioethics, 2007, p. 22). This has been conceived as the least ethically problematic branch of health promotion: 'if the strategy is successful, it is because the individuals or groups found it useful or persuasive and chose to act on it' (Gardner, 2014, p. 31).

This apparently innocuous principle, however, is not without problems. During the Ebola outbreak of 2015 in Guinea, Liberia, and Sierra Leone, traditional funeral rites and burial practices, including washing of the dead, were identified as a factor in the spread of the Ebola virus. International teams provided transparent information alongside education and interventions such as mass cremation (see Maxman, 2015). These strategies raised both practical and ethical concerns. For example, a subsequent collection of informant narratives stressed that alienation of populations may have led to non-compliance with the recommended measures, which may have increased the risk of virus transmission (Pellecchia et al., 2015). In this case, information was necessary, but biomedical information alone – about how to avoid the disease – was often of limited relevance. People had other pressing priorities such as caring for the sick and burying the dead in accordance with longstanding rituals.

Additionally, the transparent provision of public health information – though necessary and appropriately motivated – can erode meaningful cultural practices. It can leave people in an uncertain place with regards to practices that provide for a shared community and identity. And it can alienate the community so that it refuses to engage with the interventions. This applies to a paradigmatically biomedical problem such as Ebola as well as to the prevention of depression. In relation to our case study, the Public Health Team's information relies on categories of distress and their scientifically established causes and risk factors that differ markedly from those of the community. The community are being told that they need to reconsider aspects of gender roles and marriage practices as a way of preventing some of the risk factors for depression, i.e. the community is being asked to change its culture in order to reduce the incidence of depression.

5.2 Bringing about change: health beliefs and culture

The Public Health Team seek change in those aspects of gender roles and marriage practices that constitute risk factors for depression. In order to do so they could design messages of possible alternatives and publicise these through advertising campaigns and through home visits by health workers and promoters. Family Planning campaigns in Egypt in the 1990s proceeded in similar terms. In a country where it was common for rural families to have up to ten children, and where a big family was highly prized, the key message was the health, economic, and social benefits that would befall families if they limited their progeny to two or three. Campaigns of this sort, we want to argue, aim to change the culture and not just health beliefs.

The expression health beliefs is self-explanatory. What is apposite for us here is the distinction between it and the broader cultural context. For the purpose of this discussion, we understand culture as

> *socially acquired and shared symbols, meanings and significances that structure experience, behaviour, interpretation and social interaction; culture 'orients people in their ways of feeling, thinking, and being in the world' (Jenkins & Barrett, 2004). (Rashed, Bingham, Poole, Sanati, & van Staden, 2018, p. 73)*

The frameworks of understanding distress and disability posited by the community in Dakhla and by the Global Mental Health movement are examples of different cultures in the sense we intend. Spirit possession is not an isolated set of beliefs about illness, but is connected to people's orientation to the world in domains such as personal identity, morality, agency, and cosmology. And so with the biopsychosocial, naturalistic, secular understanding of distress and disability advanced by Global Mental Health. Within each cultural framework, there are beliefs specifically about health. In Dakhla, for example, a person could be judged by a healer to have incorrect health beliefs about spirit possession; he might attribute to spirits maladies that they cannot cause, as determined through expert, local knowledge. And in England, for example, a person could be judged by medical opinion to have incorrect health beliefs about vaccinations, such as with the MMR/autism controversy. In either case, a public health campaign can aim to correct these beliefs (about spirits and about vaccinations) without having to change people's overall cultural orientation.

Other times, the attempt to change health beliefs can intentionally or unintentionally overstep the mark and become an attempt to change the culture. This is particularly clear with anti-obesity campaigns and partly explains the ethical controversies that surround them (e.g., Couch, Fried,

& Komesaroff, 2018; Fry, 2012). On the face of it, anti-obesity campaigns seek to change health beliefs relating to BMI and health. Those who object to the campaigns argue that they advance a particular bodily norm and justify it by appeal to health. In doing so, the objectors continue, these campaigns might impact the broader culture by stigmatising individuals who do not fit that norm. So, what started with an attempt to change health beliefs can, in some cases, end up changing, and having a negative effect upon, the culture.

In light of the distinction between health beliefs and culture, and returning to the case study under consideration in this chapter, does the Public Health Team want to change the health beliefs of the people of Dakhla or their culture? First, it does not make sense to talk about changing health beliefs outside of a shared cultural context, for health beliefs (whether correct or incorrect) arise and acquire their sense from that context. For example, in conversation with those who argued (wrongly) that the MMR vaccine causes autism, we can assume, at the least, a range of shared biological, psychological, and behavioural concepts, and a scientific understanding of experiment and evidence. Where such assumptions cannot be made, it is hard to see how we can discuss the matter with others, say by pointing out that their interpretation of the evidence was incorrect or that the experimental design was flawed. And relating to the matter at hand, as we have seen in Sections 2 and 3, the community and the Team have different views on the nature, causes, risk factors, and prevention of depression, views that reflect differences in culture. Second, in aiming to change how people conceive of gender roles and marriage, the Team is targeting social practices that go deep in the community's worldview and are connected to religious, moral, and social commitments. On these two points, we can see that what the Team is really targeting here is the culture of the community. Unlike anti-obesity campaigns which start with health beliefs and might end up having a negative impact on the culture, the public health prevention campaign we have been describing targets the culture from the start. However, aiming to change a community's culture due to an interest in their wellbeing is a different prospect to changing a set of health beliefs; it cannot be approached with the ethical mindset of a public health intervention - as changing health beliefs might - but with the mindset of an intercultural encounter. One reason behind this is the importance of culture to identity.

Like religion, gender, and sexual orientation, culture can be understood as a component of individual identity, the latter being a person's understanding of who he is. Such an understanding is constitutive of the person in the sense that it provides him with an ethical and agentic orientation in the world — when that orientation is lost, we talk about an identity crisis. When a person endorses the various components of his identity, he becomes deeply committed to them in a way that precludes thinking of such components as garments that one can remove and change at will. This point is recognised in political-philosophical discussions on cultural rights (see Rashed, 2019). Initial liberal justifications for cultural rights centred around the argument that culture provides a context of choice for its members, a range of options from where people can select what is valuable and worth pursuing. In the absence of a secure cultural context, the capacity for autonomy cannot be properly exercised (see Kymlicka, 1991, 1995, 2001). This argument implies that as long as people are provided with *any* cultural context (in which they can be integrated), they would be able to exercise their autonomy. However, responses to Kymlicka point out that it is not cultural belonging *per se* that matters but belonging to a *particular* culture. Taylor (1994b, pp. 259—260) writes that 'for the people concerned, their way of life is a good worth preserving; indeed, it is something invaluable and irreplaceable, not just in the absence of an alternative, but even if alternatives are available'. Margalit and Halbertal (1994, p. 503) note that 'an individual has a right to *his* or *her* culture' and not just to any culture, a right that derives from the interest individuals have in preserving the sources of their *identity*, the things that give meaning to their lives. Similarly, Gans (1998, p. 164) writes that people have an interest in their culture because it is a 'component of their identity':

> *They have an interest in adhering to it as it is the interest of people to adhere to every component of their identity they wish to adhere to (their sex, sexual orientation, religion), even though it may not serve other interests they have.*

If we accept this view, we can appreciate the seriousness of attempting to change aspects of people's culture. The connection between culture and identity explains, in part, the resistance often shown to attempts to change cultural beliefs and practices. It also clarifies the harms that can arise where such change is imposed rapidly from without, as opposed to gradually and organically from within. Individually, there is the risk of loss of orientation, identity crises, and failed acculturation, all of which can result in further psychological harms. Collectively, communities that undergo rapid change and are disenfranchised of their culture can experience harms such as breakdown in communication between generations as well as destabilisation and, possibly, violent conflict.

All of this, however, does not deny the fact that sometimes, certain aspects of culture cause significant and unnecessary harm, and that efforts to change these practices are required. The issue is how to approach such efforts, given the problems we have just outlined. Here we return to what we mentioned earlier, which is that the Team ought to approach the situation with the mindset of an intercultural encounter and not a public health intervention. This requires, first, that there is a certain parity in the interaction, which a typical public health campaign might not have; both parties have something to learn from each other, even if it is the Team who approached the community with ideas about health improvement. Second, the Team needs to make its concerns clear and to point out what it thinks might help, but at the same it has to be open to correction on these matters subject to further engagement with the community. Engagement here needs to be substantial and long-term, ideally of an ethnographic nature, as demonstrated in this chapter. Writing in the context of an intercultural encounter where the understanding and evaluation of cultural practices are at stake, Taylor (1994a, p. 67) draws on a Gadamerian metaphor to describe the ideal of such encounters:

> *What has to happen is what Gadamer has called a 'fusion of horizons'. We learn to move in a broader horizon, within which what we have formerly taken for granted as the background to valuation can be situated as one possibility alongside the different background of the formerly unfamiliar culture. The 'fusion of horizons' operates through our developing new vocabularies of comparison, by means of which we can articulate these contrasts.*

When this is successful, and some overlap is achieved, a basis for creative cooperation becomes possible. An example of this can be found in a controversial area of cross-cultural health promotion: campaigns to end the practice of female genital mutilation/cutting (FGM/C). In some regions, health promoters have been able to support communities in identifying ways to retain core traditional values while abolishing harmful practices. Here, an intervention much celebrated and promoted by international women's rights organisations is alternative rites of passage, which seeks to avert the practice of FGM/C while respecting local values. In this case, the rite of passage is conceived as having fundamental importance, not the practice of FGM/C itself. Accordingly, interventions aim to facilitate events and ceremonies for girls as they reach the age at which they might otherwise undergo

FGM/C.[11] It is this sort of collaborative creativity and flexibility that is afforded by an encounter conducted as an *intercultural* encounter. Public health prevention campaigns that seek to change a community's culture, ought to be conducted in this manner.

6. Individual interventions

In Section 3 we identified several risk and protective factors of relevance to depression at the level of the individual. These include: 'depressogenic cognitions' such as the tendency for the negative appraisal of events and of oneself; the extent of a person's skills at problem-solving and conflict management; the person's ability to deal with stressful life events. A Global Mental Health approach concerned with the primary prevention of depression recommends psychological interventions modelled on Cognitive Behavioural and Interpersonal approaches. As cited earlier, there is some evidence that such approaches reduce the incidence of future depressive episodes when delivered in schools to children and adolescents. Returning to the case of Dakhla, from the perspective of the Public Health Team, psychological approaches of this sort are important interventions that can be rolled out in schools and community centres. From the discussion in the previous two sections, we can anticipate that the Team may run into problems in trying to impose a framework radically different to that of the local community. In the present example, we consider a possibility for avoiding this through collaboration with the local community. This depends on being able to adapt the intervention to the local context: Is it possible to find an overlap?

Consider an intervention designed between the Public Health Team and traditional healers of Dakhla. In Sections 2.2 and 2.3 we outlined the causes, risk factors, and preventive strategies for the category of *marad nafsi* (mental illness) as understood in Dakhla. One of the causes, from the emic perspective, is spiritual deviance; the risk factor is living an irreligious life; and the preventive strategy is grounded in Islamic cultural psychology. It requires education in the Islamic virtues, the religious meaning of suffering, and the life of the Prophet Mohammed who was regarded as a model of virtuous

[11] See, for example Amref Health Africa's campaign (online: https://www.amrefcanada.org/get-involved/stories/the-maasai-alternative-rituals-to-fgm/), and CIFORD's campaign (online: https://child.org/news/alternative-rites-of-passage).

behaviour. And, in light of this, prevention requires training of the self to adopt the qualities of the content/peaceful self and hence to contain the impulses of the inciting self, the dominance of the latter being the source of psychological maladies. Such an intervention, therefore, has a pedagogic and a practical component. The practical component requires of the person to recall, reinterpret, and recalibrate key beliefs, values, and emotions. In this sense, a possible intervention from the community's perspective shares features with common psychological interventions such as those proposed by Global Mental Health, for example cognitive restructuring and Cognitive Behavioural approaches more generally. The pedagogic component provides the content for the practical component. The overlap between the Team and the community would then consist in combining the general framework of evidence-based psychological interventions with an Islamic understanding of the self and the connections with spirituality and wellbeing.[12]

The feasibility of such a collaboration depends on establishing a functioning partnership, mutual respect, and other practicalities that have been well-described in guidance in the HIV/AIDS arena (Kayombo et al., 2007; King, 2005). Of particular relevance is the extent to which those with a biomedical perspective are able to engage with a spiritual dimension to an intervention. For example, it has been noted that physicians collaborating with traditional healers in some contexts preferred herbalists to spiritual healers. This preference has been attributed to the belief that 'herbs could be subjected to rigorous empirical methods of analysis as opposed to the use of spirits which are too abstract and subjective and therefore devoid of scientific objectivity' (Mokgobi, 2013).

This brings us to a core ethical challenge posed by collaboration across cultures: to what extent are both parties prepared to compromise their value system? A challenge for traditional healers collaborating with the Public Health Team is that the latter's view of depression might be seen to undermine their fundamentally spiritual understanding. Similarly, the Public Health Team are required to implement a programme that includes spiritual and ideological components that may not coincide with their own.

A related ethical issue arose in campaigns to end the practice of FGM/C, which we referred to in Section 5.2. Alternative rites of passage ceremonies

[12] Within academic and clinical psychology, there have been a number of attempts to develop cognitive behavioural and psychological principles from an Islamic perspective (see Hamdan, 2008; Haque, Khan, Keshavarzi, & Rothman, 2016; Husain & Hodge, 2016; Yaacob, 2013).

aim to minimise harm while respecting local culture. Yet, such ceremonies are not ethically unproblematic for external facilitators and funders. For example, in Kenya, this 'hybridised cultural assemblage' was observed to promote particular concepts of citizenship, morality and virtue, heavily underpinned by certain interpretations of Christianity:

> *In teaching girls to be good wives and mothers, to fear God, and to obey their parents, husbands and elders (in mimicry of traditional instruction), there is a risk of outmoded gender roles being reinforced that may not fit these girls for life as empowered 21st century women.*
>
> ***(Hughes, 2018, pp. 285–286).***

None of which is to determine the cost-benefit analysis in this case, but to illustrate that attempts to initiate change through collaboration across cultures may require of both parties a difficult compromise. Likewise, in the mental health case under consideration, the compromise on behalf of the Public Health Team might consist in promoting virtues and values (of a religious nature) that they might not agree with, and which go against notions they subscribe to, such as liberal notions of autonomy. Such ethical discomfort may be amplified by the related issue that psychological preventative interventions are normally targeted at children and adolescents who might not have made up their mind about their religious views, yet will be instilled with a particular worldview that can limit their future options.

A second ethical problem that can arise in the case under consideration has to do with the piecemeal nature of the suggested collaborative intervention. Risk factors for depression, as we discussed in Section 3, traverse a range of causal levels and have an impact across the life-course. To address them properly, a comprehensive prevention campaign is required. Yet as we have seen in Section 5, there are difficulties with actioning such a campaign since it requires of the community to change its culture and not just a few health beliefs. Accordingly, what the Team is left with is the potential for collaboration outlined here, a useful endeavour no doubt, but one that is likely to be of limited efficacy in preventing depression. Given this, care must be taken that the community is not misled by an overestimation of what such an intervention can achieve.

7. Conclusion

Our description of the Dakhla Oasis of Egypt illustrates the need for public mental health ethics to accommodate a detailed, contextualised understanding of the types of distress that are targeted for intervention. In our example, the different visions for the primary prevention of depression between the community and public health practitioners include differences in conceptualisation at every level: the description of the condition, its causes, the risk factors, and possible interventions to prevent the condition. We asked how a public health team can provide ethical evidence-based interventions across these worldviews. Targeting deeply held beliefs or valued social practices is often tantamount to targeting the culture itself. Given the close ties between culture and identity, we need to take seriously this aspect of the exchange. Yet, for a public health team to accommodate the local culture by adapting their intervention to fit local beliefs and practices (as seen in our discussion of individual-level interventions), they might face a difficult compromise with regards to their own value system. Notwithstanding the good intentions of the public health practitioner, we argue that public health ethics, narrowly conceived, cannot suffice to regulate this encounter. Targeting culture requires the approach not of an ethical public health intervention, but of an intercultural encounter. Accordingly, if the movement of Global Mental Health is to address the accusation that it imposes norms of health and ideas about illness on other communities, then its practitioners might have to rethink the nature of what they are doing.

References

Barrett, R. (2004). Kurt Schneider in Borneo: Do first Rank symptoms apply to the iban? In J. Jenkins, & R. Barrett (Eds.), *Schizophrenia, culture and subjectivity* (pp. 87–109). Cambridge: Cambridge University Press.

Bingham, R., & Banner, N. (2014). The definition of mental disorder: Evolving but dysfunctional? *Journal of Medical Ethics, 40*, 537–542.

Bolton, D. (2008). *What is mental disorder? An essay in philosophy, science and values*. Oxford: Oxford University Press.

Bolton, D. (2013). What is mental illness. In K. W. M. Fulford, M. Davies, R. Gipps, G. Graham, J. Sadler, G. Stanghellini, & T. Thornton (Eds.), *The oxford handbook of philosophy and psychiatry* (pp. 434–450). Oxford: Oxford University Press.

Boorse, C. (2011). Concepts of health and disease. In F. Gifford (Ed.), *Philosophy of medicine* (pp. 13–64). Amsterdam: Elsevier.

Collins, P., Patel, V., Joestl, S., March, D., Insel, T., & Daar, A. (2011). Grand challenges in global mental health. *Nature, 475*, 27–30.

Couch, D., Fried, A., & Komesaroff, P. (2018). Public health and obesity prevention campaigns - a case study and critical discussion. *Communication Research and Practice, 4*(2), 149–166.

Fabrega, H. (1989). On the significance of an anthropological approach to schizophrenia. *Psychiatry, 52*, 45–64.

Fernando, S. (2011). A 'global' mental health program or markets for big pharma? *Open Mind, 22*.

Fry, C. (2012). Ethical issues in obesity interventions for populations. *NSW Public Health Bulletin, 23*(5–6), 116–119.
Gans, C. (1998). Nationalism and immigration. *Ethical Theory and Moral Practice, 1*(2), 159–180.
Gardner, J. (2014). Ethical issues in public health promotion. *South African Journal of Bioethics and Law, 7*(1), 30–33.
Hamdan, A. (2008). Cognitive restructuring: An islamic perspective. *Journal of Muslim Mental Health, 3*, 99–116.
Haque, A., Khan, F., Keshavarzi, H., & Rothman, A. (2016). Integrating Islamic traditions in modern psychology: Research trends in last ten years. *Journal of Muslim Mental Health, 10*(1), 75–100.
Harris, M. (1968). *The rise of anthropological theory: A history of theories of culture*. New York: Thomas Y. Crowell Co.
Headland, T., Pike, K., & Harris, M. (1990). Emics and etics: The insider/outsider debate. *The Journal of American Folklore, 105*.
Hetrick, S., Cox, G., Witt, K., Bir, J., & Merry, S. (2016). Cognitive behavioural therapy (Cbt), third-wave cbt and interpersonal therapy (ipt) based interventions for preventing depression in children and adolescents. *The Cochrane Collaboration*. https://doi.org/10.1002/14651858.CD003380.pub4.
Hopper, K. (1991). Some old questions for the new cross-cultural psychiatry. *Medical Anthropology Quarterly, 5*(4), 299–330.
Hopper, K. (2004). Interrogating the meaning of culture in the WHO international studies of schizophrenia. In J. Jenkins, & R. Barrett (Eds.), *Schizophrenia, culture and subjectivity* (pp. 62–86). Cambridge: Cambridge University Press.
Horton, R. (2007). Launching a new movement for mental health. *The Lancet, 370*(9590), 806.
Horwitz, A., & Wakefield, J. (2007). *The loss of sadness: How psychiatry transformed normal sorrow into depressive disorder*. Oxford: Oxford University Press.
Hughes, L. (2018). Alternative Rites of Passage: Faith, rights, and performance in FGM/C abandonment campaigns in Kenya. *African Studies, 77*, 274–292.
Husain, A., & Hodge, D. (2016). Islamically modified cognitive behavioural therapy: Enhancing outcomes by increasing the cultural congruence of cognitive behavioural therapy self-statements. *International Social Work, 59*(3), 393–405.
Jablensky, A. (2016). Psychiatric classifications: Validity and utility. *World Psychiatry, 15*(1), 26–31.
Jenkins, J., & Barrett, R. (2004). Introduction. In J. Jenkins, & R. Barrett (Eds.), *Schizophrenia, culture and subjectivity* (pp. 1–28). Cambridge: Cambridge University Press.
Kayombo, E., Uiso, F., Mbwambo, Z., Mahunnah, R., Moshi, M., & Mgonda, Y. (2007). Experience of initiating collaboration of traditional healers in managing HIV and AIDS in Tanzania. *Journal of Ethnobiology and Ethnomedicine, 3*, 6.
Kendell, R., & Jablensky, A. (2003). Distinguishing between the validity and utility of psychiatric diagnoses. *American Journal of Psychiatry, 160*(1), 4–12.
Kinderman, P., Read, J., Moncrieff, J., & Bentall, R. (2013). Drop the Language of Disorder. *Evidence Based Mental Health, 16*, 2–3.
King, R. (2005). *Collaboration with traditional healers on prevention and care in sub saharan Africa: A practical guideline for programs*. Geneva: UNAIDS.
Kingma, E. (2013). Naturalist accounts of mental disorder. In K. W. M. Fulford, M. Davies, R. Gipps, G. Graham, J. Sadler, G. Stanghellini, & T. Thornton (Eds.), *The oxford handbook of philosophy and psychiatry* (pp. 363–384). Oxford: Oxford University Press.
Kirmayer, L. (2006). Beyond the "new cross-cultural psychiatry": Cultural biology, discursive psychology and the ironies of globalization. *Transcultural Psychiatry, 43*(1), 126–144.
Kleinman, A. (1987). Anthropology and psychiatry. *British Journal of Psychiatry, 151*, 447–454.

Kymlicka, W. (1991). *Liberalism, community, and culture*. Oxford: Oxford University Press.

Kymlicka, W. (1995). *Multicultural citizenship: A liberal theory of minority rights*. Oxford: Oxford University Press.

Kymlicka, W. (2001). *Politics in the vernacular: Nationalism, multiculturalism, and citizenship*. Oxford: Oxford University Press.

Littlewood, R. (1990). From categories to contexts : A decade of the 'new cross-cultural psychiatry'. *The British Journal of Psychiatry, 156*(3), 308–327.

Margalit, A., & Halbertal, M. (1994). Liberalism and the right to culture. *Social Research, 61*(3), 491–510.

Maxman, A. (2015). How the fight against Ebola tested a culture's traditions. *National Geographic*. https://news.nationalgeographic.com/2015/01/150130-ebola-virus-outbreak-epidemic-sierra-leone-funerals/.

Mills, C., & Davar, B. (2016). A local critique of global mental health. In S. Grech, & K. Soldatic (Eds.), *Disability in the Global South*. Springer. https://doi.org/10.1007/978-3-319-42488-0_28.

Mokgobi, M. G. (2013). Towards integration of traditional healing and western healing: Is this a remote possibility? *The African Journal for Physical, Health Education, Recreation and Dance, 1*, 47–57.

Nuffield Council on Bioethics. (2007). *Public health: Ethical issues*. Cambridge, UK: Cambridge Publishers Ltd.

Oliver, I., & Lewis, D. (2009). Public trust is necessary to protect the population from threats to public health. *Journal of Public Health, 31*(4), 468–469.

Patel, V., Collins, P., Copeland, J., Kakuma, R., Katontoka, S., Lamichhane, J., … Skeen, S. (2011). The movement for global mental health. *BJPsych, 198*, 88–90.

Patel, V., & Prince, M. (2010). Global mental health: A new global health field comes of age. *JAMA, 303*(19), 1976–1977.

Patel, V., & Saxena, S. (2014). Transforming lives, enhancing communities – innovations in global mental health. *The New England Journal of Medicine, 370*(6), 498–501.

Pellecchia, U., Crestani, R., Decroo, T., Van den Bergh, R., & Al-Kourdi, Y. (2015). Social consequences of Ebola containment measures in Liberia. *PLoS One, 10*(12).

Peterson, I., Barry, M., Lund, C., & Bhana, A. (2014). Mental health promotion and the prevention of mental disorders. In V. Patel, H. Mina, A. Cohen, & M. Prince (Eds.), *Global mental health: Principles and practice*. Oxford: Oxford University Press. https://doi.org/10.1093/med/9780199920181.003.0011.

Prevention of depression and bipolar disorder. In Evans, D., Foa, E., Gur, R., Hendin, H., O'Brien, C., Seligman, M., & Walsh, B. (Eds.), *Treating and preventing adolescent mental health disorders*, (2012). Oxford: Oxford University Press. https://doi.org/10.1093/9780195173642.003.0004.

Prince, M., Patel, V., Saxena, S., Maj, M., Maselko, J., Phillips, M., & Rahman, A. (2007). No health without mental health. *The Lancet, 370*, 859–877.

Public Health Leadership Society. (2002). *Principles of the ethical practice of public health*. https://www.apha.org/-/media/files/pdf/membergroups/ethics/ethics_brochure.ashx.

Radden, J. (2018). Public mental health and prevention. *Public Health Ethics, 11*(2), 126–138.

Rashed, M. A. (2012). *Subjectivity, society, and the experts: Discourses of madness*. PhD Thesis. College London: University.

Rashed, M. A. (2013a). Psychiatric judgements across cultural contexts: Relativist, clinical-ethnographic, and universalist-scientific perspectives. *Journal of Medicine and Philosophy, 38*(2), 128–148.

Rashed, M. A. (2013b). Culture, salience, and psychiatric diagnosis: Exploring the concept of cultural congruence and its practical application. *Philosophy, Ethics and Humanities in Medicine, 8*(5), 1–12.

Rashed, M. A. (2015). Islamic perspectives on psychiatric ethics. In J. Sadler, C. W. van Staden, & K. W. M. Fulford (Eds.), *Vol. 1. The oxford handbook of psychiatric ethics* (pp. 495–519). Oxford: Oxford University Press.
Rashed, M. A. (2019). *Madness and the demand for recognition: A philosophical inquiry into identity and mental health activism.* Oxford: Oxford University Press.
Rashed, M. A., & Bingham, R. (2014). Can psychiatry distinguish social deviance from mental disorder? *Philosophy, Psychiatry and Psychology, 21*(3), 243–255.
Rashed, M. A., Bingham, R., Poole, N., Sanati, A., & van Staden, W. (2018). Debate: The concept of culture has outlived its usefulness for psychiatry. *BJPsych Bulletin, 42*(2), 72–76.
Saxena, S., Jane-Llopis, E., & Hosman, C. (2006). Prevention of mental and behavioural disorders: Implications for policy and practice. *World Psychiatry, 5*(1), 5–14.
Saxena, S., Thornicroft, G., Knapp, M., & Whiteford, H. (2007). Resources for mental health: Scarcity, inequity, and inefficiency. *The Lancet, 370*, 878–889.
Simon, J. (2011). Medical ontology. In F. Gifford (Ed.), *Philosophy of medicine* (pp. 65–114). Amsterdam: Elsevier.
Summerfield, D. (2008). How scientifically valid is the knowledge Base of global mental health? *BMJ, 336*, 992–994.
Summerfield, D. (2012). Afterword: Against 'global mental health'. *Transcultural Psychiatry, 49*(3–4), 519–530.
Summerfield, D. (2013). 'Global mental health' is an oxymoron and medical imperialism. *BMJ, 346*. https://doi.org/10.1136/bmj.f3509.
Taylor, C. (1994a). The politics of recognition. In A. Gutmann (Ed.), *Multiculturalism: Examining the politics of recognition* (pp. 25–73). New Jersey: Princeton University Press.
Taylor, C. (1994b). Can liberalism be communitarian? *Critical Review, 8*(2), 257–262.
Timimi, S. (2014). No more psychiatric labels: Why formal psychiatric diagnostic systems should be abolished. *International Journal of Clinical and Health Psychology, 14*, 208–215.
Upshur, R. (2002). Principles for the justification of public health intervention. *Canadian Journal of Public Health, 93*(2), 101–103.
Varga, S. (2015). *Naturalism, interpretation, and mental disorder.* Oxford: Oxford University Press.
Whitley, R. (2015). Global mental health: Concepts, conflicts, and controversies. *Epidemiology and Psychiatric Sciences, 24*, 285–291.
World Health Organisation (WHO). (2004). *Prevention of mental disorders: Effective interventions and policy options*. Geneva: World Health Organisation.
World Health Organisation (WHO). (2008). *Outbreak communication planning guide*. https://www.who.int/ihr/elibrary/WHOOutbreakCommsPlanngGuide.pdf.
Yaacob, N. (2013). Cognitive therapy approach from islamic psycho-spiritual conception. *Procedia – Social and Behavioural Sciences, 97*, 182–187.
Zachar, P. (2015). Psychiatric disorders: Natural kinds made by the world or practical kinds made by us? *World Psychiatry, 14*(3), 288–290.

CHAPTER FIVE

Ethical considerations for conducting cross-cultural biological psychiatry and prevention research on depression among adolescents in low- and middle-income countries

Kamal Gautam[a], Gloria A. Pedersen[b], Syed Shabab Wahid[b] and Brandon A. Kohrt[b,*]

[a]Transcultural Psychosocial Organization Nepal, Kathmandu, Nepal

[b]Division of Global Mental Health, George Washington University, Washington DC, United States

*Corresponding author: E-mail: bkohrt@gwu.edu

Contents

1. Introduction

Today we take for granted that medical tests can inform us about our risk of heart disease, diabetes, and cancer even before we have the disorders.

Developments in Neuroethics and Bioethics, Volume 2
ISSN 2589-2959
https://doi.org/10.1016/bs.dnb.2019.05.001

These tests help us decide when and what preventive measures should be taken to reduce the risk of future disease. In mental health, the science of biological predictors of future mental illness is in its infancy. Whereas scientific research for early markers of Alzheimer's disease and other neurodegenerative conditions are well funded and supported, the idea of testing children and adolescents for future risk of mental illness is fraught with ethical and moral questions. This is especially true in low- and middle-income countries (LMICs) where there are increasing restrictions on biological psychiatry research with children and adolescents. Given the history of exploitative research and extraction of biological samples from LMIC to high-income countries (HICs), the growing limitations are not surprising. However, this sets up a challenge for improving public health in LMIC. The majority of depression and other psychiatric conditions have their onset during adolescence and the majority of the world's youth live in LMIC (Whiteford, Ferrari, & Degenhardt, 2016). Youth in LMIC bear more than 90% of the total global disability adjusted life years attributed to depression (Kieling et al., 2019) see Fig. 1. Thus, if the majority of youth live in LMIC and depression is a condition starting in adolescence, how will the science of

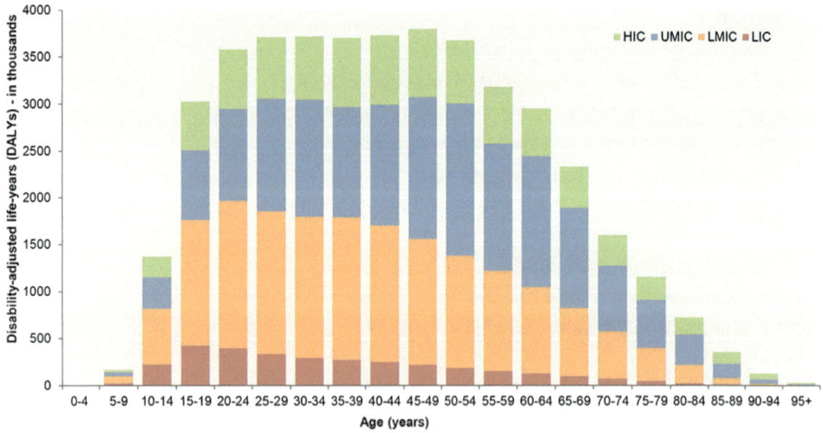

Fig. 1 Burden of depressive disorders in terms of total disability-adjusted life-years (DALYs) across the lifespan. *HIC*, high-income country; *LIC*, low-income country; *LMIC*, lower-middle-income country; *UMIC*, upper-middle-income country. *Data extracted from the Global Burden of Disease Study 2017. Figure from Kieling, C., Adewuya, A., Fisher, H. L., Karmacharya, R., Kohrt, B. A., Swartz, J. R., & Mondelli, V. (2019). Identifying depression early in adolescence.* The Lancet Child & Adolescent Health, 3*(4), 211–213. https://doi.org/10.1016/S2352-4642(19)30059-8, used with permission.*

prevention be advanced if biological psychiatry research is not conducted in LMIC settings?

The goal of this chapter is to review the policies and practices related to biological research with children and adolescents in LMIC. Reviewing these materials is vital to developing an informed agenda for prevention research that will be relevant beyond the 10% of youth who live in North America, Western Europe, and other high resource settings. Ultimately, adolescents in LMIC bear the greatest burden of untreated depression and suffer the greatest exposure to environmental and social risk factors such as poverty, violence, natural disasters, and *unavailability* of psychological treatments (Belfer, 2008; Erskine et al., 2015; Kieling et al., 2011; Leckman & Leventhal, 2008; WHO, 2014). In these low resource settings, children and adolescents have even less access to care compared to adults: 159 per 100,000 youth have available mental health services compared to 664 per 100,000 adults in LMICs (Morris et al., 2011). Proper identification of risk and protective factors, including biological risk factors, can be used to inform early interventions (Kieling et al., 2011). More research on multiple causal mechanisms—biological, social, and other pathways—for the onset of depression is desperately needed to create better targeted prevention strategies (Cuijpers, Beekman, & Renynolds, 2012).

1.1 The role of biological psychiatry research

Research on biological processes in combination with other risk factors (e.g., environmental and social) for depression is important for accurate assessment and can support effective, targeted prevention and treatment (Hankin, 2015). The addition of biomarkers in combination with other risk and protective factors has been shown invaluable in other areas of healthcare, such as with cardiovascular disease, improving risk prediction, prevention, and targeted treatment (Wang et al., 2017; WHO, 2007). Research on biological mechanisms in association with depressive symptoms is beginning to show similar potential.

In HIC, biological psychiatry research has been accumulating evidence of biological mechanisms and biomarkers in-line with depression and depressive symptoms through methods such as neuroimaging, genomic studies, psychoneuroendocrine research on inflammation, and stem cell models of neurogenesis. A recent review and meta-analysis concluded that history of childhood trauma is associated with increased inflammation with dysregulation found to last through adulthood resulting in heightened vulnerability to mental health disorders such as depression (Baumeister,

Akhtar, Ciufolini, Pariante, & Mondelli, 2016). In neuroimaging studies, reward-related ventral striatum dysfunction is associated with depressive symptomology (Hanson, Hariri, & Williamson, 2015). Heightened amygdala response to threat was found to be associated with the development of depression and anxiety symptoms in adolescents and young adults (Swartz, Hariri, & Williamson, 2017; Swartz, Williamson, & Hariri, 2015). Similarly to what is done in cardiovascular screening, including biomarkers with other risk and protective factors for the onset of depression gives the potential to enhance the predictive ability of depression screening tools, allowing increased accuracy in identifying adolescents at risk for depression and giving way for opportunities to intervene and prevent early on.

1.2 The need for biological psychiatry research in LMIC

Though biological psychiatry research is promising, the vast majority of all biological studies of depression have been limited to high-income settings. This prevents the field from being able to conclude that these are universal changes in depression. These associations may be processes that only occur in high-resource contexts. Biological and psychological profiles vary from context and settings, with populations in Western Educated Industrialised Rich Democratic (WEIRD) societies (e.g., North America, Western Europe) being the outlier when compared with low-resource settings in which the majority of the world's adolescents live (Henrich, Heine, & Norenzayan, 2010). This presents ethical issues when considering application of prevention and treatment interventions informed by HIC and implemented in LMIC, a typical practice in the global mental health field due to lack of resources, unavailability of psychological care and shortage of evidence-based research for targeted prevention in low-resource settings.

Currently, it is a crucial public health need to understand the full scope for risk and protective factors to better inform prevention strategies. Biological psychiatry research alongside other research methodologies can inform context and culturally specific targeted prevention strategies early on in life (Ormel, Cuijpers, Jorm, & Schoevers, 2019). In line with global trends, adolescents in Nepal are at an increased risk for developing depression and subsequently are burdened with higher rates of suicide mortality. Though research is limited, young age and depression are currently the most consistently reported risk factors for suicide among girls in Nepal (Marahatta, Samuel, Sharma, Dixit, & Shrestha, 2017). In 2012, the World Health Organization (WHO) ranked Nepal as seventh highest in the world for suicide rates, with an approximated age-standardised rate of 25 suicide deaths per 100,000 population—in contrast, the United States has 14 suicide deaths per 100,000 (WHO, 2014). However, accurate suicide rates are not available because completed suicide deaths in Nepal are underestimated relying mostly on police case reports and specific Nepali population studies (Marahatta et al., 2017).

In attempt to bring clarity to suicide rate estimates in Nepal, a mixed-methods psychological autopsy case-series study was conducted; among 302 rural and urban police cases between 2013 and 2015, half of all female suicide completions were under the age of 25. Moreover, the greatest burden was among those in rural areas, where 90% of females and 50% of males completed suicide before 25 years of age, in contrast, in Kathmandu 50% of female suicides and 30% of male suicides occurred before age 25 (Hagaman, Khadka, Lohani, & Kohrt, 2017). Such dire health inequities demand research for evidencing multiple determinants of adolescent depression in Nepal (Tol, 2015).

Epidemiological research in Nepal has identified some environmental and social risk and protective factors that contribute to distress and depression, e.g., political conflict, caste-based disparities, earthquake, urban habitation, domestic violence, poverty, bullying, family, peers (Kohrt et al., 2009, 2012; Risal, Manandhar, Linde, Steiner, & Holen, 2016; Upadhaya et al., 2018). However, biological psychiatry research about adolescent depression and suicide is limited—but promising. For example, a biomarker study conducted with Nepali child soldiers substantiates the potential biomarkers have for predicting risk and resilience to mental disorders and the possible opportunities for improving prevention and detection of depression (Kohrt, Worthman, et al., 2016). Nepali child soldiers who experienced high degrees of trauma and associated distress showed proinflammatory gene expression associated with the activation of the conserved transcriptional response to adversity (CTRA) gene complex. In other words, due to their high rates of fight or flight responses, the soldiers' gene transcription of inflammation was increased while their transcription of antiviral defense decreased. However, of these child soldiers with reported post-traumatic stress disorder who also reported having psychological resilience showed relatively healthy gene expression—their molecular profiles were comparable to those of civilian youth in Nepal not affected by post-traumatic stress disorder (see Fig. 2).

Another reason that research across populations is needed is because populations have different frequencies of genetic haplotypes associated

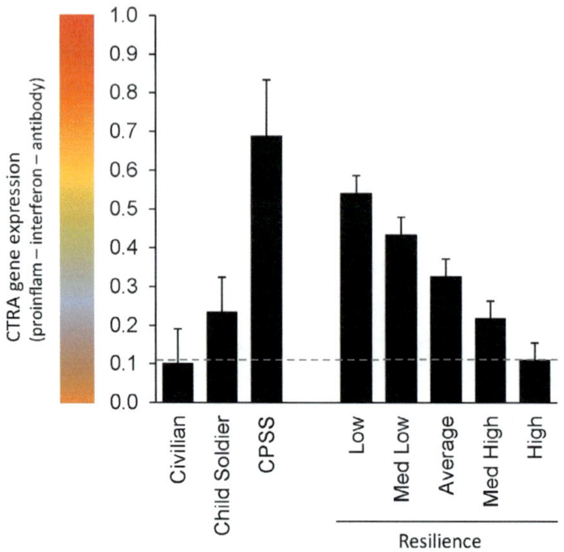

Fig. 2 Conserved transcriptional response to adversity (CTRA) gene activation among Nepali civilian children and former child soldiers. Higher CTRA activation reflects greater proinflammatory activity and lower antiviral activity, which are risk factors for poor future physical and mental health. *From Kohrt, B. A., Worthman, C. M., Adhikari, R. P., Luitel, N. P., Arevalo, J. M. G., Ma, J., … Cole, S. W. (2016). Psychological resilience and the gene regulatory impact of posttraumatic stress in Nepali child soldiers.* Proceedings of the National Academy of Sciences, 113*(29), 8156–8161. https://doi.org/10.1073/pnas.1601301113, used with permission.*

with risk. Although child maltreatment is a risk factor for depression, there are certain genetic profiles that are associated with greater prevalence of depression in the context of child maltreatment. The corticotropin releasing hormone receptor (CRHR1) is associated with greater risk of depression. Fig. 3 demonstrates the population prevalence of different CRHR1 haplotypes (0 copies of TAT, 1 copy of TAT, and 2 copies of TAT). In most studies conducted in North America, Western Europe, and New Zealand with populations of European and African descent, 0 TAT copies in the presence of child maltreatment was associated with greater risk of depression. However, in one of the European studies (Germany) and the Japanese study, the greater the number of TAT copies, the greater the risk of depression. This is important in Asia where having multiple TAT copies is the dominant haplotype, demonstrated by both the Japanese and Nepali samples. Therefore, it is crucial to have biological psychiatry research with different populations around the world.

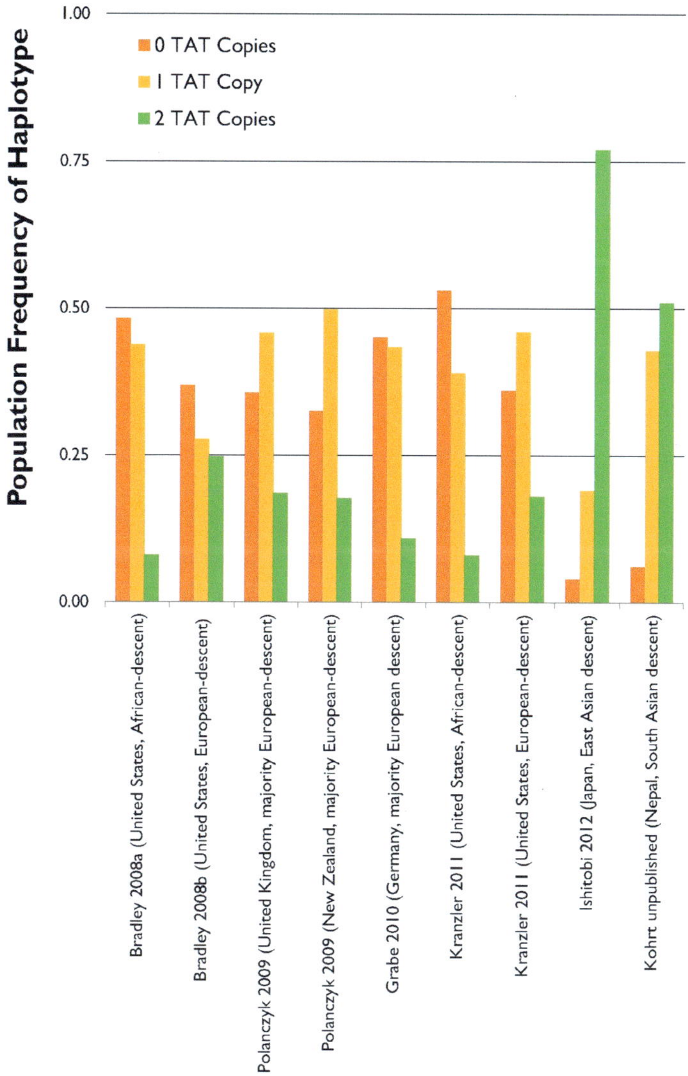

Fig. 3 Prevalence of corticotropin releasing hormone receptor (CRHR1) haplotypes by country and population. The prevalence of 2 TAT copies is more frequent in the Asian (Japanese and Nepali) populations and has been associated with greater risk of depression in some studies. *Data from Bradley, R. G., Binder, E. B., Epstein, M. P., Tang, Y., Nair, H. P., Liu, W., … Ressler, K. J. (2008). Influence of child abuse on adult depression: Moderation by the corticotropin-releasing hormone receptor gene.* Archives of General Psychiatry, *65(2), 190–200; Grabe, H. J., Schwahn, C., Appel, K., Mahler, J., Schulz, A., Spitzer, C., … Völzke, H. (2010). Childhood maltreatment, the corticotropin-releasing hormone receptor gene and adult depression in the general population.* American Journal of Medical Genetics Part B: Neuropsychiatric Genetics, 153B*(8), 1483–1493. https://doi.org/10.1002/ajmg.b.31131; Ishitobi, Y., Nakayama, S., Yamaguchi, K., Kanehisa, M., Higuma, H.,*

1.3 Ethical groundwork for biological psychiatry research in Nepal

The future for biological psychiatry research in Nepal looks hopeful, as there are some studies underway such as the Study of Health Outcome after Trauma (SHOT), which is looking at differences in health outcomes and biochemical and genetic markers between victims with single and multiple traumas in Kathmandu, Nepal (Koirala, Thapa, Hauff, & Ohja, 2016). Additionally, the Identification of Depression Early in Adolescence (IDEA) study is currently ongoing, researching the feasibility and acceptability of biological psychiatry research specifically for the onset of depression in adolescents in Kathmandu (Kieling et al., 2019).

Although capacity for this research is building, national ethical guidelines for biological psychiatry research have yet to be established in Nepal (NHRC, 2011). Biomedical research introduces important ethical considerations such as consent and assent on storage, proprietorship, and usage of biological samples. Guidelines for these ethical provisions have been addressed on a global scale but have not been tailored to a national level—particularly when working with vulnerable populations such as adolescents experiencing depression.

As Nepal builds its capacity to conduct biological psychiatry research on depression in adolescents, the feasibility of implementing specific ethical guidelines at the national level must be addressed. In this chapter, we take the opportunity to address important ethical considerations for conducting biological psychiatry research with adolescents as follows in three sections:

1. **Global Ethical Principles and their Implications in LMICs:** A review of global ethical principles for conducting human-participatory

Maruyama, Y., … Akiyoshi, J. (2012). Association of CRHR1 and CRHR2 with major depressive disorder and panic disorder in a Japanese population. American Journal of Medical Genetics Part B: Neuropsychiatric Genetics, 159B*(4), 429–436. https://doi.org/10.1002/ajmg.b.32046; Kranzler, H. R., Feinn, R., Nelson, E. C., Covault, J., Anton, R. F., Farrer, L., & Gelernter, J. (2011). A CRHR1 haplotype moderates the effect of adverse childhood experiences on lifetime risk of major depressive episode in African-American women.* American Journal of Medical Genetics. Part B, Neuropsychiatric Genetics: The Official Publication of the International Society of Psychiatric Genetics, 156B*(8), 960–968. https://doi.org/10.1002/ajmg.b.31243; Polanczyk, G., Caspi, A., Williams, B., Price, T. S., Danese, A., Sugden, K., … Moffitt, T. E. (2009). Protective effect of CRHR1 gene variants on the development of adult depression following childhood maltreatment: Replication and extension.* Archives of General Psychiatry, 66*(9), 978–985. Nepal data from Kohrt (unpublished). Original figure produced by authors.*

research and their implications in LMICs, with highlights on cultural translation, privacy and data protection risks, and specific ethical concerns when working with children and youth (e.g., consent and assent)

2. **Global Biological Research Policies:** Current global ethical guidelines for conducting biological research with humans
3. **International standards on a national scale: Implementing ethical guidelines for biological research in Nepal:** Ethical considerations of implementation for such guidelines on a national scale in a low-resource setting such as Nepal

2. Global ethical principles and their implications in LMICs

In order to set the stage for including core ethical principles for biological psychiatry research in Nepal, we widen our scope to first discuss the ethical tenets of beneficence/non-maleficence, justice or fairness, and respect of persons or autonomy, as identified by the Belmont Report, issued in 1978 (National Commission for the Protection of Human Subjects of Biomedical and Behavioral Research, 1978) and subsequent guidance from the Council for International Organizations of Medical Sciences (2002). We examine these global ethical principles in the context of LMICs in order to consider the implications of applying such core ethical guidelines in varying conditions. We offer recommendations for incorporating such principles within the challenges of culturally diverse global contexts and then narrow our scope of specific guidelines for Nepal in Section 3.

Before discussing these tenets, it is important to acknowledge that the meanings of well-being, beneficence, justice, and autonomy are subjective to cultural and contextual forces. It is important to seek clarity on how these terms are perceived and used in local settings (DuBois, 2005). For example, a form of autonomy for adolescents in the United States is the ability to emancipate from their birth or given families, in which (if granted), they are deemed legally as adults and competent to make a variety of decisions on their own behalf including giving consent for research or medical treatments (English, Bass, Boyle, & Eshragh, 2010). However, within the USA, the basis for granting emancipation, age for allowing emancipation, and certain legal rights included or excluded within emancipation vary across state lines. Similarly, definition of a minor and statutory age for consent varies across the globe (Library of Congress, 2019).

2.1 Culturally appropriate and locally supported research

Any human-participatory research conducted must minimise harm and ensure that benefits outweigh risks. Risks and benefits should be contextually assessed with a systemic consideration of sociocultural and environmental factors involving community stakeholder agreements and partnerships and avoiding assumptions of risk and benefits imported from an HIC setting (Ruiz-Casares, 2014). Research must also be novel, contribute to scientific knowledge, and ultimately be useful. Any research agenda proposed in LMICs should be weighed against these criteria, and not rely solely on the premise that because so little research has been previously conducted, any subsequent research is automatically useful. Additionally, it is paramount that the cultural validity of assessment measures is ensured, in other words, the mental disorder intended to be measured should capture and reflect the local social, cultural, and moral values, and should not be dominated by values of the community in which the tool was derived, values which are typically Western-dominated (Gureje & Stein, 2014). Instruments utilised in global mental health (GMH) research should be standardised to the populations and contexts that the research is being conducted in (Kleinman, 1977; Kohrt et al., 2011; Kohrt, Luitel, Acharya, & Jordans, 2016). Using tools such as the Composite International Diagnostic Interview (CIDI), Mini International Neuropsychiatric Inventory (MINI), and Patient Health Questionnaire for Adolescents (PHQ-As) without local validation can produce misleading information resulting in misuse of resources and potential harm to members of the population. This invokes the adage that *bad data are worse than no data.*

Potential mechanisms to address these concerns can include local stakeholder collaboration in determining research priorities salient for local communities and policy. Additionally, the development and use of culturally valid instruments through formative research (e.g., qualitative and quantitative methods to evaluate assessment and healthcare needs, acceptability, feasibility, and suitable translations) should precede further inquiry. Ethics surrounding biological psychiatry research should be conducted in adherence to both local and international ethical standards. In the absence of local guidelines, adherence to guidelines devised by international agencies of research governance should act as a minimum threshold, with active engagement efforts between international and local stakeholders to develop local research governance capacity and ethical policy.

When conducting research in LMICs, long-term societal gains may not necessarily be apparent or important to participants in the short-term. Communities may very well be unfamiliar with participant rights in research studies, and initial engagement in a research process may cause therapeutic misconceptions: impressions that healthcare services are automatically available based on participation in a study. It is the researcher's responsibility that populations with limited research literacy be given all of the necessary information for making informed decisions about study participation.

If descriptions of the aims of the research do not incorporate sufficient local terminology, participants cannot provide adequately informed consent (Summerfield, 2008). In that scenario, a signature on a consent form is indicative of unintended, or potentially, soft coercion. The concept of the signature has been debated within the lens of contextual relevance with considerations of possible deleterious implications of such 'proof' of participation (Ruiz-Casares, 2014). Researchers should determine whether an individual has the mental capacity to make the decision to participate in research and consider the social, cultural and community structures and dynamics that inform an individual's decision to provide consent (Hindmarch, Hotopf, & Owen, 2013). Additionally, criteria have been developed which offer the standards to which informed consent procedures should adhere (Fisher et al., 2002). These include considering language preference and proficiency, communication channels utilised in the local community, and the developmental experiences of respondents. The use of professionally trained interpreters is vital. There have also been calls for prospective, empirical studies of informed consent processes in LMICs so that GMH research practices are rooted in evidence-based ethics (Roth, 2003). This approach to evidence-informed ethics can be expanded to include all aspects of cultural adaption and implementation of ethical frameworks for research in LMICs as well.

The cultural and contextual challenges of conducting research in LMICs can be addressed through multiple avenues. A good place to begin would be to form collaborations with local communities by their inclusion in research advisory boards. Research teams can actively implement awareness and capacity building activities to raise the research-literacy level of communities and sensitise local stakeholders on the full spectrum of implications in the participation in research. Crosby and Grodin (2002) recommend seeking out and working closely with culturally-competent advocates through all phases of research implementation to ensure continuity of community involvement and oversight. Formative research and cultural feasibility studies can be conducted to educate and inform researchers and participants about relevant cultural factors in local communities and to contextualise the implementation within a larger community and national research agenda (Coreil et al., 1998). For example, if conducting biological psychiatry research with adolescents in Nepal, a youth advisory board including representative Nepali adolescents for ongoing guidance and advocacy of such research should be established.

2.2 Privacy and confidentiality

Privacy and confidentiality issues are salient considering mechanisms of stigma that operate across LMICs. The highest adherence to ethical precautions in protecting privacy and confidentiality has to inform all research practice, including the use of de-identifiers, vigilance in limiting access to data to only essential personnel, and ensuring appropriate secure storage—details of which are discussed later in this chapter. When dealing with small communities with strong ties, retaining privacy is paramount to protect participants by preventing accidental disclosure of mental health status.

Ultimately the ethical scaffolding of research is the responsibility of the staff who implement the study. Ethical conduct of researchers depends on their personal integrity and ethical training provided to them. Core skills need to be ingrained into researchers in their ethical training. This includes education on national and international ethical and human rights frameworks and procedures, cultural competency and retaining awareness of local customs, applicable local and international legal frameworks and considerations, research and critical thinking skills, and the sociocultural, economic, and political context of the research site (Shah, Nodell, Montano, Behrens, & Zunt, 2011). Researchers should link training with a constant exploration of their assumptions about research questions, methodology, and interpretation of results for the local cultural context (Fisher et al., 2002).

2.3 Research involving minors

The issues surrounding ethics and cultural sensitivity in GMH research are magnified when concerning children and minors. Conducting research with children was historically restricted due to concerns over the protection of a vulnerable population from unnecessary harm (Code, 1949). However, it is ethically imperative that the ultimate objective of global child health research aspires to achieve equity in child health, so that all children can attain health at the highest standard, regardless of their nationality, socioeconomic status, race, or religion (Roth, 2003). Considering the principle of beneficence, the non-inclusion of children in research ultimately produces a detrimental effect, and over time the inclusion of children in paediatric research has been strongly advocated for, and has flourished in some settings (National Council on Bioethics in Human Research, 1992; Roth, 2003). This extends to the application of biological psychiatry research with children and adolescence in Nepal, where we can repeat our earlier question: *if the majority of youth live in LMIC and depression is a condition starting in adolescence, how will the science of prevention be advanced if biological psychiatry research is not conducted in LMIC settings?*

When considering the dynamics of conducting GMH research with children, it is important to first consider that the concept of what delineates childhood from adulthood varies considerably among and within LMICs and in contrast with HICs (Kohrt & Maharjan, 2009). Although the United Nations *Convention of the Rights of the Child* defined children to include "all those under 18 years of age," (United Nations, 1989), it is not unusual for female children in their upper teenage years to be married and have children of their own. In LMICs, it may be common and expected for young children to take on economic responsibility and contribute to their family's livelihood. Similarly, other circumstances such as global changes in demography, livelihood, social media, and moral values influences the concept and age of what might qualify as an adolescent. United Nations has therefore updated the global definition of adolescence to account for these varying conditions and covers the broad range of 10–24 years of age (UNDESA, 2010). Accordingly, balancing international guidelines and procedures with local normative and societal discourse without compromising ethical research standards is essential. Age of maturity for informed consent and assent within Nepal is discussed further in Section 3.

The debate on the ethics of conducting research with children and youth in LMICs is comprised of several key considerations: determining the thresholds of acceptable risk; balancing risks and benefits; confidentiality; the role of parents and guardians; and the relationship between therapeutic and non-therapeutic research (McIntosh et al., 2000). Therapeutic research is that which provides treatment, for example, studying the therapeutic effect of a cancer treatment in cancer patients. Non-therapeutic research is that which does not provide treatment for a given condition and studies the factors and impacts of a given condition, for example, studying and collecting tissue specimen from cancer patients. Foremost, amongst these considerations for therapeutic research is determining an inclusive and culturally sensitive approach to the process of obtaining consent as the autonomy of children and youth in LMICs is often governed through societal and familial structures that differ substantially from high-income settings. According to the UN *Convention on the Rights of the Child*, "State parties shall assure to the child who is capable of forming his or her own views the right to express those views freely in all matters affecting the child, the views of the child being given due weight in accordance with the age and maturity of the child," (United Nations, 1989). The term 'child' denotes a relationship of stewardship connected to a parent or guardian. Accordingly, the consent and assent process of children's participation in a research study should be mediated through a process of consent involving the parent/guardian and the assent of the child (Morrow & Richards, 2007).

There have been variations in the way that assent has been defined in different ethical guidelines. The United States Code of Federal Regulations defines assent as "a child's affirmative agreement to participate in research," (CFR, 2018). The United Kingdom's Royal College of Paediatrics and Medical Research Council understand assent to involve a sense of acquiescence (McIntosh et al., 2000; UK Medical Research Council, 2004). Another organization that has developed ethical guidelines is the Council for International Organizations of Medical Sciences (CIOMSs); they define assent as a "knowing agreement" (2002). The International Association of Child and Adolescent Psychiatry and Allied Professions (IACAPAP) established the Declaration of Berlin in 2004 (revised in Melbourne in 2006) for guidance on consent and assent, including in the specific case of pharmacological trials with children and adolescents (Sondheimer & Rey, 2012). An excerpt of the Declaration of Berlin is provided in Box 1.

LMICs, including Bangladesh, South Africa, Thailand, and Malaysia have also indicated specific procedures of obtaining assent when conducting research with children and adolescents (Cheah & Parker, 2014). Recommendations for specific procedures of obtaining assent in biological psychiatry research in Nepal are made in Section 4.

Assent and parental/guardian consent procedures in LMICs are not free from their own set of challenges. Research issues include lack of formal education, asymmetry of power dynamics between children and parents, generational variation in knowledge and access to resources between children and parents (with children often having access to more diverse resources, e.g., social media and formal education), lack of familiarity with medical research, complex familial structures with multiple or no easily

Box 1 ***Principles of Ethics in Child and Adolescent Mental Health*** **developed by the International Association for Child and Adolescent Psychiatry and Allied Professions (IACAPAP).**

- The participation of human subjects in clinical research is for the purpose of promoting health.
- The well-being of the research participant takes priority over the interests of science and society.
- Research with adolescent human subjects should be based on a clearly written proposal that is approved by an independent ethics committee comprising of parents, guardians, and the law.
- Participation is voluntary without pressure, penalty or loss of benefit with discontinuation.
- Assent must be obtained from children and adolescents who may not have the capacity to give informed consent to participate in research.[1]
- Assent should take into account the age, maturity and psychological state of the child involved. If the child is unable to give assent, then proxy consent of a parent or legal guardian is required.
- The informed consent process requires written information about the purposes, duration and procedures of the research. It should include a description of the foreseeable risks and discomforts involved, and of the benefits to the subject expected as a result of the research. Alternative treatments should be discussed.
- New pharmaceuticals must be properly investigated for children and efficacy empirically established before they are widely used.
- The results of clinical trials should be available to the public even when the trial fails to establish effectiveness empirically, following the adage: 'No clinical trial is finished until the data are made available'.

[1] See Section 2.3 Research involving minors.

Adapted from Declaration of Berlin, Sondheimer, A., & Rey, J. (2012). Ethics and international child and adolescent psychiatry. In J. Rey (Ed.), IACAPAP e-textbook of child and adolescent mental health. *Geneva: International Association for Child and Adolescent Psychiatry and Allied Professions.*

identifiable stewardship relationship between a guardian and a child, issues with documentation and signatures, lack of specific guidelines in research in emergencies and for acute medical conditions requiring urgent action, and factors in assent and consent procedures (Cheah & Parker, 2014) (see case example in Box 2). As a way forward to resolving these issues, and in the spirit of autonomy espoused in the UN *Convention on the Rights of the Child*, children who are competent should consent for themselves and

Box 2 A grey area in consent and assent: Emergency or urgent situation in LMIC

In severe cases that could lead to mortality (e.g., severe malaria), a waiver of consent for treatment for the child may be granted by a legally accepted representative. In low-income settings where such an individual may not exist or be difficult to notify in a timely way, assent from the child may be deemed acceptable. Specific guidelines for these situations are lacking and should be addressed.

both parents' consent and children's assent should be obtained (Cheah & Parker, 2014). Competence should be ascertained on considering the capabilities of parents and children to understand the full implications of the information provided to them by researchers and the ability to make and communicate a decision based on that information.

The discussion and recommendations presented here can be analysed through the unique cultural context of different LMICs and adjusted accordingly to reflect what works best and achieves the highest good while minimising harm. Lessons learned in one LMIC may inform research, interventions, and policy in other LMICs, provided that appropriate considerations are taken regarding cultural heterogeneity. This can be facilitated through dialogue between stakeholders of LMICs in international forums. Existing international institutions engaged in GMH research can act as facilitators for this dialogue. Special considerations for children and minors must remain paramount to advance the evidence base for their well-being while balancing the need for protecting this vulnerable population. Ultimately, the application of ethical and human rights in GMH research fundamentally relies on collaboration among stakeholders, both locally and internationally, and prioritising adaptation for local cultural dynamics and considerations from the conceptualisation, through to the mobilisation and dissemination of research.

3. Global biological research policies

In this section, we outline specific recommendations for adaptation of global ethical policies for conducting **biological research in low-resource settings**. 1982, CIOMS in collaboration with WHO, produced *International Ethical Guidelines for Health-related Research Involving Humans.*

Fig. 4 Word cloud produced with text of *International Ethical Guidelines for Health-related Research Involving Humans* (CIOMS, 2016). *Original figure produced by authors.*

CIOMS ethical guidelines have recently been updated (CIOMS, 2016), to support the global application of ethical principles when conducting scientific research with humans (see Fig. 4). The 2016 CIOMS update specifically addresses the growth in translational biomedical research taking place in low-resource settings (p. ix). Details are included to support necessary modifications, for example when working with more vulnerable groups, CIOMS highlights the importance to obtain consent and assent for youth under age 18 (Guideline 17, pp. 64–68). In addition, local cultural modifications (e.g., with age of maturity for consent) should be made. Moreover, specifications are outlined for biological data collection and mechanisms to authorise data mining, as well as the importance of settings establishing local standards and guidelines for storage and usage of samples while fostering capacity building and community engagement (Guideline 12, pp. 46–52).

Related to what has been discussed in Section 2.3 *Research involving minors,* CIOMs recommends that each country considering to adapt CIOMS ethical guidelines, particularly those in low-resource settings, should establish local standards for age of maturity, rather than a fixed universal definition of age-based adulthood in order to carefully determine country specific consent versus assent. Then, for children and adolescents who may have given *informed assent* with their guardian to participate in biological research below the age of maturity, a mechanism should be in place so that once they reach the local age of maturity, they are given the opportunity to re-consent, that is give informed consent without need for a parent or guardian. This allows the participant to decide if their biological samples can continue to be stored and used or destroyed (p. 45). For example, in a hypothetical scenario, a blood sample has been collected from an adolescent with depression at the age of 15 for analysis of serum cortisol. At age 15 years, the consent was obtained from the parent and assent from the adolescent for analysis and storage up to 10 years. When the adolescent turns 18 years of age, she can provide consent for continued storage and analysis or require that the sample be destroyed. Based on this approach, an adolescent participating in biological psychiatry research—especially in instances where genetic samples have been collected – has the right to opt out from the continued storage and usage of the sample in the future.

The Nuffield Council on Bioethics explores ethical issues in biology and medicine. In their report *Children and Clinical Research: Ethical Issues*, they address management of biological material (e.g., cells and tissues), which require important considerations including technical, scientific, organisational, ethical, and legal guidance (Nuffield Council on Bioethics, 2015). If collection and storage of biological samples (referred to as a biobank) is taking place in any country, then a governance system must be implemented that follows the principle of accountability and maintenance of good stewardship of stored biological materials (CIOMS, 2016, p. 43), and ensure that the rights and welfare of the individuals from whom the biological materials are collected are not adversely affected (p. 43).

Once a source of funding or certain donor has been obtained to support biological research, the researchers must then obtain approval from their local and national ethical research review board which will review the ethical standards, fit of scope, and ongoing research conduct for the proposed science. Researchers should submit a clear and thorough study protocol outlining the purpose, importance, and overall risks and benefits for conducting biological research and include details for the use of the samples (e.g., type, time(s) of collection, purpose), role of the donor, and overall proposed impact and applicability to multiple beneficiaries such as immediate participants, their communities, and the greater society (CIOMS, 2016, p. 43). Timing of collection, storage and analysis of samples may vary across settings, resources, and determined research purposes. The biological research protocol must detail also how the researchers plan to attain both *specific* and *broad* informed consent from the human research participants prior to collection (CIOMS, 2016, p. 42).

Two types of informed consent are applicable depending on the nature of intended use of biological samples in research. If the future use of the biological sample is determined at the time of sample collection, a specific informed consent should be taken from the parents or guardians and assent from the adolescent. It involves explaining the predetermined use of the biological samples in language the participant understands delivered both verbally and in written form (CIOMS, 2016, p. 43). For example, if a saliva sample has been collected from an adolescent for a study involving quantification of RNA and the analysis is predetermined, a specific informed consent/assent must be taken. However, it is possible for biological samples to be useful for other forms of analyses and purposes which may not be known at the time of collection. A broad informed consent can be taken in such cases encompassing the range of future uses taking into consideration that this broad consent must be explicit and is not a blanket consent for use of biological material without any restriction (of note, critiques of broad consent are provided below). To limit potential abuse and of samples and avoid any violation of trust between the researcher and the participant, this type of consent should outline certain limitations on the future use of biological material (Beauchamp, 2011). Broad consent must specify rules on access to the sample, foreseeable use of the sample, the possibility of unsolicited findings and methods of dealing with them and allow for re-consent of the participant. The provision of broad informed consent must be ensured by the supervising research ethics committee (CIOMS, p. 43).

The consent process is ongoing, and if at any time during the study the participant is deemed incapable of informed consent (e.g., signals some form of incapacitation that objectively prevents her from understanding the outlined research processes) or decides they would no longer be included in the study, she must be provided with an opportunity to withdraw from the study (CIOMS, p. 43). Similarly, if the human biological samples remain residual after the primary research purpose is met, a specific informed consent or broad informed consent must be taken for use in future undefined research or an informed opt-out procedure needs to be followed wherein the individual must be aware of its existence, provided with sufficient information, and given a genuine opportunity to object from continued use of their samples (CIOMS, p. 43).

Biobank governance must maintain the confidentiality of the participants whose biological samples have been stored through anonymization or coding of the material and the key to the code must remain with the custodian (CIOMs, p. 44). Similarly, emerging consensus says that results from analysis of biological samples with analytical validity, clinical significance and actionability must be returned or disclosed to research participants (CIOMS, p. 45). International guidelines mandate that the transfer of human biological materials must be covered by Material Transfer Agreement (MTA) which specifically mentions documentation and retrieval procedure, the range and duration of use of the sample (CIOMS, p. 45). A plan for appropriate transfer or disposal of the stored biological materials must be made in collaboration with the local health authorities if the biobank is being closed (CIOMS, p. 45).

4. International standards on a national scale: implementing ethical guidelines for biological psychiatry research in Nepal

Now that the foundation for adaptation of biological research ethical policies in low-resource settings has been set, we take the opportunity to tailor recommendations for implementing such ethical standards for biological psychiatry research in Nepal.

The Nepal Health Research Council (NHRC) is the key governing body in health research in Nepal and is responsible for reviewing, approving, and conducting ongoing monitoring of all health research in Nepal. NHRC formulated the *National Ethical Guidelines for Health Research in Nepal and Standard Operating Procedures* in 2011 to develop ethical and scientific standards for conducting health research in Nepal (NHRC, 2011). The national guidelines are based on the Belmont Report (1978), Helsinki Declaration (World Medical Association, 2002), and CIOMS Guidelines (2016). Currently, there is no separate guideline for biomedical research *conducted* in Nepal. The current NHRC guidelines address basic considerations for consent, proprietorship, and capacity building that have implications for biological research:

1. Informed consent must be obtained from all intended participants of any health research study;
2. If biological samples are transferred abroad for storage and analysis, a memorandum of understanding has to be signed by the donor/sponsor and the NHRC which clearly defines the purpose for the transfer, the

material that is being transferred, ownership of intellectual property rights, and provisions for privacy protection;

3. International biological transfers will only be permitted for the reasons originally stated in the research proposal;
4. Biological research must be sensitive to the need and existing culture and social norms of the communities where it will be carried out;
5. Any externally sponsored research should demonstrate provisions for capacity building and strengthening that field of research, with mandatory provision for the study to have at least one co-investigator from Nepal;
6. Any research must include potential benefits and minimal to no harm to the participants; and
7. Participation must be voluntary, without coercion, and confidential.

As biological psychiatry research capacity builds in Nepal, it is strongly encouraged that Nepal immediately build upon these minimal guidelines and create and implement a national, comprehensive strategy for conducting biological psychiatry research to ensure ethical practice that protects and benefits Nepali researchers, research participants, and the broader communities throughout the country. A limited resource setting, Nepal lacks trained human resources and infrastructure for collection, storage and transfer of biological samples for some types of laboratory analyses. Therefore, it is important that guidelines established by the NHRC address the role of international collaborations in this domain.

The following ethical pitfalls and potential harms may result if national guidelines for conducting biological psychiatry research in Nepal are not created:

a. Ambiguity regarding ethical practices to be followed by biological researchers during studies;
b. Increased probability of inconsistent approval standards by the national review board;
c. Obstacles to capacity building of Nepali researchers in the domain of biological research;
d. Difficulty with regulation of international partners conducting biological research in Nepal;
e. Ambiguity on biological psychiatry research in collaboration with non-public sector international partners, especially private multi-national pharmaceutical companies;
f. Potential for coercion or deception of the study participants, data manipulation, and ultimately increased harm to participants;

g. Elicitation of trafficking of human biological samples;
h. Risk of potential harm to minors who are incapable of providing informed consent; and,
i. Adolescents and children may get excluded from biological research and the potential benefits on their outcomes because of a lack of clear policies.

Clear guidelines will facilitate equitable international collaborations, partnerships, and expertise in the country to expedite human resource development and infrastructure growth to conduct independent biological psychiatry research with adolescents in the long run. To date, lack of biomedical research guidelines has led to an ambiguity amongst the ethical review board as well as researchers in conducting and regulating biological research in the country. One important step could be the endorsement of CIOMS *International Ethical Guidelines for Health-related Research Involving Humans* and the Nuffield Council on Bioethics report on *Children and Clinical Research: Ethical Issues*, with a notation of what recommendations would not be applicable or require adaptation in Nepal. A number of factors need to be considered while developing or adapting the available guidelines on biological psychiatry research in Nepal.

First, context plays a significant role in ethical considerations in research. Nepal is a low-income country with cultural and ethnic variations and differences in beliefs, traditions and practices. Therefore, within-country variation needs to be taken into account rather than treating Nepal as a single homogeneous culture. The ethical guidelines on biological psychiatry research should be sensitive enough to specifically focus on these contextual factors. A large number of studies in Nepal are conducted through international grants and funding. The ethical guideline on research should specifically ascertain that the sponsor, researchers, and relevant public health authority conducting research must ensure that the research is responsive to the health needs or priorities of the communities or populations where the research will be conducted (c.f., CIOMS, p. 3).

Second, biological psychiatry research in Nepal should always be paired with capacity building in terms of human expertise, laboratory infrastructure, and analytic software. Human capacities, resources, and infrastructures are determining factors in conducting research with a specific population. Biological psychiatry research requires specific laboratory setup for collection of samples, storage, quality assurance and analysis of samples. Nepal being a developing country has limited resources and infrastructure, and the sophisticated laboratory facilities are limited to few laboratories owned by private and non-governmental organizations. Similarly, analysing biomarkers in psychiatric disorders often require expertise in analysis on the part of laboratory personnel. Therefore, the ethics on biological psychiatry research with adolescents in Nepal should consider capacity building of biological researchers in conducting research as well as laboratory analysis ultimately leading to feasibility of biological psychiatry research in the long term. Additionally, the ethics guidelines should also consider allowing researchers to conduct biological research with adolescents through transfer of the biological samples to concerned laboratories with adequate capacity to analyse the samples until the time local capacity is sufficiently established.

Third, biological research studies should ensure that there are also short-term benefits to participants. The ethics guidelines should also ensure obligations from the sponsors and international researchers that the intervention developed or evidence generated is beneficial to the community where the research was carried out and should also assist in building local health system capacity (c.f., CIOMS, p. 3). Similarly, the guidelines should also ensure an overall fair distribution of the benefits and burdens of the research and additional benefits such as investments in the local health infrastructure should be provided to the population or community in a low resource setting like Nepal. An example of short-term benefits may be that blood collection studies for genetic analyses also include point-of-care information such as haematocrit, haemoglobin A1c, and lipid levels. There are a number of technologies that can provide immediate information for participants and their health workers. Point-of-care information must not be unduly influential in the sense that participants are lured into participation in a genetic study by receiving other health-related information. This coercion can be mitigated by having point-of-care biomarkers available throughout the beneficiary community and not only contingent upon study enrolment.

Fourth, given that Nepali adolescents often assume adult roles, especially in rural regions, a combined dual consent process should be explored. Specifically, in relation to children and adolescence, the NHRC should insist that adolescents can provide consent at younger ages (e.g., 16 years old) and/or between the ages of 16 and 18 both parental and adolescent consent are be required. Along these lines, the NHRC should establish policies regarding age of maturity in which participants providing sample before reaching age 18 years can then decide if the materials can be retained and continued in analysis after reaching 18 years of age. Given that the majority of psychiatric disorders begin in adolescence, it is essential to include this age group, and special considerations should be made appropriate to Nepal's cultures and context.

Fifth, a policy of waivers governing the use of biological research conducted in routine clinical care needs to be established. NHRC should consider the concept of waivers of informed consent and specify the conditions wherein the privilege applies. A waiver of consent allows researchers to conduct studies without obtaining fully informed consent for reasons of practicality. For example, patients in clinical settings can consent to samples being used without requiring that additional consent being provided for every analysis. A number of high-income countries use this model to evaluate clinical services and public health performance without requiring approval for every analysis. However, this approach could also be abused if clinical services shared genetic data or other biomarkers with pharmaceutical companies, insurance companies, and other institutions without participant's consent. It is possible for the NHRC to waive the requirement of informed consent on use of samples if the research is deemed to have significant social value, cannot be done without the waiver, and poses no more than minimal risks to participants or to the group (c.f., CIOMS, pp. 37–38). However, the waivers of informed consent require justification and approval from the ethical review board and the researchers and research ethics committees must seek to preserve as much of the informed consent process as possible.

5. Conclusion

As efforts to increase capacity for mental health research in LMIC grow across the globe, there are considerable reasons for Nepal and other countries to view these developments with both optimism and caution. To support valuable Nepali research across multiple disciplines and gather much needed evidence on the development of depression in adolescents, specific biological research ethical guidelines must be implemented. Important and sensitive research should proceed within the principles of a thorough ethical framework of the highest standard, with special considerations for children and adolescents, and be tailored for the context and specific needs of local Nepali beneficiaries.

Acknowledgments

We would like to thank Jyoti Bohara for support in gathering evidence for this chapter. This chapter was inspired by the IDEA project, and we thank the lead investigators, Valeria Mondelli, Christian Kieling, Helen Fisher, Johnna Swartz, Abiodun Adewuya and Rakesh

Karmacharya, for their support. We also thank the editors, Kelso Cratsley and Jennifer Radden, for their feedback for revisions.

Funding

MQ Transforming Mental Health Charity, Brighter Futures grant named "Identification of Depression Early in Adolescence", Grant #MQBF/1 IDEA. Additional support was provided by the UK Medical Research Council (MC_PC_MR/R019460/1) and the Academy of Medical Sciences (GCRFNG\100281). BAK is supported by the U.S. National Institute of Mental Health (K01MH104310, R21MH111280).

References

Baumeister, D., Akhtar, R., Ciufolini, S., Pariante, C. M., & Mondelli, V. (2016). Childhood trauma and adulthood inflammation: A meta-analysis of peripheral C-reactive protein, interleukin-6 and tumour necrosis factor-alpha. *Mol Psychiatry, 21*(5), 642–649. https://doi.org/10.1038/mp.2015.67.

Beauchamp, T. L. (2011). Informed consent: Its history, meaning, and present challenges. *Cambridge Quarterly of Healthcare Ethics, 20*(4), 515–523. https://doi.org/10.1017/S0963180111000259.

Belfer, M. L. (2008). Child and adolescent mental disorders: The magnitude of the problem across the globe. *Journal of Child Psychology & Psychiatry & Allied Disciplines, 49*(3), 226–236.

Bradley, R. G., Binder, E. B., Epstein, M. P., Tang, Y., Nair, H. P., Liu, W., ... Ressler, K. J. (2008). Influence of child abuse on adult depression: Moderation by the corticotropin-releasing hormone receptor gene. *Archives of General Psychiatry, 65*(2), 190–200.

CFR. (2018). *50.55 Requirement for permission by parents or guardians and for assent by children.* Retrieved from Washington DC. https://www.accessdata.fda.gov/scripts/cdrh/cfdocs/cfcfr/CFRSearch.cfm?fr=50.55

Cheah, P. Y., & Parker, M. (2014). Consent and assent in paediatric research in low-income settings. *BMC Medical Ethics, 15*(1), 22. https://doi.org/10.1186/1472-6939-15-22.

CIOMS. (2016). International ethical guidelines for health-related research involving humans. In *C. Council for international organizations of medical sciences.*

Coreil, J., Losikoff, P., Pincu, R., Mayard, G., Ruff, A. J., Hausler, H. P., ... Halsey, N. A. (1998). Cultural feasibility studies in preparation for clinical trials to reduce maternal-infant HIV transmission in Haiti. *AIDS Educ Prev, 10*(1), 46–62.

Council for International Organizations of Medical Sciences. (2002). *International ethical guidelines for biomedical research involving human subjects.* Retrieved from Geneva, Switzerland.

Crosby, S., & Grodin, M. A. (2002). The forum. The challenges of cross-cultural research in the international setting. *Ethics Behav, 12*(4), 371–372, 376–378.

Cuijpers, P., Beekman, A. T. F., & Renynolds, C. F. I. (2012). Preventing depression: A global priority. *JAMA, 307*(10), 1033–1034.

DuBois, J. (2005). *Decision-making capacity and surrogate decision making: Ethics in mental health research.* Retrieved from: http://www.emhr.net.

English, A., Bass, L., Boyle, A. D., & Eshragh, F. (2010). In C. f. A. H. t. Law (Ed.), *State minor consent laws: A summary.*

Erskine, H. E., Moffitt, T. E., Copeland, W. E., Costello, E. J., Ferrari, A. J., Patton, G., ... Scott, J. G. (2015). A heavy burden on young minds: The global burden of mental and substance use disorders in children and youth. *Psychological Medicine, 45*(07), 1551–1563. https://doi.org/10.1017/S0033291714002888.

Fisher, C. B., Hoagwood, K., Boyce, C., Duster, T., Frank, D. A., Grisso, T., ... Zayas, L. H. (2002). Research ethics for mental health science involving ethnic minority children and youths. *American Psychologist, 57*(12), 1024–1040.

Grabe, H. J., Schwahn, C., Appel, K., Mahler, J., Schulz, A., Spitzer, C., ... Völzke, H. (2010). Childhood maltreatment, the corticotropin-releasing hormone receptor gene and adult depression in the general population. *American Journal of Medical Genetics Part B: Neuropsychiatric Genetics, 153B*(8), 1483–1493. https://doi.org/10.1002/ajmg.b.31131.

Gureje, O., & Stein, D. J. (2014). Disorders, diagnosis, and classification. In *Global mental health: Principles and practice*. Oxford University Press.

Hagaman, A. K., Khadka, S., Lohani, S., & Kohrt, B. (2017). Suicide in Nepal: A modified psychological autopsy investigation from randomly selected police cases between 2013 and 2015. *Social Psychiatry and Psychiatric Epidemiology*. https://doi.org/10.1007/s00127-017-1433-6.

Hankin, B. L. (2015). Depression from childhood through adolescence: Risk mechanisms across multiple systems and levels of analysis. *Current Opinion in Psychology, 4*, 13–20. https://doi.org/10.1016/j.copsyc.2015.01.003.

Hanson, J. L., Hariri, A. R., & Williamson, D. E. (2015). Blunted ventral striatum development in adolescence reflects emotional neglect and predicts depressive symptoms. *Biological Psychiatry, 78*(9), 598–605. https://doi.org/10.1016/j.biopsych.2015.05.010.

Henrich, J., Heine, S. J., & Norenzayan, A. (2010). The weirdest people in the world? *Behavioral & Brain Sciences, 33*(2–3), 61–83. https://doi.org/10.1017/S0140525X0999152X. discussion 83–135.

Hindmarch, T., Hotopf, M., & Owen, G. S. (2013). Depression and decision-making capacity for treatment or research: A systematic review. *BMC Medical Ethics, 14*, 14–54. https://doi.org/10.1186/1472-6939-14-54.

Ishitobi, Y., Nakayama, S., Yamaguchi, K., Kanehisa, M., Higuma, H., Maruyama, Y., ... Akiyoshi, J. (2012). Association of CRHR1 and CRHR2 with major depressive disorder and panic disorder in a Japanese population. *American Journal of Medical Genetics Part B: Neuropsychiatric Genetics, 159B*(4), 429–436. https://doi.org/10.1002/ajmg.b.32046.

Kieling, C., Adewuya, A., Fisher, H. L., Karmacharya, R., Kohrt, B. A., Swartz, J. R., & Mondelli, V. (2019). Identifying depression early in adolescence. *The Lancet, Child & Adolescent Health, 3*(4), 211–213. https://doi.org/10.1016/S2352-4642(19)30059-8.

Kieling, C., Baker-Henningham, H., Belfer, M., Conti, G., Ertem, I., Omigbodun, O., ... Rahman, A. (2011). Child and adolescent mental health worldwide: Evidence for action. *The Lancet, 378*(9801), 1515–1525.

Kleinman, A. M. (1977). Depression, somatization and the "new cross-cultural psychiatry". *Social Science & Medicine (1967), 11*(1), 3–9. https://doi.org/10.1016/0037-7856(77)90138-X.

Kohrt, B. A., Hruschka, D. J., Worthman, C. M., Kunz, R. D., Baldwin, J. L., Upadhaya, N., ... Nepal, M. K. (2012). Political violence and mental health in Nepal: Prospective study. *British Journal of Psychiatry, 201*(4), 268–275. https://doi.org/10.1192/bjp.bp.111.096222.

Kohrt, B. A., Jordans, M. J., Tol, W. A., Luitel, N. P., Maharjan, S. M., & Upadhaya, N. (2011). Validation of cross-cultural child mental health and psychosocial research instruments: Adapting the depression self-rating scale and child PTSD symptom scale in Nepal. *BMC Psychiatry, 11*(1), e127.

Kohrt, B. A., Luitel, N. P., Acharya, P., & Jordans, M. J. D. (2016). Detection of depression in low resource settings: Validation of the patient health Questionnaire (PHQ-9) and cultural concepts of distress in Nepal. *BMC Psychiatry, 16*(1), e58. https://doi.org/10.1186/s12888-016-0768-y.

Kohrt, B. A., & Maharjan, S. M. (2009). When a child is no longer a child: Nepali ethnopsychology of child development and violence. *Studies in Nepali History and Society, 14*(1), 107–142.

Kohrt, B. A., Speckman, R. A., Kunz, R. D., Baldwin, J. L., Upadhaya, N., Acharya, N. R., ... Worthman, C. M. (2009). Culture in psychiatric epidemiology: Using ethnography and multiple mediator models to assess the relationship of caste with depression and anxiety in Nepal. *Annals of Human Biology, 36*(3), 261–280.

Kohrt, B. A., Worthman, C. M., Adhikari, R. P., Luitel, N. P., Arevalo, J. M. G., Ma, J., ... Cole, S. W. (2016). Psychological resilience and the gene regulatory impact of posttraumatic stress in Nepali child soldiers. *Proceedings of the National Academy of Sciences, 113*(29), 8156–8161. https://doi.org/10.1073/pnas.1601301113.

Koirala, R., Thapa, S. B., Hauff, E., & Ohja, S. (2016). The SHOT study: Study of health outcome after trauma. In *A comparative study of health outcome, biochemical and genetic markers between victims of single and multiple traumas in Kathmandu, Nepal (ongoing)*. Retrieved from: https://www.researchgate.net/project/The-SHOT-Study-Study-of-Health-Outcome-after-Trauma-A-Comparative-Study-of-Health-Outcome-Biochemical-and-Genetic-Markers-between-Victims-of-Single-and-Multiple-Traumas-in-Kathmandu-Nepal.

Kranzler, H. R., Feinn, R., Nelson, E. C., Covault, J., Anton, R. F., Farrer, L., & Gelernter, J. (2011). A CRHR1 haplotype moderates the effect of adverse childhood experiences on lifetime risk of major depressive episode in African-American women. *American Journal of Medical Genetics. Part B, Neuropsychiatric Genetics: The Official Publication of the International Society of Psychiatric Genetics, 156B*(8), 960–968. https://doi.org/10.1002/ajmg.b.31243.

Leckman, J. F., & Leventhal, B. L. (2008). Editorial: A global perspective on child and adolescent mental health. *Journal of Child Psychology & Psychiatry & Allied Disciplines, 49*(3), 221–225.

Library of Congress. (2019). *Children's rights: International and national laws and practices*. Retrieved from: https://www.loc.gov/law/help/child-rights/index.php.

Marahatta, K., Samuel, R., Sharma, P. R., Dixit, L., & Shrestha, B. R. (2017). Suicide burden and prevention in Nepal: The need for a national strategy. *WHO South-East Asia Journal of Public Health, 6*(1).

McIntosh, N., Bates, P., Brykczynska, G., Dunstan, G., Goldman, A., Harvey, D., ... Shelley, P. (2000). Guidelines for the ethical conduct of medical research involving children. Royal College of paediatrics, child health: Ethics advisory committee. *Arch Dis Child, 82*(2), 177–182.

Morris, J., Belfer, M., Daniels, A., Flisher, A., Ville, L., Lora, A., & Saxena, S. (2011). Treated prevalence of and mental health services received by children and adolescents in 42 low-and-middle-income countries. *J Child Psychol Psychiatry, 52*(12), 1239–1246. https://doi.org/10.1111/j.1469-7610.2011.02409.x.

Morrow, V., & Richards, M. (2007). The ethics of social research with children: An overview. *Children & Society, 10.*

National Commission for the Protection of Human Subjects of Biomedical and Behavioral Research. (1978). *The Belmont report: Ethical principles and guidelines for the protection of human subjects of research.*

National Council on Bioethics in Human Research. (1992). *Report on research involving children*. Retrieved from Ottawa, Canada.

NHRC. (2011). *National ethical guidelines for health research in Nepal and standard operating procedures*. Ramash Path, Kathmandu, Nepal: Nepal Health Research Council (NHRC).

Nuffield Council on Bioethics. (2015). *Children and clinical research: Ethical issues*. London, UK, Retrieved from: http://nuffieldbioethics.org/wp-content/uploads/Children-and-clinical-research-full-report.pdf.

Ormel, J., Cuijpers, P., Jorm, A. F., & Schoevers, R. (2019). Prevention of depression will only succeed when it is structurally embedded and targets big determinants. *World Psychiatry : Official Journal of the World Psychiatric Association (WPA), 18*(1), 111–112. https://doi.org/10.1002/wps.20580.

Polanczyk, G., Caspi, A., Williams, B., Price, T. S., Danese, A., Sugden, K., … Moffitt, T. E. (2009). Protective effect of CRHR1 gene variants on the development of adult depression following childhood maltreatment: Replication and extension. *Archives of General Psychiatry, 66*(9), 978–985.

Risal, A., Manandhar, K., Linde, M., Steiner, T. J., & Holen, A. (2016). Anxiety and depression in Nepal: Prevalence, comorbidity and associations. *BMC Psychiatry, 16*(1), 1–9. https://doi.org/10.1186/s12888-016-0810-0.

Roth, D. (2003). An ethics-based approach to global child health research. *Paediatrics & Child Health, 8*(2), 67–71.

Ruiz-Casares, M. (2014). Research ethics in global mental health: Advancing culturally responsive mental health research. *Transcultural Psychiatry, 51*(6), 790–805. https://doi.org/10.1177/1363461514527491.

Shah, S. K., Nodell, B., Montano, S. M., Behrens, C., & Zunt, J. R. (2011). Clinical research and global health: Mentoring the next generation of health care students. *Glob Public Health, 6*(3), 234–246. https://doi.org/10.1080/17441692.2010.494248.

Sondheimer, A., & Rey, J. (2012). Ethics and international child and adolescent psychiatry. In J. Rey (Ed.), *IACAPAP e-textbook of child and adolescent mental health*. Geneva: International Association for Child and Adolescent Psychiatry and Allied Professions.

Summerfield, D. (2008). How scientifically valid is the knowledge base of global mental health? *BMJ, 336*(7651), 992–994.

Swartz, J. R., Hariri, A. R., & Williamson, D. E. (2017). An epigenetic mechanism links socioeconomic status to changes in depression-related brain function in high-risk adolescents. *Molecular Psychiatry, 22*(2), 209–214. https://doi.org/10.1038/mp.2016.82.

Swartz, J. R., Williamson, D. E., & Hariri, A. R. (2015). Developmental change in amygdala reactivity during adolescence: Effects of family history of depression and stressful life events. *The American Journal of Psychiatry, 172*(3), 276–283. https://doi.org/10.1176/appi.ajp.2014.14020195.

Tol, W. A. (2015). Stemming the tide: Promoting mental health and preventing mental disorders in low- and middle-income countries. *Global Mental Health, 2*. https://doi.org/10.1017/gmh.2015.9.

Trials of war criminals before the nuremberg military tribunals under control Council law No. 10.(1949).

UK Medical Research Council. (2004). *MRC Ethics Guide: Medical research involving children*. Retrieved from: https://mrc.ukri.org/documents/pdf/medical-research-involving-children/.

UNDESA. (2010). *Definition of youth*. Retrieved from: https://www.un.org/esa/socdev/documents/youth/fact-sheets/youth-definition.pdf.

United Nations. (1989). *United nation conventions on the rights of child*. United Nation Conventions on the Rights of Child.

Upadhaya, N., Tize, C., Adhikari, R., Gurung, D., Pokhrel, R., Maharjen, S., & Reis, R. (2018). Geographies of adolescent distress: A need for a community-based psychosocial care system in Nepal. *Intervention*. https://doi.org/10.4103/INTV.INTV_21_18.

Wang, J., Tan, G.-J., Han, L.-N., Bai, Y.-Y., He, M., & Liu, H.-B. (2017). Novel biomarkers for cardiovascular risk prediction. *Journal of Geriatric Cardiology: JGC, 14*(2), 135–150. https://doi.org/10.11909/j.issn.1671-5411.2017.02.008.

Whiteford, H., Ferrari, A., & Degenhardt, L. (2016). Global burden of disease studies: Implications for mental and substance use disorders. *Health Affairs, 35*(6), 1114–1120. https://doi.org/10.1377/hlthaff.2016.0082.

WHO. (2007). *Prevention of Cardiovascular Disease: Guidelines for assessment and management of cardiovascular risk.*

WHO. (2014). *Preventing suicide: A global imperative.* Retrieved from Geneva.

World Medical Association. (2002). *World medical association declaration of Helsinki ethical principles for medical research involving human subjects.* Retrieved from Washington, DC.

CHAPTER SIX

International and cross-cultural parenting research and intervention ethics

Rebecca S. Hock[a,b,*], Elizabeth J. Levey[a,b,c], Christine E. Cooper-Vince[a,b] and Benjamin L. Harris[d]

[a]The Chester M. Pierce, MD Division of Global Psychiatry, Massachusetts General Hospital, Boston, MA, United States
[b]Department of Psychiatry, Harvard Medical School, Boston, MA, United States
[c]Department of Psychiatry, University of Illinois College of Medicine, Chicago, IL, United States
[d]A.M. Dogliotti College of Medicine, University of Liberia, Monrovia, Liberia
*Corresponding author: E-mail: rhock@mgh.harvard.edu

Contents

1. Introduction

In an increasingly globalized world, a greater amount of public health research and intervention work involves international collaborations, as well as researchers working across cultures and contexts. A search for literature on "cultural adaptation" from 1990 to 2017 found a 291.8% increase in citations in Google Scholar when comparing number of publications between 1990 and 1999 with publications between 2010 and 2017 and a 792.8% increase in PsychINFO search engines comparing the same time periods (Bernal & Adames, 2017).

Developments in Neuroethics and Bioethics, Volume 2
ISSN 2589-2959
https://doi.org/10.1016/bs.dnb.2019.06.001

Parenting research and interventions have also increased rapidly over time, including work applied internationally and cross-culturally (Knerr, Gardner, & Cluver, 2013; Morelli et al., 2018). Parenting interventions span a variety of goals, including promoting infant nutrition and health (Black & Aboud, 2011), supporting children's cognitive and social-emotional development (Page, Wilhelm, Gamble, & Card, 2010), preventing and treating emotional and behavioral disorders in children (Comer, Chow, Chan, Cooper-Vince, & Wilson, 2013), fostering secure attachment (Lyons-Ruth & Spielman, 2004), and preventing maltreatment (Levey et al., 2017a), among other goals.

As is the case with public health work generally, there are many benefits to international and cross-cultural applications of parenting work, and also potential harm that can inadvertently be caused, even with the best of intentions. In this chapter, we explore some of the considerations regarding benefits and risks related to the international and cross-cultural applications of parenting work.

We begin by introducing philosophical concepts that can help frame the ethical issues at stake. Next, we discuss some of the benefits of parenting research and interventions, as well as the importance of increasing their availability and accessibility for underserved populations globally. This increased representation in research and interventions must also include greater representation in the research process itself and involve mutually-beneficial partnerships when non-local collaborators are involved. We then discuss potential ethical challenges and how to navigate them. Finally, we provide a chapter synthesis, including recommendations. Some previously published guidelines for ethical conduct in international and cross-cultural collaborations, as well as global mental health work in particular are presented, as well as examples drawn from local ethics codes. This chapter explores some of the ethical issues involved in global applications of parenting work. While we cannot address every ethical issue that may arise in this complex field, we hope it may stimulate reflection and further study.

2. Philosophical concepts to be explored

One common framework used for the analysis of ethics in medicine and human subjects research involves four core principles. Three of these were initially articulated in the Belmont Report of Ethical Principles and

Guidelines for the Protection of Human Subjects Research (United States National Commission for the Protection of Human Subjects of Biomedical and Behavioral Research, 1978), and a fourth was later added to this model (Beauchamp, 2007; Beauchamp & Childress, 2008). The four principles are as follows: (1) Beneficence; (2) Justice; (3) Respect for autonomy; and (4) Nonmaleficence.

This framework is central to research ethical oversights employed by institutional review boards (IRBs). Although this framework was developed in the United States and does not represent the full range of ethical models conceived by all cultures and worldviews, we use it as a starting point to organize our discussion of the ethics in the design and implementation of parenting research and interventions cross-culturally. We additionally draw in thoughts from frameworks of ethics for global health and refer to some examples of ethical models drawn from more local contexts.

3. Applied parenting research benefits

3.1 Beneficence

"The principle of beneficence refers to a moral obligation to act for the benefit of others."

(Beauchamp, 2007)

Beneficence is a core value behind the rationale for parenting research and interventions. Children need positive relationships with caregivers and freedom from maltreatment to foster their healthy development (Ainsworth, Blehar, Waters, & Wall, 2015; Bowlby, 1988; Freud & Burlingham, 1943). The early years of life have been found to not only influence health and well-being during that time, but also to have significant long-term implications for health and well-being throughout the life course, and even into the next generation (Galler & Rabinowitz, 2014). Unfortunately, exposure to early life adversities is prevalent. Data from the National Survey of Children's Exposure to Violence in the US found that at least one in four children has experienced neglect and/or abuse (including physical, emotional, and sexual) in their lives and one in seven children has experienced it in the past year (Finkelhor, Turner, Shattuck, & Hamby, 2015). Over 1 billion children worldwide experience violence of this nature each year (Hillis, Mercy, Amobi, & Kress, 2016).

The far-reaching consequences of early life experiences are well-documented. Studies have found that experiencing maltreatment in childhood greatly increases the likelihood of a range of mental health problems in adulthood, including depression (Chapman et al., 2004), substance use disorders (Banducci, Hoffman, Lejuez, & Koenen, 2014), suicidal behavior (Dube et al., 2001), and post-traumatic stress disorder (Frans, Rimmo, Aberg, & Fredrikson, 2005). Early life abuse and neglect are also significant risk factors for a range of physical health conditions and chronic diseases in adulthood, including cancer, heart disease, liver disease, chronic lung disease, obesity, high cholesterol, high blood pressure, and high levels of C-reactive protein (Danese et al., 2009; Felitti et al., 1998; Gilbert et al., 2015). In addition to mental and physical health outcomes, early life relationships set the tenor for later relationships and interpersonal interactions. People who survived abuse in childhood are more likely to experience intimate partner violence (Bensley, Van Eenwyk, & Wynkoop Simmons, 2003; Coid et al., 2001; Schaaf & McCanne, 1998) and their children are also more likely to suffer from abuse (Berlin, Appleyard, & Dodge, 2011; Madigan et al., 2014).

The mechanisms underlying associations between early life experiences and later outcomes are multifold and are still in the process of being fully delineated, but there is evidence that these mechanisms include both biological and psychosocial processes. These experiences can be so powerful as to have a lasting influence on development, including brain structure and function, as well as hormone regulation and response to stressful situations (McEwen, 2013; Shonkoff et al., 2012; Teicher et al., 2003).

From a positive perspective, parenting behavior and attachment can be influenced via intervention work, which represents a key pathway to preventing negative health consequences later in life (Cooper et al., 2009; DuMont et al., 2008; Fearon et al., 2017; Olds, Henderson, & Kitzman, 1994). Given the importance of early life experiences and relationships with caregivers, interventions that help target these experiences are key pathways to improving individual and population health.

Parenting interventions have been found to have the potential to help with a range of issues, including feeding difficulties, physical, cognitive, and social development, emotional and behavioral disorders, and to help buffer children against the effects of early-life adversities (Cluver et al., 2018; Comer et al., 2013; Puffer et al., 2015; Shonkoff et al., 2012; Tomlinson et al., 2016). Not only do these interventions help the health and development of children, but they have also been found to help improve the mental health of the caregiver (Olds et al., 2004), and can have preventive effects that endure long-term (Fearon et al., 2017; Prinz, Sanders,

Shapiro, Whitaker, & Lutzker, 2009). There are positive societal implications of parenting interventions because early life experiences and relationships have been found to influence health, mental health, and interpersonal interactions throughout life.

Summary and recommendations

Given the extensive research documenting the importance of the early environment in child development, maximizing positive relationships with caregivers in childhood and preventing abuse should be key targets to improve the mental health of children, families, and populations. Etiologies of many mental and other health disorders can begin in the earliest stages of development and thus prevention of this type of suffering may be achieved from focusing on the early environment. To truly improve public mental health, prevention is essential.

3.2 Justice

> *This principle concerns the question of "who ought to receive the benefits of research and bear its burdens?" (Belmont Report, 1978) An aspect of this pertains to striving for "fairness in distribution."*

There is a tremendous global burden of trauma. The prevalence of childhood abuse is estimated at 36% globally (Euser, Alink, Stoltenborgh, Bakermans-Kranenberg, & van, 2015). In countries affected by war and civil strife, many young children do not have the safe, stable environments they need for healthy growth and development. War can also break down the social fabric and institutions that can support recovery from trauma (Betancourt, McBain, Newnham, & Brennan, 2014). While there is a global need for research and interventions supporting children and families, the majority of parenting research has historically been conducted in a narrow segment of the world's population (Arnett, 2008; Henrich, Heine, & Norenzayan, 2010a,b; Morelli et al., 2018). Eighty-five percent of the world's population lives in low- and middle-income countries, but only 1% of research regarding mental health interventions has been done in low-income countries and 10% from middle-income countries (Patel et al., 2007; Saxena, Paraje, Sharan, Karam, & Sadana, 2006). Additionally, there is a large treatment gap for mental health services globally (Eaton et al., 2011). Psychological findings derived from one population may not be generalizable to all populations (Henrich et al., 2010a, b). There is need for further research so that all populations may benefit.

Programs that teach parents new childrearing skills are increasingly being recognized as important on a global level, but there is still further need for research. In order to ensure widespread and just benefits of programs and interventions, research and appropriate methodology require upscaling. Given the unique assets, cultural factors, and resource constraints that influence parenting across cultural and socioeconomic contexts, there is a need for further studies to inform evidence-based development and adaptation of parenting interventions for currently underserved populations. A review of parenting interventions in low- and middle-income countries showed a need for more evidence to be built in this area (Knerr et al., 2013).

The concept of justice should include the understanding that researchers working with underserved populations they are not members of must work with local collaborators in partnerships that are mutually beneficial (Shuchman, Wondimagegn, Pain, & Alem, 2014). These partnerships are essential for generating research that is responsive to local health needs, as well as ensuring that the research process itself is beneficial for the advancement of careers and opportunities of local scientists and other collaborators. A few guidelines to ensure that these relationships are mutually beneficial include that partnerships should be long-term, involve intellectual and technical contributions from all parties including authorship on papers, and fair practices regarding ownership of the data (Ng et al., 2016). Researchers coming from high-income settings and working with underserved populations should be aware of historical and present-day power dynamics among partners from different backgrounds and countries, so as to avoid situations where there is an unfair balance of benefits and risks or burdens resulting from these relationships (Emanuel, Wendler, Killen, & Grady, 2004; Ng et al., 2016). Two-way capacity building can help ensure that further training opportunities are available to all parties involved and increase the potential for local mental health needs to be met. It is also important to engage community members beyond local scientists, and community-based participatory research (CBPR) techniques that involve population members throughout the research process—from formulating the research question to disseminating the results—can help with this process (Viswanathan et al., 2004). Implementation and scale up of these interventions are important ways to ensure that populations benefit from study findings (Ng et al., 2016).

The concept of justice also considers who ought to bear the burdens involved in research and whether populations will reap benefits, such as improved health, services, or knowledge, from any burdens involved. One common criticism often cited by local communities regarding non-local researchers is that the communities are not always informed about the results of the research. While the researcher benefits personally and professionally, when the results are not shared, the local community does not have the opportunity to benefit from the knowledge acquired. Thus, researchers should be mindful of how to share study findings with the local community in culturally appropriate and sensitive ways so that the results can be actively utilized for community benefit. Implementation and scale up of these interventions are important ways to ensure that populations benefit from study findings.

An example of how partnerships can be utilized to conduct research that will inform local interventions is as follows. The Republic of Liberia experienced civil war and political unrest from 1989 until 2003. Two of the authors of this chapter (BH and EL) have conducted research in Liberia regarding the mental health needs of the population generally, and children, specifically (Borba et al., 2016; Levey et al., 2013). Dr Benjamin Harris is an Associate Professor of Psychiatry and Chairman of the Department of Psychiatry at A.M. Dogliotti Medical College at the University of Libera in Monrovia, Liberia and the only child psychiatrist practicing in Liberia. He was the lead World Health Organization (WHO) consultant to the development of Liberia's National Mental Health Policy, adopted by Parliament in 2009. A rapid needs assessment survey was carried out in 2009 by the Liberia Ministry of Health and Social Welfare in collaboration with researchers from Massachusetts General Hospital to inform the development of the policy. Key informants, including parents, teachers, clinicians, and other stakeholders, were interviewed to ascertain their perceptions of the needs of youth in their communities, in order to design interventions that were tailored to the cultural, social and political context (Levey et al., 2013).

In a follow up study, qualitative interviews were conducted with youth. These interviews underscored the importance of early caregiving relationships for the development of resilience (Levey et al., 2016). There were an estimated 340,000 orphans in Libera, 18% of the total child population of the country (LISGIS, 2009). Young people who had lost a parent or other primary caregiver described the significant impact this loss had on their psychosocial and emotional health and ability to function (Levey et al., 2017b). These findings helped to focus Dr Harris's work on the needs of the most vulnerable youth.

Summary and recommendations

Parenting research and intervention work can lead to significant benefits for children and families, and this work is greatly needed. To work towards justice, scientific progress in child development and parenting research should include focus on engaging traditionally underserved populations. Collaborative partnerships with local communities, researchers, and other stakeholders are essential for ensuring that research is responsive to local needs, acceptable, and ethical. It is also important to consider and communicate how study findings may benefit populations that incur any burdens related to the research.

4. Applied parenting research risks

4.1 Respect for autonomy

This principle relates to having the freedom to make one's own choices without being subjected to controlling influences.

> *"To respect an autonomous agent is to recognize with due appreciation that person's capacities and perspectives, including his or her right to hold certain views, to make certain choices, and to take certain actions based on personal values and beliefs ... As a negative obligation, autonomous actions should not be subjected to controlling constraints by others. As a positive obligation, this principle requires both respectful treatment in disclosing information and actions that foster autonomous decision making."*
>
> ***(Beauchamp, 2010, p. 37).***

Autonomy can be thought of in terms of the autonomy of individuals but also in terms of the autonomy of a group of people, such as a community, a cultural group, or a population (Kymlicka, 2008, pp. 11–41). The concept of autonomy frequently enters ethical conversations regarding human subjects research in terms of respect for the autonomy of the individual. For instance, violations would be lack of informed consent or putting participants in a situation where they do not feel fully comfortable declining to participate (e.g., if they fear negative repercussions). At its core, the concept of autonomy involves respect for the values and goals of another being and allowing them to make informed decisions free of coercion. Similarly, considering autonomy on a group level necessitates respecting the values of another entity and striving to avoid any pressures related to power asymmetries.

In accordance with this idea of the importance of group and cultural autonomy, many ethical frameworks emphasize the importance of respect for local culture, history, and self-determination. Accordingly, a common thread in many codes of ethics includes the mandate for "… the need for consultation with local communities and respect and appreciation for local knowledge …" (Morelli et al., 2018). One example comes from "The African Charter on the Welfare and Rights of the Child" from the "Organization of African Unity" (African Commission on Human and Peoples' Rights, n.d.), which includes a statement on the importance of respect for "the virtues of their cultural heritage, historical background, and the values of the African civilization." In terms of raising children, this document states the importance of including in children's education "the preservation and strengthening of positive African morals, traditional values and cultures." Another example statement, which speaks to the importance of local voices being centered, is included in the 2017 San Code of Research Ethics, from the South African San Institute saying that "We require respect for our culture, which also includes our history. We have certain sensitivities that are not known by others" (South Africa San Institute, 2017, p. 1).

Respect for communities, cultures, and their value systems is especially important when considering parenting interventions given that belief and value systems are embedded within behavior change interventions and can promote the altering of people's "views, practices, social norms, and/or lifestyles" (Guttman, 2017, p. 2; Morelli et al., 2018). Meanwhile, parenting is an important means by which groups can transmit cultural values to the next generation (Prioste, Narciso, Goncalves, & Pereira, 2015). In addition, parenting practices can be a means of socializing children to adaptively fit into their society and local context. Thus, teaching different ways of child rearing derived from other cultural contexts, could potentially impact culture and society in ways that are unforeseen and may be undesirable.

Contextual and cultural awareness are important, and a first step is to recognize that parenting interventions developed in one context may come with certain assumptions that may or may not apply universally. A recently published article on ethics related to cross-cultural applications of parenting interventions (Morelli et al., 2018) provides examples of some child-rearing assumptions commonly held by people with Western lifestyles, which may not apply universally. These assumptions include that there is a primary caregiver, often the mother, with whom the child spends most of their time, that this primary caregiver interacts with the child in certain ways, such as face-to-face with regular talking and mutual gaze in order to show sensitive and responsive care, and that children should develop into seeing themselves as independent and self-contained. Accordingly, many Western-developed parenting interventions include a focus on "exclusive and dyadic exchanges with the child that rely on distal senses," often focused specifically on the mother and the child (Morelli et al., 2018, p. 10).

The article goes on to explain that these assumptions do not necessarily apply to all cultures and worldviews. For instance, extended family may play a greater role in infant care in a way that shapes attachment formation (e.g., Morelli & Tronick, 1992; Weisner et al., 1977), there may be more varied caregivers, including other children, attending to the child besides principally the mother, sensitivity and responsiveness may be expressed in different types of ways, and there may be more of a focus on the self as interrelated with others and embedded in a family, community, or other network such that dyadic relating is not as highly emphasized (Markus & Kitayama, 2010; Morelli et al., 2018).

These different practices may be ways of socializing the child to fit into the community and communicating certain values. For instance, a study comparing developmental stories among three groups—the Matsigenka in the Peruvian Amazon, Samoans, and middle-class families in Los Angeles, USA—found that even subtle practices, such as whether the child is routinely physically oriented outward or toward caregivers, may be tied to larger values and have implications for child development. The authors "observed Matsigenka and Samoan caregivers routinely orienting infants and young children outward to notice other people engaged in ongoing practical activities and artifacts used in these activities. US infants and young children instead are routinely positioned face-to-face with caregivers who accommodate them and to child-oriented artifacts (Ochs & Schieffelin, 1984)" (Ochs & Izquierdo, 2009, p. 407). This led the authors to conclude that Matsigenka and Samoan children's body orientation allowed for the early development of social awareness, social responsiveness, and self-reliance in everyday activities (Ochs & Izquierdo, 2009). These types of developmental factors could have even broader implications for the society at large, such as the extent to which values of individualism, collectivism, or another approach are emphasized.

Respect for communities necessitates awareness of cultural factors that impact how research and interventions are experienced. We emphasize the importance of partnering with and centering the voices of local community members, including non-scientists, and also acceptance of the potential for a diversity of valid ways of parenting. Qualitative research is also important for ensuring that culture and context are taken into account and respected. In addition, we recognize that autonomy of a group is not entirely straightforward as there is the question of who can be said to represent the community and there can be tensions between the autonomy of individuals in the community and the community as a whole. Acknowledging autonomy-related risks of behavior change interventions and being mindful of power dynamics, including historical dynamics of colonialism and imperialism, are important to the ethical conduct of this work. Partnering with local researchers and communities in respectful, mutually-beneficial ways, keeping in mind one's own biases, centering interventions on the needs and desires of the community, and respecting the potential for different ways of parenting are some ways to help mitigate this risk.

By virtue of their humanity, children, in addition to their parents, are also rights holders. Conferring rights on children means recognizing their moral equality with adults, thereby underscoring the moral worth of all human beings, irrespective of their situation (Kaime, 2009). There is overwhelming evidence that abuse and neglect have negative consequences for child development. There is also evidence that positive parenting has benefits. The challenge is that in order to be effective, interventions need to teach parents what to do, rather than simply telling them what not to do. Parenting involves many complex behavior patterns. When identifying effective parenting, it is nearly impossible to identify which individual behaviors are essential to bringing about positive outcomes. Thus, Western-based interventions applied in other contexts may fail to recognize locally-based beneficial parenting practices and encourage supplanting of behaviors that may in fact serve a function in that context.

Summary and recommendations

The concept of autonomy is related to respect and freedom from coercion and can apply to both individuals and groups. There is complexity to the concept of autonomy on a group level. Respect for communities, including culture and context is important and can be shown by engaging in formative qualitative work, partnering with and centering the voices of local researchers and non-researchers, and being open to a diversity of ways of parenting so long as they are not causing harm. In cases of suspected harm, reliance on local expertise is essential to understand and respond to these events in a culturally informed manner. Research and interventions should be grounded in knowledge of parenting in the local context and researchers should be open to diverse views of good parenting.

4.2 Nonmaleficence

> *"Physicians have long avowed that they are obligated to avoid doing harm to their patients. Among the most quoted principles in the history of codes of health care ethics is the maxim primum non nocere: 'Above all, do no harm'."*
>
> ***(Beauchamp, 2010, p. 38).***

In addition to awareness of the principle of respect for autonomy, there can be additional potential hazards when applying parenting work, and mental health research more generally, to broader cultures and contexts than those in which they were initially developed. One example is that, if culture and context are not taken into account, there is the potential to pathologize or stigmatize behaviors, practices, and ways of behaving that are merely different ways of going about parenting and other ways of life. When working with communities that are already vulnerable and may be faced with poverty, there can be a focus on deprivation and deficiencies (Morelli et al., 2018; Raz, 2013), and accordingly some examples of parenting research in the US have been experienced as stigmatizing or lead to further stigma (Raz, 2013). Relatedly, there is additionally the danger of placing responsibility for poverty and other conditions on individual parents if it is assumed that their parenting is part of the problem.

Given these hazards, it is important to consider how to go about the process of research and intervention work in a sensitive and thoughtful way and to be careful about potentially stigmatizing language or ways in which research and intervention work could lead to inadvertent harm. We advocate supporting caregivers in coping and parenting via a "strengths-based" approach that emphasizes self-determination and strengths (Puffer, Pian, & Sikkema, 2013; Zimmerman, 2013), engaging parents themselves in identifying treatment targets, engaging the local community beginning with qualitative assessments, and partnering with local scientists and non-scientists. Change agents should additionally be aware of biases they may bring into research based on their own backgrounds and lenses that can affect how they interpret situations.

It would be a mistake to wait to engage in parenting work until larger social inequalities can be solved. However, it is important not to lose sight of the need to advocate for social justice and address social inequalities while engaging in parenting work, as well as being cautious about stigmatizing the people one aims to help. In addition, we should recognize that structural inequalities contribute to family dynamics—for instance chronic poverty, stressors, and trauma can be risk factors for maltreatment (Belsky, 1984; Goodman et al., 2011; Meinck, Cluver, & Boyes, 2015; Taylor, Guterman, Lee, & Rathouz, 2009)—and those structural inequalities also need to be addressed.

Multiple approaches have been developed that can be used to identify potentially useful evidence-based interventions and adapt them to a new cultural context. One example of a framework for adapting evidence-based interventions that can help enact some of these principles is the ADAPT-ITT approach (Wingood & DiClemente, 2008). This model was initially developed for adapting evidence-based interventions (EBIs) for HIV, but is relevant to the adaptation of EBIs in other fields of study, too, such as parenting interventions. This model outlines eight sequential phases to be followed in the adaptation process: Assessment, Decision, Administration, Production, Topical experts, Integration, Training, and Testing. Per our recommendations, this approach begins with an assessment phase involving qualitative research such as focus groups, elicitation interviews, or needs assessments with the local population. In addition, members of the local target population and local experts are involved throughout the entire intervention adaptation process. There is a recognition of the need to balance fidelity to core elements with adaptation to local needs and context. This is one model that can be helpful and additional models may also be sought out to help maximize benefits and mitigate potential risks. A second example is the employment of a community-based participatory research (CBPR) approach, in which the community members are centrally engaged throughout the entire research process (Viswanathan et al., 2004).

Summary and recommendations

There is great potential for parenting research to help children, families, and communities. As with all public health work, there is also the potential to inadvertently cause harm. Some hazards to be aware of include the potential to pathologize behaviors when culture and context are not taken into account, the potential to stigmatize vulnerable populations, as well as the potential to lose sight of larger social conditions, such as poverty, when focused on the goal of behavior change among individuals. We recommend being mindful of these potential hazards and considering them when designing and conducting studies. Centering the voices of local researchers and non-researchers will also help to create research that maximizes benefits and minimizes potential harms. Finally, in working with underserved communities, we must not lose sight of larger social inequalities that affect children and families and must continue to work toward a more just society.

5. Conclusion

We began this chapter providing some key information supporting the value and importance of parenting research and intervention work. The early environment, and particularly relationships with caregivers, are fundamental to child growth and development and have long-term implications for physical and mental health as well as interpersonal relationships throughout life. Parenting interventions can help minimize early life adversities and increase nurturing environments, which have far-reaching implications for individuals as well as society at large. We next described the current state of affairs which is that many children and families worldwide could benefit from research and intervention work supporting parenting, but historically the majority of this research has taken place in a small segment of the world's population. The concept of justice mandates that scientific progress should benefit everyone and not just a small segment of the population. To truly be just and beneficial, the entire scientific process should also include central involvement and intellectual influence from people around the globe, especially those most affected by a given public health concern. In this vein, we provided some thoughts on respectful and mutually-beneficial collaborations when non-local collaborators are involved, which is a key message of this chapter. Justice also includes the weighing of burdens and benefits that research entails. This is an important consideration, especially with respect to the outcome and dissemination of research findings so that they may benefit the population that engaged in the research, as well as the scaling up of interventions found to be effective.

We next introduced some potential risks when applying parenting interventions developed in one culture and context for use in others. These include potential violations of autonomy on the individual or group level as well as the potential to inadvertently pathologize, stigmatize, or lose sight of larger social inequalities, among other potential risks. Acknowledging these risks and being mindful of power dynamics, including historical dynamics of colonialism and imperialism are important to the ethical conduct of this work. As such, it is recommended to partner with local researchers and communities in respectful, mutually-beneficial ways, keep in mind one's own biases, center interventions on the needs and desires of the community, and respect the potential for different ways of parenting.

We believe it is important to reflect on these issues and stimulate discussion. We also acknowledge the limitations of the review and recommendations provided in this chapter. First, the ethical guidelines we have presented to frame the chapter are a useful starting point but also have been critiqued as not being comprehensive (Friesen, Kearns, Redman, & Caplan, 2017). In addition, this framework does not provide guidance on what to do when several of these values conflict—for instance, autonomy and justice can be in conflict in some situations. Several proposed places the Belmont Report falls short include the lack of an explicit mandate for *transparency* in research, which calls for trust and openness with research. Transparency is important for research integrity since it helps increase accountability for keeping promises and being truthful. In addition, the report addresses harms to individual persons but not to communities. We have tried to address this issue by including in our section on *respect for persons* the notion of respect for the autonomy of communities.

In the context of partnerships that are mutually beneficial and respectful with local researchers and community members, there is a greater likelihood of conducting culturally valid research that truly serves the community, addresses the problems that are important to them, and centers their values, priorities, and interpretations of the research. There is a greater likelihood of beneficial, ethical parenting research when a range of cultural and other perspectives are represented, including from researchers and non-researcher community members and leaderships. This approach may not wholly solve the ethical issues we have raised, but we argue that the *process* of the research is central to navigating these concerns.

Acknowledgements

Author time for this work was supported by grant numbers T32MH116140 (Christine Cooper[HYPHEN]Vince; PI: David Henderson) and K23MH115169 (Elizabeth Levey; PI: Elizabeth Levey) from the National Institutes of Mental Health (NIMH) and R01HD060986 (Rebecca Hock; PI: Janina Galler) from the National Institute of Health (NIH) Eunice Kennedy Shriver National Institute of Child Health & Human Development (NICHD).

References

African Commission on Human Peoples' Rights. (n.d.). *African charter on the rights welfare of the child*. Retrieved from http://www.achpr.org/instruments/child/#preamble.

Ainsworth, M. D., Blehar, M. C., Waters, E., & Wall, S. (2015). *Patterns of attachment: A psychological study of the strange situation*. New York, NY: Psychology Press.

Arnett, J. J. (2008). The neglected 95%: Why American psychology needs to become less American. *American Psychologist, 63*, 602–614. https://doi.org/10.1037/0003-066X.63.7.602.

Banducci, A. N., Hoffman, E. M., Lejuez, C. W., & Koenen, K. C. (2014). The impact of childhood abuse on inpatient substance users: Specific links with risky sex, aggression, and emotion dysregulation. *Child Abuse and Neglect, 38*, 928–938. https://doi.org/10.1016/j.chiabu.2013.12.007.

Beauchamp, T. L. (2007). The 'four principles' approach to health care ethics. In R. E. Ashcroft, A. Dawson, & H. Draper (Eds.), *Principles of health care ethics*. John McMillan.

Beauchamp, T. L. (2010). *Standing on principles: Collected essays*. New York, NY: Oxford University Press.

Beauchamp, T. L., & Childress, J. F. (2008). *Principles of biomedical ethics* (6th ed.). New York, NY: Oxford University Press.

Belsky, J. (1984). The determinants of parenting: A process model. *Child Development, 55*, 83–96.

Bensley, L., Van Eenwyk, J., & Wynkoop Simmons, K. (2003). Childhood family violence history and women's risk for intimate partner violence and poor health. *American Journal of Preventive Medicine, 25*, 38–44.

Berlin, L. J., Appleyard, K., & Dodge, K. A. (2011). Intergenerational continuity in child maltreatment: Mediating mechanisms and implications for prevention. *Child Development, 82*, 162–176. https://doi.org/10.1111/j.1467-8624.2010.01547.x.

Bernal, G., & Adames, C. (2017). Cultural adaptations: Conceptual, ethical, contextual, and methodological issues for working with ethnocultural and majority-world populations. *Prevention Science, 18*, 681–688.

Betancourt, T. S., McBain, R., Newnham, E. A., & Brennan, R. T. (2014). Context matters: Community characteristics and mental health among war-affected youth in Sierra Leone. *Journal of Child Psychology and Psychiatry, 55*, 217–226. https://doi.org/10.1111/jcpp.12131.

Black, M. M., & Aboud, F. E. (2011). Responsive feeding is embedded in a theoretical framework of responsive parenting. *The Journal of Nutrition, 141*, 490–494. https://doi.org/10.3945/jn.110.129973.

Borba, C. P., Ng, L. C., Stevenson, A., Vesga-Lopez, O., Harris, B. L., Parnarouskis, L., … Henderson, D. C. (2016). A mental health needs assessment of children and adolescents in post-conflict Liberia: Results from a quantitative key-informant survey. *International Journal of Culture and Mental Health, 9*, 56–70.

Bowlby, J. (1988). *A secure base: Parent-child attachment and healthy human development*. London: Basic Books.

Chapman, D. P., Whitfield, C. L., Felitti, V. J., Dube, S. R., Edwards, V. J., & Anda, R. F. (2004). Adverse childhood experiences and the risk of depressive disorders in adulthood. *Journal of Affective Disorders, 82*, 217–225. https://doi.org/10.1016/j.jad.2003.12.013.

Cluver, L. D., Meinck, F., Steinert, J. I., Shenderovich, Y., Doubt, J., Herrero Romero, R., … Gardner, F. (2018). Parenting for lifelong health: A pragmatic cluster randomised controlled trial of a non-commercialised parenting programme for adolescents and their families in South Africa. *BMJ Global Health, 3*, e000539. https://doi.org/10.1136/bmjgh-2017-000539.

Coid, J., Petruckevitch, A., Feder, G., Chung, W., Richardson, J., & Moorey, S. (2001). Relation between childhood sexual and physical abuse and risk of revictimization in women: A cross-sectional survey. *Lancet, 358*, 450–454.

Comer, J. S., Chow, C., Chan, P. T., Cooper-Vince, C., & Wilson, L. A. S. (2013). Psychosocial treatment efficacy for disruptive behavior problems in very young children: A meta-analytic examination. *J Am Acad Child Adolesc Psychiatry, 52*, 26–36. https://doi.org/10.1016/j.jaac.2012.10.001.

Cooper, P. J., Tomlinson, M., Swartz, L., Landman, M., Molteno, C., Stein, A., … Murray, L. (2009). Improving quality of mother-infant relationship and infant attachment in socioeconomically deprived community in South Africa: Randomised controlled trial. *BMJ, 338*, b974. https://doi.org/10.1136/bmj.b974.

Danese, A., Moffitt, T. E., Harrington, H., Milne, B. J., Polanczyk, G., Pariante, C. M., … Caspi, A. (2009). Adverse childhood experiences and adult risk factors for age-related disease: depression, inflammation, and clustering of metabolic risk markers. *Arch Pediatr Adolesc Med, 163*, 1135–1143. https://doi.org/10.1001/archpediatrics.2009.214.

Dube, S. R., Anda, R. F., Felitti, V. J., Chapman, D. P., Williamson, D. F., & Giles, W. H. (2001). Childhood abuse, household dysfunction, and the risk of attempted suicide throughout the lifespan: Findings from the adverse childhood experiences study. *JAMA, 286*, 3089–3096.

DuMont, K., Mitchell-Herzfeld, S., Greene, R., Lee, E., Lowenfels, A., Rodriguez, M., & Dorabawila, V. (2008). *Child Abuse and Neglect, 32*, 295–315. https://doi.org/10.1016/j.chiabu.2007.07.007.

Eaton, J., McCay, L., Semrau, M., Chatterjee, S., Baingana, F., Araya, R., … Saxena, S. (2011). Scale up of services for mental health in low-income and middle-income countries. *Lancet, 378*, 1592–1603. https://doi.org/10.1016/S0140-6736(11)60891-X.

Emanuel, E. J., Wendler, D., Killen, J., & Grady, C. (2004). What makes clinical research in developing countries ethical? The benchmarks of ethical research. *The Journal of Infectious Diseases, 189*, 930–937.

Euser, S., Alink, L. R., Stoltenborgh, M., Bakermans-Kranenberg, M. J., & van, I. M. H. (2015). A gloomy picture: A meta-analysis of randomized controlled trials reveals disappointing effectiveness of programs aiming at preventing child maltreatment. *BMC Public Health, 15*, 1068. https://doi.org/10.1186/s12889-015-2387-9.

Fearon, R. M. P., Tomlinson, M., Kumsta, R., Skeen, S., Murray, L., Cooper, P. J., & Morgan, B. (2017). Poverty, early care, and stress reactivity in adolescence: Findings from a prospective, longitudinal study in South Africa. *Dev. Psychopathol., 29*, 449–464. https://doi.org/10.1017/S0954579417000104.

Felitti, V., Anda, R., Nordenberg, D., Williamson, D. F., Spitz, A. M., Edwards, V., … Marks, J. S. (1998). Relationship of childhood abuse and household dysfunction to many of the leading causes of death in adults. The Adverse Childhood Experiences (ACE) Study. *American Journal of Prevention Medicine, 14*, 245–258.

Finkelhor, D., Turner, H. A., Shattuck, A., & Hamby, S. L. (2015). Prevalence of childhood exposure to violence, crime, and abuse: Results from the national survey of children's exposure to violence. *JAMA Pediatrics, 169*, 746–754.

Frans, O., Rimmo, P. A., Aberg, L., & Fredrikson, M. (2005). Trauma exposure and post-traumatic stress disorder in the general population. *Acta Psychiatr Scand, 111*, 291–299. https://doi.org/10.1111/j.1600-0447.2004.00463.x.

Freud, A. B., & Burlingham, D. T. (1943). In P. R. Lehrman (Ed.), *War and children*. New York, NY: Ernst Willard.

Friesen, P., Kearns, L., Redman, B., & Caplan, A. L. (2017). Rethinking the Belmont report? *The American Journal of Bioethics, 17*, 15–21.

Galler, J., & Rabinowitz, D. G. (2014). The intergenerational effects of early adversity. *Progress in Molecular Biology and Translational Science, 128*, 177–198.

Gilbert, L. K., Breiding, M. J., Merrick, M. T., Thompson, W. W., Ford, D. C., Dhingra, S. S., & Parks, S. E. (2015). Childhood adversity and adult chronic disease: An update from ten states and the District of Columbia, 2010. *American Journal of Prevention Medicine, 48*, 345–349.

Goodman, S. H., Rouse, M. H., Connell, A. M., Broth, M. R., Hall, C. M., & Heyward, D. (2011). Maternal depression and child psychopathology: A meta-analytic review. *Clin. Child Fam. Psychol. Rev., 14*, 1–27. https://doi.org/10.1007/s10567-010-0080-1.

Guttman, N. (2017). Ethical issues in health promotion and communication interventions. *Oxford Research Encyclopedia of Communication*. Retrieved from http://communication.oxfordre.com/view/10.1093/acrefore/9780190228613.001.0001/acrefore-9780190228613-e-118.

Henrich, Heine, & Norenzayan. (2010a). Most people are not WEIRD. *Nature, 466*, 29.

Henrich, Heine, & Norenzayan. (2010b). The weirdest people in the world? *Behavioral and Brain Sciences, 33*, 61–83.

Hillis, S., Mercy, J., Amobi, A., & Kress, H. (2016). Global prevalence of past-year violence against children: A systematic review and minimum estimates. *Pediatrics, 137*. https://doi.org/10.1542/peds.2015-4079.

Kaime, T. (2009). The foundations of rights in the African Charter on the Rights and Welfare of the Child: A historical and philosophical account. *African Journal of Legal Studies, 3*, 120–136.

Knerr, W., Gardner, F., & Cluver, L. (2013). Improving positive parenting skills and reducing harsh and abusive parenting in low- and middle-income countries: A systematic review. *Prevention Science, 14*, 352–363. https://doi.org/10.1007/s11121-012-0314-1.

Kymlicka, W. (2008). The evolving basis of European norms of minority rights: Rights to culture, participation and autonomy. *The protection of Minorities in the wider Europe*.

Levey, E. J., Borba, C. P. C., Harris, B. L., Carney, J. R., Dominguez, S., Wang, E. K. S., … Henderson, D. C. (2013). Assessment of the needs of vulnerable youth populations in post-conflict Liberia. *African Journal of Psychiatry, 16*, 349–355.

Levey, E. J., Oppenheim, C. E., Lange, B. C., Plasky, N. S., Harris, B. L., Kekulah, I., … Borba, C. P. (2016). A qualitative analysis of factors impacting resilience among youth in post-conflict Liberia. *Child and Adolescent Psychiatry and Mental Health, 10*(26). https://doi.org/10.1186/s13034-016-0114-7.

Levey, E. J., Gelaye, B., Bain, P., Rondon, M. B., Borba, C. P., Henderson, D. C., & Williams, M. A. (2017a). A systematic review of randomized controlled trials of interventions designed to decrease child abuse in high-risk families. *Child Abuse and Neglect, 65*, 48–59. https://doi.org/10.1080/17450128.2016.1262978.

Levey, E. J., Oppenheim, C. E., Lange, B. C., Plasky, N. S., Harris, B. L., Lekpeh, G. G., … Borba, C. P. (2017b). A qualitative analysis of parental loss and family separation among youth in post-conflict Liberia. *Vulnerable Children and Youth Studies, 12*, 1–16.

LISGIS. (2009). *2008 population and Housing Census*. Retrieved from http://www.lisgis.net/page_info.php?7d5f44532cbfc489b8db9e12e44eb820=MzQy.

Lyons-Ruth, K., & Spielman, E. (2004). Disorganized infant attachment strategies and helpless-fearful profiles of parenting: Integrating attachment research with clinical intervention. *Infant Mental Health Journal, 25*, 318–335.

Madigan, S., Wade, M., Plamondon, A., Vaillancourt, K., Jenkins, J. M., Shouldice, M., & Benoit, D. (2014). Course of depression and anxiety symptoms during the transition to parenthood for female adolescents with histories of victimization. *Child Abuse and Neglect, 38*, 1160–1170. https://doi.org/10.1016/j.chiabu.2014.04.002.

Markus, H. R., & Kitayama, S. (2010). Cultures and selves: A cycle of mutual constitution. *Perspectives on Psychological Science, 5*, 420–430.

McEwen, B. S. (2013). The brain on stress: Toward an integrative approach to brain, body, and behavior. *Perspectives on Psychological Science, 8*, 673–675.

Meinck, F., Cluver, L. D., & Boyes, M. E. (2015). Household illness, poverty and physical and emotional child abuse victimization: Findings from South Africa's first prospective cohort study. *BMC Public Health*. https://doi.org/10.1186/s12889-015-1792-4.

Morelli, G., Quinn, N., Chaudhary, N., Vicedo, M., Rosabel-Coto, M., Keller, H., ... Takada, A. (2018). Ethical challenges of parenting interventions in low- to middle-income countries. *Journal of Cross-Cultural Psychology, 49*, 5–24.

Morelli, G., & Tronick, E. (1992). Male care among Efe foragers and Lese farmers. In B. Hewlett (Ed.), *The father-child relationship: Cultural and biosocial contexts* (pp. 231–261). New York, NY: Aldine de Gruyter.

Ng, L. C., Magidson, J. F., Hock, R. S., Joska, J. A., Fekadu, A., Hanlon, C., ... Henderson, D. C. (2016). Proposed training areas for global mental health researchers. *Academic Psychiatry, 40*, 679–685.

Ochs, E., & Izquierdo, C. (2009). Responsibility in childhood: Three developmental trajectories. *Ethos, 37*, 391–413.

Ochs, E., & Schieffelin, B. B. (1984). Language acquisition and socialization: Three developmental stories. In R. A. Shweder, & R. A. LeVine (Eds.), *Culture theory: Essays on mind, self, and emotion* (pp. 276–320). Cambridge: Cambridge University Press.

Olds, D. L., Henderson, C. R., Jr., & Kitzman, H. (1994). Does prenatal and infancy nurse home visitation have enduring effects on qualities of parental caregiving and child health at 25 to 50 months of life? *Pediatrics, 93*, 89–98.

Olds, D. L., Robinson, J., Petitt, L., Luckey, D. W., Holmberg, J., Ng, R. K., ... Henderson, C. R., Jr. (2004). Effects of home visits by paraprofessionals and by nurses: Age 4 follow-up results of a randomized trial. *Pediatrics, 114*, 1560–1568.

Page, M., Wilhelm, M. S., Gamble, W. C., & Card, N. A. (2010). A comparison of maternal sensitivity and verbal stimulation as unique predictors of infant social-emotional and cognitive development. *Infant Behavior and Development, 33*, 101–110.

Patel, V., Araya, R., Chatterjee, S., Chisholm, D., Cohen, A., De Silva, M., ... van Ommeren, M. (2007). Treatment and prevention of mental disorders in low-income and middle-income countries. *Lancet, 370*, 991–1005.

Prinz, R. J., Sanders, M. R., Shapiro, C. J., Whitaker, D. J., & Lutzker, J. R. (2009). Population-based prevention of child maltreatment: The U.S. Triple P system population trial. *Prevention Science, 10*, 1–12. https://doi.org/10.1007/s11121-016-0631-x.

Prioste, A., Narciso, I., Goncalves, M. M., & Pereira, C. R. (2015). Family relationships and parenting practices: A pathway to adolescents' collectivist and individualist values? *Journal of Child and Family Studies, 24*, 3258–3267.

Puffer, E. S., Green, E. P., Chase, R. M., Sim, A. L., Zayzay, J., Friis, E., ... Boone, L. (2015). Parents make the difference: A randomized-controlled trial of a parenting intervention in Liberia. *Global Mental Health, e15*. https://doi.org/10.1017/gmh.2015.12.

Puffer, E. S., Pian, J., & Sikkema, K. J. (2013). Developing a family-based prevention intervention in Rural Kenya: Challenges in conducting community-based participatory research. *Journal of Empirical Research on Human Research Ethics, 8*, 119–128. https://doi.org/10.1525/jer.2013.8.2.119.

Raz, M. (2013). *What's wrong with the poor: Psychiatry, race, and the war on poverty*. Chapel Hill: University of North Carolina Press.

Saxena, S., Paraje, G., Sharan, P., Karam, G., & Sadana, R. (2006). The 10/90 divide in mental health research: Trends over a 10-year period. *British Journal of Psychiatry, 188*, 81–82.

Schaaf, K. K., & McCanne, T. R. (1998). Relationship of childhood sexual, physical, and combined sexual and physical abuse to adult victimization and posttraumatic stress disorder. *Child Abuse and Neglect, 22*, 1119–1133.

Shonkoff, J. P., Garner, A. S., Siegel, B. S., et al. (2012). The lifelong effects of early childhood adversity and toxic stress. *Pediatrics, 129*, e232–e246. https://doi.org/10.1542/peds.2011-2663.

Shuchman, M., Wondimagegn, D., Pain, C., & Alem, A. (2014). Partnering with local scientists should be mandatory. *Nature Medicine, 20*, 12. https://doi.org/10.1038/nm0114-12.

South Africa San Institute. (2017). *San code of research ethics*. Retrieved from http://trust-project.eu/wp- content/uploads/2017/03/San-Code-of-RESEARCH-Ethics-Booklet-final.pdf.

Taylor, C. A., Guterman, N. B., Lee, S. J., & Rathouz, P. J. (2009). Intimate partner violence, maternal stress, nativity, and risk for maternal maltreatment of young children. *American Journal of Public Health, 99*, 175–183. https://doi.org/10.2105/AJPH.2007.126722.

Teicher, M. H., Anderson, S. L., Polcari, A., Anderson, C. M., Navalta, C. P., & Kim, D. M. (2003). The neurobiological consequences of early stress and childhood maltreatment. *Neuroscience & Biobehavioral Reviews, 27*, 33–44.

Tomlinson, M., Rotheram-Borus, M. J., le Roux, I. M., Youssef, M., Nelson, S. H., Scheffler, A., … Worthman, C. M. (2016). Thirty-six-month outcomes of a generalist paraprofessional perinatal home visiting intervention in South Africa on maternal health and child health and development. *Prevention Science, 17*, 937–948. https://doi.org/10.1007/s11121-016-0676-x.

United States. (1978). *The Belmont report: Ethical principles and guidelines for the protection of human subjects of research*. Bethesda, Md: The Commission.

Viswanathan, M., Ammerman, A., Eng, E., Garlehner, G., Lohr, K. N., Griffith, D., … Whitener, L. (2004). In AHRQ (Ed.), *Community-based participatory research: Assessing the evidence: Summary*. Agency for Healthcare Research and Quality.

Weisner, T., Gallimore, R., Bacon, M., Barry, I. I. I. B., Bell, C., Caiuby, S., … Rhys Williams, T. (1977). My brother's keeper: Child and sibling caretaking. *Current Anthropology, 18*, 169–190.

Wingood, G. M., & DiClemente, R. J. (2008). The ADAPT-ITT model: A novel method of adapting evidence-based HIV interventions. *JAIDS, 47*, S40–S46. https://doi.org/10.1097/QAI.0b013e3181605df1.

Zimmerman, M. A. (2013). Resiliency theory: A strengths-based approach to research and practice for adolescent health. *Health Education & Behavior, 40*, 381–383.

CHAPTER SEVEN

Conceptual and ethical problems in screening for major depressive disorder

Dany Lamothe[a] **and Mona Gupta**[b,*]

[a]Université de Sherbrooke, Centre Hospitalier Universitaire de Sherbrooke, Sherbooke, QC, Canada

[b]Département de Psychiatrie et d'Addictologie, Université de Montréal, Montréal, QC, Canada

*Corresponding author: E-mail: mona.gupta@umontreal.ca

Contents

1. Introduction

As the burden of mental illness worldwide becomes more apparent over time (WHO, 2013), healthcare facilities, professional organisations and government agencies, are implementing various strategies to improve the identification, intervention, and support of affected persons. Among these is the drive towards screening for various conditions. Screening assumes that earlier identification of disease, or risk factors for disease, will allow earlier intervention leading to a reduction in the burden of disease both to the individual and to society (Joffres et al., 2013; Siu et al., 2016). While the purpose of screening is to identify those people who are likely to have or develop

Developments in Neuroethics and Bioethics, Volume 2
ISSN 2589-2959
https://doi.org/10.1016/bs.dnb.2019.05.002

a disease, a screening test itself is not diagnostic. Instead, a separate diagnostic evaluation is required to confirm the presence of disease. Thus, the value of screening depends upon the accuracy of diagnostic testing.

Although screening and diagnosis are distinct activities, they share certain assumptions about disease, diagnosis, and diagnostic error. Using the case of screening for major depressive disorder as an example, we explore conceptual issues arising from these assumptions. We argue that the logic of screening, as developed in general medicine, requires the ability to establish the accuracy of the diagnostic procedure such that contradictory results between screening and diagnostic tests can be resolved. However, the nature of major depressive disorder[1] is such that in the current state of diagnostic practice, diagnoses cannot be established as right or wrong therefore, when contradictions arise, it is uncertain whether it is the screening test or the diagnostic test that should be relied upon. Under such circumstances, screening cannot be applied to major depressive disorder. The fact that screening for major depressive disorder has been undertaken in several national healthcare contexts (Canada, USA, and the UK) gives rise to important ethical questions which we discuss at the end of the paper.

2. What is screening?

Screening exists along a continuum of preventive activities. Commonly, these are described as primary, secondary, and tertiary prevention strategies (Tulchinsky, 2000). These levels differ in temporality meaning that they are deployed at different points in time in the natural history of a disease. Primary prevention aims to prevent the onset of disease in a healthy population. Anti-smoking campaigns are example of primary disease prevention strategies which target a risk factor (cigarette smoking) for a variety of diseases (e.g., chronic obstructive pulmonary disease, lung cancer) and aim to prevent people from developing these diseases. Secondary prevention — also known as early diagnosis - involves the identification of the disease before the appearance of symptoms or at least before people seek assistance themselves. Pap smear testing of women to detect cancerous or precancerous cervical cells is such an example. Tertiary prevention involves attempting to prevent complications of disease once the disease process is already

[1] We believe our argument applies to most other psychiatric disorders but in this paper we discuss only major depressive disorder.

underway. The recommendation that Type 1 diabetics maintain their blood sugar within a certain range aims to prevent the long-term complications of the disease but does not prevent the disease itself. Screening for major depressive disorder is a secondary prevention activity: it aims to identify people who have the disease but who have not yet come to clinical attention.

Screening for disease depends upon the concept of disease, a central component of the biomedical model. The biomedical model is the dominant paradigm that shapes medical practice. Although there is debate about how best to characterise this model, we take the following two aspects to be central: 1. that illness can be fully accounted for by changes in the biological organism and 2. that behind illness (clusters of signs and symptoms) lie pathological processes that can be conceptualised as discrete disease entities. Disease entities designate the *whole course* of the process (Hucklenbroich, 2014), from cause(s), to the pathophysiological process including the symptomatic phase and the multitude of disease outcomes. Diseases usually involve detectable pathophysiological changes that can be targets for screening: abnormal lab results values, tissue changes, system dysfunction among others. But, these do not all have direct and predictable relationships to the disease process. That is, a parameter (e.g., a lab value) may change throughout the course of the disease but it may not change in predictable ways or, the changes may not correlate to predictable changes in disease course. Ideally, a screening test should detect a target that is strongly linked to the disease outcome one wishes to prevent (Harris, Sawaya, Moyer, & Calonge, 2011).

Screening reverses the traditional direction of the clinical encounter with the healthcare system. In general, a person experiences symptom then solicits medical attention in order to find relief for these symptoms and to eliminate the cause of them where possible. Screening functions by seeking objective, measurable bodily changes early in the pathophysiological process before there are symptoms or at least before the person is bothered enough by them to seek assistance. Thus, screening is an intervention offered to people who do not perceive themselves as ill and/or are not seeking care. Because of this reversal, it is ethically important that a screening test be able to discriminate between cases and non-cases so as to ensure that people are not led to perceive themselves as ill unnecessarily and/or be exposed to painful, disruptive, or costly diagnostic procedures.

Sensitivity and specificity are concepts that provide information about the accuracy of screening tools; that is, their ability to identify only those people affected by the condition as compared to a gold standard diagnostic evaluation. A gold standard diagnostic test is the most accurate test available for detecting those people who really have the underlying disease and those who do not. Sensitivity describes the true positive rate, that is, the percentage of those who truly have the condition who get a positive result on a screening test. Sensitivity is expressed by the following formula: TP/TP+FN where TP means true positives and FN means false negatives.[2] Specificity describes the true negative rate, that is, the percentage of those with negative results on a screening test who really do not have the condition. Specificity is expressed by the formula: TN/TN + FP where TN means true negatives and FP[3] means false positives. The following table expresses these concepts.

		Disease or condition	
		+	−
Test	+	True positive (TP)	False positive (FP)
	−	False negative (FN)	True negative (TN)

Screening depends on the assumptions that a) there is a disease, b) that the gold standard diagnostic procedure can be counted on to detect it, and c) that we can know under what circumstances the diagnostic procedure is correct or is in error. That is, FN rates (as part of the sensitivity formula) and FP rates (as part of the specificity formula) require knowledge of the error rate of the diagnostic procedure. The sensitivity and specificity of a screening tool can then be assessed against this diagnostic gold standard. The next section will discuss diagnosis and diagnostic error.

3. Diagnosis and diagnostic error

According to the biomedical model, accurate diagnosis is a crucial step in treating a clinical problem as it means accurate prognostication and greater ability to prescribe treatments likely to slow down, halt or even reverse the

[2] FN is someone who gets a negative result on the test but really does have the disease according to the diagnostic gold standard.

[3] FP is someone who gets a positive result on the test but really does not have the disease according to the diagnostic gold standard.

pathophysiological process. An accurate diagnosis achieves three things: it distinguishes pathological[4] from non-pathological states of health; it identifies the distinct cause(s) and/or pathophysiological process responsible for a patient's presentation; and it allows for the prediction about the evolution of the disease over time. Symptomatic treatment (e.g., of vomiting or of pain) can be achieved without accurate diagnosis, however, this would be considered insufficient if there is an underlying pathology that could be identified and corrected, or at least attenuated, because without treating the underlying cause, the symptoms would likely continue. However, even with an accurate diagnosis, there may be no treatment available which targets the underlying pathology, in which case symptomatic treatment may be all that is available.

Laypeople and clinicians alike have a working concept of diagnostic inaccuracy or diagnostic error. If someone gets "diagnosis A" at one point in time and later this is changed to "diagnosis B", one will consider the first diagnosis to be an error. On this view, a diagnostic error has taken place in light of a later, correct diagnosis. Different authors have refined this basic conceptualisation. For example, Graber, Franklin, & Gordon (2005) proposes the following definition of diagnostic error: a diagnosis that was wrong (another diagnosis was made before the correct one), unintentionally delayed (sufficient information was available earlier), or missed (no diagnosis was made). According to Graber, Franklin, & Gordon (2005), this definition is based on a definitive confirmation of an error, such as pathognomonic physical signs, tissue analysis, or radiographic findings. Still other definitions suggest that diagnostic error must be linked to failure in the diagnostic process (Schiff et al., 2009) or missed opportunities to make the correct diagnosis (Singh, 2014). The argument behind this approach is that diagnostic errors must be errors of process rather than simply of outcome. In fact, a recent conceptual model for diagnostic error explicitly distinguishes between diagnostic process failures and diagnostic label failures (Newman-Toker, 2014). While these definitions vary, they all presume a diagnostic gold standard test which can pick out the underlying cause or pathological process of a disease and which will be accurate all the time or almost all of the time.

[4] We acknowledge that determining what constitute a pathological state is a value-laden exercise. However, for many conditions, there is a relative social consensus concerning the disvalue of those health states which facilitates the designation of such states as pathological. This can be more contentious and for a larger number of mental disorders than for non-mental disorders where there is robust debate as to whether the experience is pathological or simply a normal variant e.g., gender dysphoria, autism spectrum disorder.

While this approach to conceptualising diagnostic accuracy and error may seem unproblematic in most areas of medicine, there are certain conditions for which it does not seem to apply. For example, migraine headaches (without aura) are diagnosed by means of operationalised diagnostic criteria rather than by reference to underlying pathology. The International Classification of Headache Disorders, third edition (Headache Classification Committee of the International Headache Society [IHS], 2018) provides explicit rules about the necessary and sufficient characteristics of the diagnosis.

Diagnostic criteria:

A. At least five attacks[1] fulfilling criteria B-D

B. Headache attacks lasting 4–72 h (untreated or unsuccessfully treated)[2,3]

C. Headache has at least two of the following four characteristics:

  1. unilateral location

  2. pulsating quality

  3. moderate or severe pain intensity

  4. aggravation by or causing avoidance of routine physical activity (e.g., walking or climbing stairs)

D. During headache at least one of the following:

  1. nausea and/or vomiting

  2. photophobia and phonophobia

E. Not better accounted for by another ICHD-3 diagnosis.

The criteria used to define the migraine syndrome are considered unambiguous (Olesen, 2006) and can be evaluated in a dichotomous fashion (present or absent). Therefore, whether a person's clinical presentation qualifies as an example of a case (caseness) requires minimal judgment on the part of clinician. Indeed, computerised diagnosis shows high rate of concordance with clinical diagnosis (Eslami et al., 2013). Even if diagnostic accuracy cannot be judged against definitive confirmation of underlying pathology, fulfillment of the diagnostic criteria can be evaluated unequivocally meaning that diagnosis can distinguish pathological from non-pathological states. Furthermore, the diagnosis permits accurate prognostication. Diagnostic accuracy and error can therefore be judged against these two criteria: is the person who receives a diagnosis of migraine headache experiencing a pathological state of health and does the diagnosis enable the accurate prediction of the course of disease over time? If yes, then the diagnosis is correct.

As a branch of medicine psychiatry is committed to the notion that mental disorders are real entities, independent of our awareness and experience of them that lead to changes in the bodies of affected persons.

Psychiatry extends this view through its commitment to the biopsychosocial model (Engel, 1980). According to this framework, mental disorders are conceptualised as the product of multiple, multilevel interacting causes therefore, illness cannot fully be accounted for by changes within the biological organism. However, after decades of research concerning the aetiologies of psychiatric disorders, no specific pathophysiological (or psychopathological) causal mechanisms of specific conditions have been identified. Instead, psychiatric syndromes are assessed by matching patients' symptoms to sets of operationalised criteria belonging to different diagnostic categories. For example, the Diagnostic and Statistical manual (DSM) produced by the American Psychiatric Association, offers a framework for psychiatric diagnosis as well as an exhaustive compendium of diagnostic categories and operationalised criteria (American Psychiatric Association, 2013). It is commonly used as the diagnostic reference manual in clinical practice in Canada and the United States. The accepted diagnostic instrument in research studies is the structured diagnostic interview based on the DSM (e.g., SCID-5-Research Version) or other such systems such as the International Classification of Disease whereas in clinical practice the accepted method is the psychiatric assessment completed by a trained clinician (see Appendix A). Many of the diagnostic categories require exclusion of specific aetiologies like substances, general medical condition or other psychiatric syndromes.

Similar to migraine headaches, major depressive disorder is also defined by operationalised criteria. For example, major depressive disorder (for adults) in the DSM-5 is defined by criteria A through E (adapted from Sadock, Sadock, & Ruiz, 2015)[5]:

A. Five (or more) of the following symptoms have been present during the same 2-week period and represent a change from previous functioning; at least one of the symptoms is either (1) depressed mood or (2) loss of interest or pleasure.

   1. Depressed mood most of the day, nearly every day, as indicated by either subjective report (e.g., feels sad, empty, hopeless) or observation made by others (e.g., appears tearful).

   2. Markedly diminished interest or pleasure in all, or almost all, activities most of the day, nearly every day (as indicated by either subjective account or observation.)

[5] These criteria reflect the Major Depressive Disorder criteria as outlined in the DSM-5.

3. Significant weight loss when not dieting or weight gain (e.g., a change of more than 5% of body weight in a month), or decrease or increase in appetite nearly every day. Insomnia or hypersomnia nearly every day.
4. Psychomotor agitation or retardation nearly every day (observable by others, not merely subjective feelings of restlessness or being slowed down).
5. Fatigue or loss of energy nearly every day.
6. Feelings of worthlessness or excessive or inappropriate guilt (which may be delusional) nearly every day (not merely self-reproach or guilt about being sick).
7. Diminished ability to think or concentrate, or indecisiveness, nearly every day (either by subjective account or as observed by others).
8. Recurrent thoughts of death (not just fear of dying), recurrent suicidal ideation without a specific plan, or a suicide attempt or a specific plan for committing suicide.

B. The symptoms cause clinically significant distress or impairment in social, occupational, or other important areas of functioning.

C. The episode is not attributable to the physiological effects of a substance or to another medical condition.

D. The occurrence of the major depressive episode is not better explained by schizoaffective disorder, schizophrenia, schizophreniform disorder, delusional disorder, or other specified and unspecified schizophrenia spectrum and other psychotic disorders.

E. There has never been a manic episode or a hypomanic episode.[6]

Again, like a diagnosis of migraine headache, a diagnosis of major depressive disorder does not point towards underlying pathology. However, unlike the migraine criteria, the category contains several symptoms, not all of which are required to receive the diagnosis. Thus, patients diagnosed with major depressive disorder may differ greatly from one another in terms of their symptoms. This within-category heterogeneity means that diagnosis is not necessarily linked to prognosis. In other words, it is not possible to know, at the time of diagnosis, how a given person's condition will evolve and whether the person will improve with specific treatments (e.g., medication and/psychotherapy), social interventions (e.g., income support

[6] Note that the C, D and E criteria are exclusion criteria meaning they indicate when a diagnosis of major depression cannot be given but they do not indicate when the diagnosis can be given.

programs), general interventions (e.g., availability for validation and support), or no intervention at all. Therefore, diagnostic accuracy and error for major depression cannot be assessed against the ability to prognosticate.

Can diagnosis for depression be judged accurate on its ability to distinguish between disordered and non-disordered, such that the diagnostic procedure clearly identifies caseness? Compared to the diagnosis of migraine headache, making a diagnosis of major depression according to the operationalised criteria proposed by the DSM-5 requires clinical judgment and contextualisation. Caseness requires evaluator judgment as to whether or not a criterion is fulfilled as well as consideration of contextual factors lying outside the criteria. However, these tasks are themselves not operationalised. For example, the companion text to the DSM notes that evaluation of criteria requires a considerable degree of clinical judgment. The authors write, "Will a psychiatrist tell a patient he is not depressed just because he fails to meet the diagnostic threshold?" (Black & Grant, 2014). This statement implies that fulfilling diagnostic criteria may not be necessary nor sufficient for diagnosis. However, what else is needed is not explained. In another DSM companion text, the Study Guide to DSM-5 (Roberts & Louie, 2015), the authors state that a clinician should make a diagnosis *in context* and that failing to do so can give rise to incorrect diagnosis. Here context is meant in two ways. First, it is taken to mean the purpose of the diagnostic interview and how this can influence what information the patient is willing to share. Second, the authors state that clinicians should be aware of all three components of the biopsychosocial model, (biological, psychological, and social) and how they interact with and contribute to a patient's clinical presentation. Clinicians are instructed to "weigh them in applying diagnostic criteria and arriving at diagnosis." (Roberts & Louie, 2015). Again, something beyond the operational criteria is required to determine whether a person has a condition that is defined by these criteria. Thus, there is a tension between the operationalised diagnostic criteria and the requirement and methods to take into consideration factors outside criteria in a manner that itself is undefined.

Philips writes that in the actual state of the classification of mental disorder, it is difficult to speak meaningfully of diagnostic errors: " … the lack of strong biomarkers in psychiatry hobbles both the construction of scientifically valid diagnosis and the assessment of error" (Phillips, 2014). Nevertheless, psychiatrists do speak of error in the sense of diagnosis being accurate and inaccurate. For example, DSM experts use the language of true positive and true negative to discuss diagnostic accuracy in psychiatry (Roberts &

Louie, 2015). According to these authors, some people are diagnosed as having a disorder when they do not have it (e.g., someone incorrectly diagnosed with bipolar disorder because the patient did not disclose that the manic episode was substance-induced (Roberts & Louie, 2015)). Others are not diagnosed when they have the disorder (e.g., a substance-related disorder that goes undetected or a post-partum major depressive episode that is mistaken for an adjustment disorder (Roberts & Louie, 2015). In these examples, the diagnostic error results from not correctly applying (or not having the information to apply) the diagnostic criteria to the clinical presentation. Indeed, studies that attempt to estimate diagnostic error is psychiatry employ this definition of error. For example, Glazier et al. report that on a written case vignette in which the DSM-V criteria for obsessive compulsive disorder (OCD) were unambiguously present, 50% of OCD cases are misdiagnosed or not diagnosed by a primary care physician (Glazier, Swing, & McGinn, 2015). Mitchell, Vaze, & Rao, 2009 compared different diagnostic/interviewing methodologies. They report a meta-analysis of studies comparing unassisted general practitioner diagnosis of depression to structured or semi-structured diagnosis and show that depression is correctly (where correctly means in accordance with DSM criteria) identified by general practitioners in 47.3 % of cases.

Wakefield introduces another usage of diagnostic error in psychiatry by referring to 'false positive diagnosis' by which he means attributing a diagnosis to someone when the person does not have a disorder (Wakefield, 2010). The diagnosis is considered erroneous not in light of the criteria themselves but because the symptoms are thought to represent a normal reaction rather than a dysfunction in internal mechanism (Wakefield, 2013). As Bolton (2007) comments, "Wakefield's argument has been that this methodology (the DSM's) in effect detaches troublesome mental states and behaviours from their context, failing to take account of whether they are 'normal' responses to adversities ... ". While there is continued debate over Wakefield's concept[7] and more generally how to differentiate disorder from non-disorder, on this view understanding of the context of symptoms is central to accurate diagnosis of mental disorder. Without

[7] Wakefield introduces the harmful dysfunction analysis as a model of disorder. In order to be a disorder, a condition must be harmful, as judged by social values, and due to a dysfunction, that is a failure in biologically designed functioning of internal mechanism. For example, he has argued against the elimination of bereavement exclusion in the diagnosis of major depressive disorder on the grounds that it does not reflect dysfunction (Wakefield, 2013).

context, the psychiatric diagnostic evaluation cannot succeed in its task of distinguishing disorder from non-disorder. In other words, disorder cannot be judged solely in accordance to criteria fulfillment because there are clinical situations in which we accept fulfillment of the criteria as reflecting diagnostic accuracy and others in which we do not. As pointed out above, some DSM experts suggest that criteria fulfillment must occur but that understanding and weighing of context is essential to the method of applying criteria. Others imply that criteria fulfillment is not necessary in all cases.

Thus, it is uncertain how to define diagnostic accuracy and error for the gold standard diagnostic evaluation for major depressive disorder and indeed for mental disorders more generally. Diagnostic accuracy and error cannot be judged against external criteria such as underlying pathology, prognosis, or the ability to distinguish normal from abnormal. And if we cannot do so, it is difficult to then assess the accuracy of a screening test. In the next section, we will apply this analysis to the specific example of screening for depression.

4. Screening for depression

Screening for major depressive disorder is achieved using a questionnaire (a screening tool). Such tools ask the person being screened whether or not she experiences some (e.g., Whooley Questions - see Appendix B (Whooley, Avins, Miranda, & Browner, 1997)) or all (e.g., PHQ-9- see Appendix C (Kroenke, Spitzer, & Williams, 2001)) of the symptoms of major depressive disorder. By contrast, the diagnostic procedure ensures that a person has at least five of the symptoms while fulfilling the remaining criteria, B through E. In other areas of medicine, the diagnostic evaluation following a positive result at the screening stage attempts to confirm the presence of the underlying entity or process causing the illness (which then permits prognostication and prediction of treatment response). However, in psychiatry there is no underlying entity or process to confirm. The diagnostic evaluation of major depression serves to confirm or disconfirm that a person's situation corresponds to the operationalised criteria for major depression.

In the absence of a method for judging accuracy of diagnosis, contradictory results between a screening tool and a diagnostic procedure pose a particular problem. True positives and true negatives are those people whose positive or negative results on the screening tool match those on the structured or expert clinical interview (see page 3). For major depression, the difference between a screening questionnaire and the diagnostic evaluation lies in the additional information sought. A structured interview such as the structured clinical interview for DSM-5 (First, Williams, Karg, & Spitzer, 2016) will ask about the B and C criteria and not just symptoms (A criteria). Meanwhile, in an expert clinical interview, the clinician will cover additional areas such as the medical history, personal history, and social circumstances (see Appendix A). When the results between a screening tool and the diagnostic interview do not match, we assume the diagnostic interview is correct and the screening tool incorrect, but on what grounds?

When a screening tool yields a positive result but the diagnostic procedure a negative result, we have an apparent false positive on the screening tool. In the context of major depressive disorder, a false positive means that a person has some but not all of the symptoms to fulfill the A criterion or fulfills the A criteria but not one or more of the B, C, D or E criteria. For example, a patient might have a positive result on a screening test (because she has 4 symptoms under criterion A) but have a negative result on the diagnostic procedure (because she has only 4 symptoms under criterion A). Yet, as DSM experts suggest, the diagnosis of depression may nevertheless be correct due to the weight placed by the diagnostician on the severity of those four symptoms, or the answers to B criterion questions or both. In other words, we cannot take for granted that the negative result from the diagnostic procedure is the right result.

When a screening tool reports a negative result but the diagnostic procedure gives a positive result, we have apparently, a false negative on the screening tool. It seems surprising that when being asked about the symptoms of depression a person would not have enough symptoms (less than 5 of 9) to meet the minimum requirement in order to obtain a positive result at screening but does have enough symptoms (at least 5 of 9) according to those very same questions asked during a structured or clinical interview. It may be that the person gives the same answers but that a diagnostician hearing the answers interprets them as indicating a higher level of pathology than indicated by the face value answers given on a screening tool. Or the person gives the same answers, but the diagnostician interprets the answers to the B criterion questions in such a manner that she sets aside the five symptom minimum for the A criterion. In either case, something outside of the criteria is being used to confirm that the person fulfills the criteria when the screening tool indicated the opposite. In such cases, we do not have access to the mechanism and therefore cannot take for granted that it yields correct results. It may also be possible that the person simply gave different answers to the screening questions as compared to the diagnostic questions. In this case, there is no standard to judge that the answers on the screening tool were wrong while those during the diagnostic interview is correct.

The logic of a screening program depends on the idea that the gold standard diagnostic procedure can be counted on to provide accurate diagnosis. Some authors (McGorry, Hickie, Yung, Pantelis, & Jackson, 2006) have developed a clinical staging model for psychiatric conditions including depression. This approach divides the condition into clinical stages based on timeline and severity of illness. According to these authors, early stages are opportunities for early intervention. However, this model still requires some sort of condition (whether a standard mental disorder category such as a major depressive disorder or a symptom cluster such as psychosis) and therefore faces the same problem we have discussed in this paper that is, the need to know when the condition is present and when that procedure is in error. As the discussion above illustrates, it is difficult to establish the grounds on which we would say that the diagnostic procedure in psychiatry is wrong and when it is right as compared to a screening tool. Any presumed greater accuracy of the diagnostic evaluation in psychiatry compared to a screening test requires recourse to something that lies outside the diagnostic criteria themselves: clinical interpretation and judgment.

There are various approaches that attempt to give a more explicit role to clinical interpretation and judgment in diagnosis, that frame clinical phenomena in their broader context and that include hypothesis-building about the possible contributors or causes should be an integral part of the diagnosis of a mental health condition. These include the biopsychosocial model (Engel, 1980); the perspectives approach, (McHugh & Slavney, 1998; Peters, Taylor, Lyketsos, & Chisolm, 2012); the Person-Centered Integrative Diagnosis (Mezzich, Snaedal, van Weel, Botbol, & Salloum, 2011); and the WPA idiographic personalised diagnostic formulation (IGDA Workgroup, WPA, 2003). Exploration of each of these proposals exceeds the scope of this chapter, and it is uncertain whether such methods will permit a robust conception of accuracy and error which could arbitrate between contradictory results from screening tests and diagnostic evaluations.

In his model-based view of science, Heinrichs (2015) offers an approach that could potentially solve this problem. He proposes to develop models, which are sets of hypotheses about the relevant elements of a clinical case and the relationships between them. These would reflect the unique clinical context of the patient and point to hypotheses about causes of the person's problems. As models are explanations, they allow for prediction of behavior and outcome. Because a person's situation is dynamic and influenced by many factors, models are always partial, selective and subject to further refinement and adjustment with time. For example, in the case of a patient who has current depressive symptoms, another medical condition that can cause depressive symptoms (such as hypothyroidism), and who is going through a difficult life event, the model developed should take into account these diverse factors and the extent to which they predict evolution of the person's condition and point to appropriate interventions. If a model does not predict outcome and/treatment response, it might be that it did not fit the phenomena we want to explain, that the factors identified were not as significant as they were considered to be, or that other unidentified factors mitigated the effect of intervention(s). Thus, a model-based approach has the potential to be comprehensive and testable. However, this type of approach requires an epistemological shift from the ideas of accuracy and error. Instead of conceptualising diagnosis as something that is true or false, a model would need to be evaluated on how well it represents a given patient and his or her evolution in time that is, a good diagnosis rather than a correct one.

5. Ethical considerations

Thombs et al. (2012) have argued against routine screening for major depressive disorder on the basis that it has not been demonstrated in randomised controlled trials to improve health outcomes for those screened compared to those not screened. These authors question the ethics of screening on the grounds that it may not lead to benefits and may even lead to harms (for example, if resources are spent on a screening program of uncertain effectiveness this could take resources away from effective programs for those already diagnosed with major depressive disorder). For these reasons, routine screening for major depressive disorder in primary care settings has been the subject of debate resulting in a recommendation against the practice in Canada when previously it had been advised (Joffres et al., 2013). We agree that screening may have no benefit, and may even be harmful, but for different reasons. Thombs et al. (2012) assume a model of early diagnosis borrowed from general medicine. We contribute to the ongoing debate about screening by arguing that this model cannot be applied to the diagnosis of mental disorders.

Screening is designed to identify two distinct subgroups of people: those who are symptomatic but have not come to clinical attention, regardless of the reason (e.g., lack of access, lack of knowledge of how to access services, fear of stigma, fear of treatment) and those who have a condition but have no symptoms. We will discuss each group in turn.

5.1 People who are suffering but who do not come to clinical attention

Screening for major depressive disorder might be a way of helping to improve people's access to care. Screening programs are, by design, intended to be applied to large numbers of people in order to identify those who have a condition that can be treated sooner than later in order to reduce their suffering. The conceptual problems we have discussed concerning diagnosing mental disorder, namely, the inability to use standard criteria of accuracy and error (distinguishing normal from abnormal, identification of underlying pathophysiology, and prognostication) make it difficult to resolve conflicting results between a screening test and a diagnostic evaluation. Logically speaking, screening does not narrow the field of possible patients who require a diagnostic intervention. In other words, in order to improve access to care, everyone screened, regardless of positive or negative result, would require a diagnostic evaluation. Screening then is an ineffective way of case-finding and represents a waste of scarce resources.

If on the other hand, we take for granted that the diagnostic procedure is more accurate, than the screening test, we are still faced with the problems inherent to the diagnostic procedure itself, namely, the difficulties involved in differentiating between normal and abnormal in the spectrum of mood states lying between normal sadness and major depressive disorder (American Psychiatric Association, 2013). Since screening seeks to identify people before they come to clinical attention, screening programs will identify a certain number of people whose experiences are in this middle zone of this spectrum and whose experiences will go on to be medicalised. Is it ethical to conceive of these problems as disorders, residing within individuals? Does it point up ethically justifiable solutions in terms of the burdens those individual will have to bear? Because it is a healthcare activity, screening for major depression invites these questions.

5.2 People who are affected by major depressive disorder but do not think they are

There is a sense in which such a category is implausible. Being depressed is a situation in which one is feeling in certain ways and is not functioning well in his or her life as a result. A person can be found to have a silent pancreatic tumour and upon being told the diagnosis respond with 'but I feel fine.' By definition, this same situation could not occur in the case of major depressive disorder. But even if there were some way in which a person could be depressed without self-identifying as such, there is a potential ethical problem in inviting him to conceive of his experiences this way. A recurring criterion in the diagnostic categories for mental disorders is the requirement that the symptoms of the condition cause "clinically significant distress or impairment in social, occupational, or other important areas of functioning." Norms about social and/or occupational dysfunction vary between individuals, groups and societies. Self-identifying as ill is ethically important in the case of mental disorder because we are otherwise telling such a person to perceive himself as dysfunctional when he does not. Further, by taking on these norms, the diagnosed individual is being asked to assume the responsibility for and costs and risks of treatment when he might instead legitimately expect society to adapt itself to his way of being. Washington argues that the evaluative nature of psychiatric diagnosis means that one must consider psychiatric symptoms in light of whether they do or do not contribute to individual flourishing (Washington, 2018). She points out that in psychiatry, failure to consider what the person considers to be mentally healthy has permitted diagnoses which we now consider to show disrespect for human diversity at best and disregard for human rights at worst. A similar line of argument has been used successfully by disability scholars and activists to expose the notion of disability itself as demeaning and discriminatory (Washington, 2018).

6. Conclusions

The goal of screening for major depression is early diagnosis and early treatment with a view to reducing the overall burden of suffering as well as the economic impact of mental illness. Achieving early diagnosis depends upon an accurate diagnostic evaluation procedure which itself requires an understanding of what constitutes a diagnostic error. Using the example of major depressive disorder, we have argued that apart from criteria-based failure there is no way to assess diagnostic error in psychiatry. Furthermore, criteria-based error is an insufficient concept of error to inform screening programs because of the tautological nature of psychiatric diagnosis: having the disorder (meeting the operationalised criteria) means meeting the operationalised criteria. Existing resources for resolving contradictions between screening tools and diagnostic evaluations are not an explicit part of the diagnostic procedure (clinical judgment and interpretation), and, to date, there are no methods for evaluating the quality of clinical judgment and interpretation in order to confirm the superiority of the gold standard. Thus, the accuracy of the diagnostic gold standard is simply assumed.

On these grounds, we have questioned the validity of screening for major depressive disorder and therefore the ethical implications of such programs. Because screening programs seek to identify people before they self-identify as ill or suffering, it is important that this be done on the basis of shared norms. Psychiatric diagnosis requires third party (clinician) interpretation and judgment to determine whether a person's condition is considered to be a disorder or not. Thus, screening large numbers of people for major depression risks pathologizing and medicalising circumstances which those people may view very differently.

We think a conceptually and ethically sound public health approach involves ensuring suffering people have access to an evaluation whose goal is to develop a context-dependent understanding of their problems and orienting them towards services best suited to those problems (including treatment for major depressive disorder where necessary). Instead of borrowing from the public health approach more suited to the conceptual apparatus of general medicine, such an approach reflects more accurately the current model (biopsychosocial) of mental disorders. Ultimately such an approach must itself be evaluated — perhaps not using the concepts of accuracy and error — but on how well it can portray the patient's reality, whether it can be point up appropriate solutions, and to what extent it can reduce the burden of suffering.

Acknowledgment

The authors would like to thank the reviewers for their constructive comments on the paper as well as participants of the 2016 annual meeting of the Association for the Advancement of Philosophy of Psychiatry (AAPP) for comments on an earlier version of this paper.

Appendix A. Psychiatric Assessment

Adapted from parts of an initial psychiatric interview (Sadock et al., 2015)

I. Identifying data
II. Source and reliability
III. Chief complaint
IV. Present illness
V. Past psychiatric history
VI. Substance use/abuse
VII. Past medical history
VIII. Family history
IX. Developmental and social history
X. Review of systems
XI. Mental status examination
XII. Physical examination
XIII. Formulation
XIV. DSM-5 diagnoses
XV. Treatment plan

Appendix B. Whooley screening questions for depression

Adapted from Whooley et al. (1997)

1. "During the past month, have you often been bothered by feeling down, depressed, or hopeless?" (yes/no)
2. "During the past month, have you often been bothered by little interest or pleasure in doing things?" (yes/no)

Appendix C. Screening questions from the Patient Health Questionnaire (PHQ-9)

Adapted from Kroenke et al. (2001)

1. Over the last 2 weeks, how often have you been bothered by any of the following problems? Read each item carefully, and circle your response.

a. Little interest or pleasure in doing things

Not at all	Several days	More than half the days	Nearly every day

b. Feeling down, depressed, or hopeless

Not at all	Several days	More than half the days	Nearly every day

c. Trouble falling asleep, staying asleep, or sleeping too much

Not at all	Several days	More than half the days	Nearly every day

d. Feeling tired or having little energy

Not at all	Several days	More than half the days	Nearly every day

e. Poor appetite or overeating

Not at all	Several days	More than half the days	Nearly every day

f. Feeling bad about yourself, feeling that you are a failure, or feeling that you have let yourself or your family down

Not at all	Several days	More than half the days	Nearly every day

g. Trouble concentrating on things such as reading the newspaper or watching television

Not at all	Several days	More than half the days	Nearly every day

h. Moving or speaking so slowly that other people could have noticed. Or being so fidgety or restless that you have been moving around a lot more than usual

Not at all	Several days	More than half the days	Nearly every day

i. Thinking that you would be better off dead or that you want to hurt yourself in some way

Not at all	Several days	More than half the days	Nearly every day

2. If you checked off any problem on this questionnaire so far, how difficult have these problems made it for you to do your work, take care of things at home, or get along with other people?

Not difficult at all	Somewhat difficult	Very difficult	Extremely difficult

References

American Psychiatric Association. (2013). *Diagnostic and statistical manual of mental disorders: DSM-5* (5th ed.). Washington, DC: American Psychiatric Publishing.

Black, D. W., & Grant, J. E. (2014). *DSM-5 guidebook : The essential companion to the diagnostic and statistical manual of mental disorders* (5th ed.). Washington, DC : American Psychiatric Association.

Bolton, D. (2007). The usefulness of Wakefield's definition for the diagnostic manuals. *World Psychiatry, 6*(3), 164–165.

Engel, G. L. (1980). The clinical application of the biopsychosocial model. *The American Journal of Psychiatry, 137*(5), 535–544. https://doi.org/10.1176/ajp.137.5.535.

Eslami, V., Rouhani-Esfahani, S., Hafezi-Nejad, N., Refaeian, F., Abdi, S., & Togha, M. (2013). A computerized expert system for diagnosing primary headache based on International Classification of Headache Disorder (ICHD-II). *SpringerPlus, 2*(1), 199. https://doi.org/10.1186/2193-1801-2-199.

First, M., Williams, J., Karg, R., & Spitzer, R. (2016). *Structured clinical interview for DSM-5 disorders, clinician version (SCID-5-CV)*. Arlington, VA: American Psychiatric Association.

Glazier, K., Swing, M., & McGinn, L. K. (2015). Half of obsessive-compulsive disorder cases misdiagnosed: Vignette-based survey of primary care physicians. *The Journal of Clinical Psychiatry, 76*(6), e761–e767. https://doi.org/10.4088/JCP.14m09110.

Graber, M. L., Franklin, N., & Gordon, R. (2005). Diagnostic error in internal medicine. *Archives of Internal Medicine, 165*(13), 1493–1499. https://doi.org/10.1001/archinte.165.13.1493.

Harris, R., Sawaya, G. F., Moyer, V. A., & Calonge, N. (2011). Reconsidering the criteria for evaluating proposed screening programs: Reflections from 4 current and former members of the U.S. Preventive services task force. *Epidemiologic Reviews, 33*, 20–35. https://doi.org/10.1093/epirev/mxr005.

Headache Classification Committee of the International Headache Society (IHS). (2018). The international classification of headache disorders, 3rd edition. *Cephalalgia, 38*(1), 1–211. https://doi.org/10.1177/0333102417738202.

Heinrichs, D. W. (2015). Model-based science and the ethics of ongoing treatment negotiation. In *The oxford handbook of psychiatric ethics*. https://doi.org/10.1093/oxfordhb/9780198732372.013.26.

Hucklenbroich, P. (2014). "Disease entity" as the key theoretical concept of medicine. *The Journal of Medicine and Philosophy, 39*(6), 609–633. https://doi.org/10.1093/jmp/jhu040.

IGDA Workgroup, WPA. (2003). IGDA. 8: Idiographic (personalised) diagnostic formulation. *The British Journal of Psychiatry. Supplement, 45*, s55–s57.

Joffres, M., Jaramillo, A., Dickinson, J., Lewin, G., Pottie, K., Shaw, E., … Canadian Task Force on Preventive Health Care. (2013). Recommendations on screening for depression in adults. *CMAJ: Canadian Medical Association Journal = Journal de l'Association Medicale Canadienne, 185*(9), 775–782. https://doi.org/10.1503/cmaj.130403.

Kroenke, K., Spitzer, R. L., & Williams, J. B. W. (2001). The PHQ-9. *Journal of General Internal Medicine, 16*(9), 606–613. https://doi.org/10.1046/j.1525-1497.2001.016009606.x.

McGorry, P. D., Hickie, I. B., Yung, A. R., Pantelis, C., & Jackson, H. J. (2006). Clinical staging of psychiatric disorders: A heuristic framework for choosing earlier, safer and more effective interventions. *The Australian and New Zealand Journal of Psychiatry, 40*(8), 616–622. https://doi.org/10.1080/j.1440-1614.2006.01860.x.

McHugh, P. R., & Slavney, P. R. (1998). *The perspectives of psychiatry* (2nd ed.). Baltimore: Johns Hopkins University Press.

Mezzich, J. E., Snaedal, J., van Weel, C., Botbol, M., & Salloum, I. (2011). Introduction to person-centred medicine: From concepts to practice. *Journal of Evaluation in Clinical Practice, 17*(2), 330–332. https://doi.org/10.1111/j.1365-2753.2010.01606.x.

Mitchell, A. J., Vaze, A., & Rao, S. (2009). Clinical diagnosis of depression in primary care: A meta-analysis. *Lancet (London, England), 374*(9690), 609–619. https://doi.org/10.1016/S0140-6736(09)60879-5.

Newman-Toker, D. E. (2014). A unified conceptual model for diagnostic errors: Underdiagnosis, overdiagnosis, and misdiagnosis. *Diagnosis (Berlin, Germany), 1*(1), 43–48. https://doi.org/10.1515/dx-2013-0027.

Olesen, J. (2006). *The headaches* (p. c2006). Philadelphia, PA : Lippincott Williams & Wilkins (Ressource électronique RB 128).

Peters, M. E., Taylor, J., Lyketsos, C. G., & Chisolm, M. S. (2012). Beyond the DSM: The perspectives of psychiatry approach to patients. *The Primary Care Companion for CNS Disorders, 14*(1). https://doi.org/10.4088/PCC.11m01233.

Phillips, J. (2014). Detecting diagnostic error in psychiatry. *Diagnosis (Berlin, Germany), 1*(1), 75–78. https://doi.org/10.1515/dx-2013-0032.

Roberts, L. W., & Louie, A. K. (2015). *Study guide to DSM-5* (1st ed.). Washington, DC: American Psychiatric Publishing.

Sadock, B. J., Sadock, V. A., & Ruiz, P. (2015). *Kaplan & Sadock's synopsis of psychiatry : Behavioral sciences/clinical psychiatry*. Philadelphia: Wolters Kluwer.

Schiff, G. D., Hasan, O., Kim, S., Abrams, R., Cosby, K., Lambert, B. L., ... McNutt, R. A. (2009). Diagnostic error in medicine: Analysis of 583 physician-reported errors. *Archives of Internal Medicine, 169*(20), 1881–1887. https://doi.org/10.1001/archinternmed.2009.333.

Singh, H. (2014). Editorial: Helping health care organizations to define diagnostic errors as missed opportunities in diagnosis. *Joint Commission Journal on Quality and Patient Safety, 40*(3), 99–101.

Siu, A. L., US Preventive Services Task Force (USPSTF), Bibbins-Domingo, K., Grossman, D. C., Baumann, L. C., Davidson, K. W., ... Pignone, M. P. (2016). Screening for depression in adults: US preventive services task force recommendation statement. *JAMA, 315*(4), 380–387. https://doi.org/10.1001/jama.2015.18392.

Thombs, B. D., Coyne, J. C., Cuijpers, P., de Jonge, P., Gilbody, S., Ioannidis, J. P. A., ... Ziegelstein, R. C. (2012). Rethinking recommendations for screening for depression in primary care. *CMAJ : Canadian Medical Association Journal, 184*(4), 413–418. https://doi.org/10.1503/cmaj.111035.

Tulchinsky. (2000). *The new public health*. San Diego, CA: Academic Press.

Wakefield, J. C. (2010). Misdiagnosing normality: Psychiatry's failure to address the problem of false positive diagnoses of mental disorder in a changing professional environment. *Journal of Mental Health (Abingdon, England), 19*(4), 337–351. https://doi.org/10.3109/09638237.2010.492418.

Wakefield, J. C. (2013). The DSM-5 debate over the bereavement exclusion: Psychiatric diagnosis and the future of empirically supported treatment. *Clinical Psychology Review, 33*(7), 825–845. https://doi.org/10.1016/j.cpr.2013.03.007.

Washington, N. (2018). Contextualism as a solution to paternalism in psychiatric practice. *Philosophy, Psychiatry, & Psychology, 25*(4), 235–243. https://doi.org/10.1353/ppp.2018.0034.

WHO. (2013). *Investing in mental health: Evidence for action*. Available at: https://www.who.int/mental_health/publications/financing/investing_in_mh_2013/en/.

Whooley, M. A., Avins, A. L., Miranda, J., & Browner, W. S. (1997). Case-finding instruments for depression. Two questions are as good as many. *Journal of General Internal Medicine, 12*(7), 439–445.

CHAPTER EIGHT

Machine learning and suicide prevention: considering context as a guide to ethical design

Phoebe Friesen[a,*] and Katie O'Leary[b]
[a]Ethox Centre, University of Oxford, Oxford, United Kingdom
[b]The Information School, University of Washington, Seattle WA, United States
*Corresponding author: E-mail: phoebe.friesen@ndph.ox.ac.uk

Contents

1. Introduction

As data-driven health research continues to offer remarkable advances in terms of the detection and prediction of many health conditions, those concerned with the extraordinarily difficult task of predicting suicide have taken note. Within suicidology, decades of research have yielded few accurate predictors. Having a history of self-injurious thoughts and behaviours is commonly held to be the best predictor of a suicide attempt, and yet a recent systematic review concluded that such a history only provides an improvement in prediction marginally above chance (Ribeiro et al., 2016). Given the enormous data sets that are increasingly available in electronic health records and online communities, the interest in developing tools for

Developments in Neuroethics and Bioethics, Volume 2
ISSN 2589-2959
https://doi.org/10.1016/bs.dnb.2019.04.006

predicting and intervening on suicide attempts has been taken up by traditional health care organisations and technology companies alike. Here, we use the lens of patient and user expectations to examine the ethical complexities that can arise when personal data and artificial intelligence (AI) are harnessed for suicide prevention.[1] In order to do so, we examine two cases in which similar technologies are developed in divergent contexts. In an analysis of these cases, we draw out the ways in which medical, social, online and offline contexts can shape expectations of suicide prevention and interventions, and consider the ethical implications of these expectations for the design of these novel technologies.

In the following section, we introduce the work of Helen Nissenbaum regarding the importance of expectations for grounding the ethical treatment of personal data in online contexts. Next, in section three we introduce two cases that will form the backdrop of our ethical analysis, both of which involve the use of personal data in the development of machine learning techniques for suicide prevention. In section four, we examine the ways in which expectations related to one's intended audience, data collection, use and protections, as well as the nature of the intervention differ substantially in these cases. Finally, we end with an examination of the ways in which these expectations can underwrite the ethical development and implementation of suicide prevention techniques, and offer recommendations for researchers and technologists working in the design space of machine learning systems for suicide prevention.

2. Expectations and privacy

In her development of a theory of online privacy, Helen Nissenbaum highlights the importance of considering user expectations when developing privacy protections for personal information online (Nissenbaum, 2009, 2011). Nissenbaum critiques the ubiquitous notice-and-consent model, which involves merely notifying and receiving consent from individuals with regards to the use of their personal data, arguing that this model is

[1] It is worth defining both artificial intelligence and machine learning at the outset, as both terms arise frequently in this context. Artificial intelligence (AI) is the study and development of automated methods for solving complex tasks requiring human-level intelligence, such as decision-making, visual perception, learning, and natural language processing. Machine learning is a core subarea of AI which involves using data and experience to develop and improve upon these methods through the use of inferences and patterns, rather than explicit instructions.

unable to adequately respect individual privacy online. One of the many issues with the model, she points out, is that it treats all online spaces as equal when it comes to informational norms. The notice-and-consent model suggests, falsely, that online 'is a distinctive venue, sphere, place, or space defined by the technological infrastructures and protocols of the Net, for which a single set of privacy rules can, or ought to, be crafted' (Nissenbaum, 2011, p. 38). Instead, Nissenbaum argues that our expectations related to online privacy are governed by 'context-relative informational norms', which when violated, result in grounds for complaint (Nissenbaum, 2011, p. 33). Just as one would not expect one's final grade to be announced to the class by one's teacher, it should not be posted publicly online. Similarly, in the context of health care:

> *Patients expect their physicians to keep personal medical information confidential, yet they accept that it might be shared with specialists as needed. Patients' expectations would be breached and they would likely be shocked and dismayed if they learned that their physicians had sold the information to a marketing company. In this event, we would say that informational norms for the health care context had been violated.*
>
> ***(Nissenbaum, 2011, p. 33).***

Rather than emphasising the need to develop a novel account of privacy to deal with online spaces, Nissenbaum points out that we ought to rely on existing 'systems of informational norms that have evolved to accommodate diverse legitimate interests as well as general moral and political principles and context-specific purposes and values' (Nissenbaum, 2011, p. 37). Once we recognise how deeply embedded these expectations and norms are within our lives, we can then develop appropriate mechanisms for adhering to them in online spaces. You probably do not expect your results from a medical test to be publicly announced within a walk-in clinic, but may not mind a friend repeating a story you told in a social setting. Similarly, online spaces that store your medical information ought to ensure that your data is kept secure, while online social platforms may build in mechanisms that make it easier to spread your contributions to other users. Such measures respect the expectations related to personal data that are carried forward from offline settings into online versions of the same context. Nissenbaum argues that to create appropriate privacy protections within online spaces, we should understand existing offline informational norms related to shopping, dating, socialising, seeking medical care, and so on, and use these expectations as the basis for the development of policies and tools that deal with personal information online.

While Nissenbaum is concerned primarily with the question of how we might respect user privacy in online settings, we hope to draw on her insight related to the importance of contextual expectations in both online and offline settings. Through an examination of the differences in expectations that appear within two settings, both of which have been used for the development of tools for suicide prevention and interventions, we illuminate the ways in which ethical tensions may track norms particular to a given context. Similarly, while Nissenbaum is interested in exploring a variety of contexts in which informational norms have evolved, we focus here on norms related to personal health information, and in particular, in relation to highly sensitive health data regarding suicidal risk. While all health data is sensitive, symptoms and confessions of suicidal behaviours are especially so, due to the potentially serious legal, social, and economic ramifications associated with mental illness (Sharac, Mccrone, Clement, & Thornicroft, 2010). Below, we introduce two cases that involve the use of algorithms to predict suicide, followed by a discussion of how expectations within them differ.

3. Two cases

Machine learning systems are developed by training machines to identify patterns in an existing corpus of data so that they can make predictions when given novel data. For example, in the context of suicide prevention, a machine can be trained on a corpus of medical records to identify patterns of symptom or demographic factors that correlate with patient suicidality. That machine can henceforth be used to predict suicidality in new patient medical records based on what it learnt from the prior corpus. While there are different forms of machine learning, we focus on such classic cases of supervised learning within this chapter.[2] Supervised learning systems are derived from training data which has been labelled by humans with the relevant categories (e.g., at risk or not at risk of suicide), which the algorithm then uses

[2] The three primary subfields of machine learning are supervised learning, unsupervised learning, and reinforcement learning. As described above, supervised learning systems take in data annotated with specific labels (e.g., a collection of emails with each labelled as spam or not spam), uncover patterns in this data, and use these patterns to predict the labels of new data (e.g., whether newly received emails are spam or not). Conversely, unsupervised learning systems take in unlabelled data (e.g., a collection of emails without spam labels) from which patterns are uncovered. In reinforcement learning, the amount of label information is intermediate between supervised and unsupervised learning; these systems take in unlabelled data but are provided weak feedback during training regarding whether a prediction is correct or not (Jordan & Mitchell, 2015).

to uncover patterns that can help predict the existence of those categories in new data. Unsurprisingly, the accuracy of this labelling can have a significant impact on the accuracy of the machine's predictions. Once trained, the machine is deployed on new data that does not contain a label for the predicted variable, and the accuracy of its predictions can be assessed in relation to the "ground truth," i.e., the human-provided label.

To demonstrate ethical tensions in using machine learning for preventing suicide, we draw on two cases of supervised learning: (1) a machine learning system supporting clinicians to identify suicidal risk based on data from electronic health records; and (2) a machine learning system operating within a peer-to-peer social network, drawing on user generated content to predict and intervene in cases of suicidal risk. In both of these cases, subsequent interventions ensue as a result of the machine learning predictions of suicidality.

(1) In the first case, a suicide prediction model was designed to help clinicians identify patients at risk for suicide (Barak-Corren et al., 2016). The model was trained on 15 years of patient data from the electronic health records of 1,728,549 patients, including demographic characteristics, diagnostic codes, laboratory results and prescribed medications. Ground truth, the basis for judging the precision and accuracy of predictions from a supervised machine learning model, was established by three clinicians, who labelled which patients in the training data set had been diagnosed with suicidal behaviour (according to the International Classification of Diseases, ICD-9) and/or had died from suicide (according to death certificates). The suicide prediction model identified nearly half of all suicides and suicidal behaviours with 90% specificity, an average of 3–4 years in advance of the suicidal behaviour, showing promise as an effective clinical decision support tool. The authors note that the tool is not meant to offer a substitute for clinical evaluation, but only to help identify patients who require screening for suicidal ideation (Barak-Corren et al., 2016). Such screening is currently far below recommended levels, which is concerning because of the important role screening can play in connecting suicidal individuals to effective interventions (O'Connor, Gaynes, Burda, Williams, & Whitlock, 2013).

(2) In the second case, a machine learning model was used to identify users at risk of suicide on Reddit, a social website containing news and forums in which content is curated by users through votes and organised according to topics of interest (Popper, 2017). The machine learning model was trained on data collected from Koko, a social networking application for people with depression and anxiety. The data, consisting of short text descriptions of Koko users' emotional problems, was labelled according to the subjective intensity of the problem by online crowd workers - people who are paid small amounts of money to complete short micro tasks online such as labelling images or transcribing audio (Paolacci & Chandler, 2014). It is noteworthy that no explicit declaration of an intention to self-harm or attempted suicide was required for an individual to be flagged as at risk by the algorithm. Once trained, the supervised learning model identified users in the Reddit forums r/depression and r/SuicideWatch that it predicted were emotionally vulnerable and could benefit from intervention. Using this prediction, an emotionally supportive artificially intelligent chatbot approached the user to offer help. The motivation for this machine learning approach was to offer protection and safety to people whose emotional distress could go unnoticed online, or worse, could lead to dangerous encounters with malicious intruders. However, when the model was let loose within the Reddit community, users were outraged at being approached by an artificial intelligence about highly sensitive personal information, with one user writing 'I feel deeply disturbed that they would use a bot to do this' (Popper, 2017).

Both of these cases relied on readily available data of patients and users, trained an algorithm to identify which features of their data predict suicide risk, refrained from acquiring informed consent with participants or users, and were built with the intention of preventing deaths by suicide and offering support to individuals in need. As will be shown below, there are also several important differences between the two cases. In the following section, we focus on three differences that illuminate ethical tensions in the design and implementation of these technologies: (1) the intended audience one is sharing health information with, (2) expectations related to data collection, protections, and use; and (3) the nature of the intervention.[3]

[3] While we do not have space to consider them in detail here, there are also important differences in these cases related to the representativeness of the samples used, the nature and breadth of the data used to represent ground truth, and the predictive accuracy of the algorithms, all of which were superior in the case developed with the use of electronic health data, as compared to the case developed in an online setting. This is consistent with a recent scoping review of research using Twitter for health data collection, which concluded that data collected from the social network is inherently difficult to verify, and methods (e.g., triangulation) ought to be built in to research to increase the possibility of verification (Zhang, Albrecht, & Scott, 2018).

4. Examining differences

4.1 Intended audience

A key difference between patients in a clinical setting and users in an online social setting, all of whom might be disclosing health information that may help predict future suicidal behaviour, is the intended audience for their disclosures. In the context of a health care setting, patients approach providers with the intention of seeking professional help for an issue they are dealing with. They may do so because they are seeking support, have a trusting relationship with their provider, and/or because they are hoping to access treatments. As Rosamond Rhodes points out, the system of medicine we have in place depends on patients making themselves very vulnerable, exposing sensitive information and body parts to clinicians that they may not be willing to share with anyone else (Rhodes, 2001). Because of the vulnerability of patients, Rhodes argues that it is crucial that clinicians both seek trust and be deserving of it, acting in the best interest of their patients and protecting their confidentiality. If such trust is violated, the system cannot continue to function, since providing good care depends on the honesty and vulnerability of patients, which is only forthcoming when trust is present (Rhodes, 2001). Therefore, when patients share sensitive health information in clinical contexts, they do so on the basis of a foundation of trust, and the knowledge that such information will be used in order to promote their best interests.

However, a number of reasons may prompt individuals to utilise informal support rather than clinical services, either in online or offline settings, when they are struggling with psychological distress. People may seek out online peer communities in which they can receive emotional support while remaining anonymous, because they feel more comfortable discussing their concerns with their peers, and/or because they are not ready to receive or interested in receiving formal services (Bernecker, Banschback, Santorelli, & Constantino, 2017; O'Leary, Bhattacharya, Munson, Wobbrock, & Pratt, 2017). In a survey of social media users, of all participants who reported having used social media for 'an emotional problem related to suicide', the majority (68%) said they had never sought professional mental health support on social media, while the remainder said they did so 'almost never' or 'occasionally/sometimes' (Robinson, Rodrigues, Fisher, Bailey, & Herrman, 2015). The same sample reported the benefits of social media to be 'its ability to provide a platform for them to express their feelings (91%), receiving emotional support from others (82%), talking to others with a similar problem (73%), and helping others (73%)' (Robinson et al., 2015, p. 30). Sometimes support from peers may be all that is desired, as well as all that is required to prevent a suicide attempt. Evidence suggests that merely talking about suicide can help to alleviate suicidal thoughts and 'reduce the pressure to take suicidal action' (Biddle et al., 2013; Eichenberg, 2008, p. 112).

Furthermore, particular groups may be even less likely to desire professional help when engaging in online settings, as a result of past experiences with institutions, medical and non-medical. For example, within Australia, Aboriginal and Torres Islander people tend to engage in help seeking through informal online routes (family, friends, and social networks) more often than formal ones (mental health professionals) as a result of intergenerational trauma, distrust of governmental agencies, and widespread stigma of help-seeking behaviours (Farrelly, 2008)[4] (Carson, Farrelly, Frazer, & Borthwick, 2015). Similarly, online participants who identify as ex-patients or survivors of experiences within the mental health system often seek and develop grassroots communities of healing and empowerment in order to provide safe alternatives to a system in which many of them have experienced trauma (DuBrul, 2014; Naslund, Aschbrenner, Marsch, & Bartels, 2016). As such, social networks are often perceived as places to escape judgement and intervention, a space antithetical to clinical settings: 'you might be too embarrassed or not want to talk to a person physically about it, whereas you can just go online and talk to someone or even look it up on the internet … It's less confronting (Daniel, Male, 19)' (Carson et al., 2015, p. 7). It is likely that the Reddit forums r/depression and r/SuicideWatch, in which Koko's chatbot was introduced in the case described above, provide such safe spaces for many individuals seeking peer support online.[5]

Disclosure requirements present another difference related to the collection of sensitive health information in clinical settings and online communities, that might shape individual decisions related to who they share sensitive information with. Many individuals are aware of the limits to confidentiality that health care providers are subject to when one shares information regarding an intention of self harm, as well as the potential implications

[4] Given the presence of these features in other Indigenous communities, it is likely that this phenomenon is not specific to Aboriginal and Torres Island people in Australia, although we limit our discussion here, drawing on (Carson et al., 2015).

[5] Of course, despite expectations, there is no guarantee that individuals will find the support they need online, or that they will not be given bad advice.

of disclosing suicidal intent, given that the risk of self harm or suicide can be, and often is, used as grounds for police and hospital intervention against a person's will. Friends, family members, and anonymous interlocutors in online communities are not subject to the same reporting requirements, however, and so may be sought out by individuals who are seeking to communicate about their suffering without worrying about their thoughts and intentions being repeated elsewhere, or who fear being subjected to forced treatment.

4.2 Data collection, use, and protection

A second significant difference between the two cases introduced above revolves around individual expectations related to the collection and use of their health data, as well as what protections apply to such data. When patients visit a health care centre, such as a hospital or clinic, they are actively seeking help, providing sensitive health data to clinicians with full awareness (and indeed, expectation) that the clinician is listening, recording, and protecting their information. In contrast, users in online settings divulging information to peers, without any awareness of the presence of a machine learning system, are much less likely to expect that sensitive information related to their health is being collected.[6] Similarly, in a clinical setting, it is not unsurprising when one's doctor calls to discuss one's test results or to remind one of the need for a screen or check up. This aligns with their role as the keeper of health data, trusted to act in their patient's best interest. In the case of a setting like Reddit, however, questions or reminders related to one's personal health seem out of place. As a sphere for sharing information, conversation, and news, nothing about the site suggests that one should anticipate mental health interventions to be built in. The expectation of being in a safe space where one is not being watched can be read from the user involved in the Reddit forums r/depression and r/suicideWatch who felt 'deeply disturbed'. Another user similarly expressed the feeling of violation and outrage present in the online community: 'Disgusting that assholes would try and take advantage of people' (Popper, 2017). A similar reaction occurred on Twitter when an app called Samaritan Rader was used to publicly offer help to Twitter users it identified as in distress, leading to upset users who felt their privacy had been invaded (Popper, 2017).

[6] Such expectations are in flux, however, as the media increasingly reports on the collection of health data in online settings (e.g., Singer, 2018), and are also likely to vary significantly across groups (e.g., depending on age, data literacy).

While users are not always aware of them, the laws and regulations protecting one's health data also differ considerably in clinical settings and online social platforms. In the United States, HIPAA, the Health Insurance Portability and Accountability Act, ensures that individuals have the right to access their personal health information (PHI), request corrections or complaints related to that information, limit who else has access to it, and know the identity of who has received reports related to that information. These protections apply only to PHI that is individually identifiable and held or transmitted by a covered entity (including health plans, clearinghouses, or providers) or its business associates (Health Insurance Portability and Accountability Act (HIPAA), 2004). In online spaces such as Reddit, however, little protections apply. A recent report by the United States Department of Health and Human Services (HHSs) noted that there are significant gaps in the protection of personal health information, especially with regards to health technologies and online spaces. The authors of the report noted that this is largely the result of current legal protections applying only to certain types of entities, while the type of data being collected and how it is used is not taken into account (U.S. Department of Health and Human Services, 2016). Somewhat paradoxically, this means that in many cases in which one is aware that one's personal health information is being collected, such data is protected, but in cases in which personal health information about oneself is being surreptitiously created from other data one provides, little to no protections exist.

Personal data protections tend to be most limited in online public spaces, which are often thought to require little research ethics oversight. As Metcalf and Crawford have argued, this rests on a faulty assumption that 'the risk to research subjects depends on what kind of data is obtained and how it is obtained, not what is done with the data after it is obtained' (Metcalf & Crawford, 2016, p. 7). However, data that may seem harmless and impersonal (e.g., one's Twitter feed, browsing history), when coupled with additional information about oneself (e.g., demographic, related internet use) and examined within a machine learning system, may reveal highly sensitive information related to one's health status.[7] Moreover, the HHS report further notes that 'individuals may have a limited or incorrect understanding of when data about their health is protected by law' (U.S. Department of Health and Human Services, 2016, p. 4).

[7] Or, for example, one's immigration status or sexual orientation.

4.3 Nature of the intervention

Another key difference between the two cases revolves around the nature of the intervention, a crucial aspect of suicide prevention that is both ethically and emotionally charged. When ethics committee members were surveyed about issues related to research involving people who might be suicidal, 57% mentioned concerns related to a duty of care and the possibility that researchers or mainstream services may uncover 'problems that could not adequately be dealt with', such as when information about services is provided to an individual who is depressed and lacks the energy to contact such resources (Lakeman & FitzGerald, 2009, p. 16). Similarly, in a survey of stakeholders related to social media and suicide prevention, all three groups of stakeholders (social media users, researchers, and organisations that use social media for suicide prevention) expressed concerns about the ability of social media site administrators to 'operate safe and effective interventions online' (Robinson et al., 2015). Because of the amount and accuracy of the data utilised by the machine learning system developed from electronic health records, the model was able to predict cases of suicidal behaviour with 90% specificity 3–4 years in advance of suicide attempts (Barak-Corren et al., 2016). This distance between the prediction and predicted event makes space for a wide variety of possible interventions well in advance of a suicide attempt.[8] Furthermore, this lengthy window of opportunity makes space for predictions to be fed back to clinicians and other health care professionals whom the patient trusts. When there is little urgency, detecting individuals at risk could lead to early support and screening, as well as the development of advance directives, and when there is greater urgency, interventions can be provided by familiar faces, such as friends, family members, or health care professionals.

Due to the nature of data taken from social media sites, however, predictions are more likely to occur just-in-time, before or during a crisis, narrowing the scope of possible interventions considerably. Users seeking help online are often reporting distress in the moment of crisis, and systems trained to predict suicidality in this setting often use only that crisis data to make predictions. Given this narrow focus, there is a short time window to react to the suicide prediction, and the responsibility for the intervention is often passed to chatbots and/or local law enforcement. As can be seen in

[8] Although it also introduces novel worries related to unnecessary treatment, patient privacy, and the desire not to know.

the responses of Reddit users discussed above, unsolicited help from chatbots is not always well received. Police officers, as well, tend to lack adequate training in how to respond appropriately to distress and many have come to resent their role as crisis responders (Lamb, Weinberger, & DeCuir, 2002; Price, 2005).[9] Calling the police to the home of a person in crisis without their consent can do more harm than good, as such calls can result in unwarranted forced treatment, the escalation of conflicts, and sustained trauma (Wise, 2017). Although it is rare, individuals who have called the police for assistance have also ended up being killed by those responding to the call (Tastrom, 2017).

The accountability of the humans-in-the-loop for intervening in cases where suicide risk is high is likely to differ in these two cases as well. Central to a clinician's role is their obligation to act in the best interests of their patients. The fiduciary relationship between a doctor and patient is characterised by entrustment, in which a vulnerable patient entrusts aspects of their wellbeing to a doctor who is then permitted to act with some discretion in order to achieve the aim of contributing to the patient's wellbeing (Richardson & Belsky, 2012). Such entrustment generates a sense of accountability in those who take on responsibility for another's wellbeing, be they a doctor, a lawyer, or a kindergarten teacher. As such, when information regarding a patient's suicide risk is fed back to a clinician as a result of a machine learning system, they are obligated to respond to that information in a way that promotes the patient's health. In contrast, the sense of accountability towards social media users felt by those developing suicide prediction and intervention technologies behind the scenes is likely to differ substantially from that of clinicians responding to data related to suicide prediction. As pointed out by Metcalf and Crawford, it is not uncommon for those working with large data sets to see themselves as 'absolved of ethical concerns', particularly when they are working with public data sets, which often appear to involve no violations of privacy (Metcalf & Crawford, 2016, p. 2). This experience of ethical detachment may be attributable, in part, to the ways in which 'data science methods create an abstract relationship between researchers and subjects, where work is being done at a distant remove from

[9] Mark Zuckerberg reported in a blog post that in 2018, Facebook's suicide prevention program sent first responders to the homes of 3,500 people who were at risk of suicide (Zuckerberg, 2018). Previously, Facebook reported that they were working with some first responders to perform wellness checks after these emergency responses (Rosen, 2017), but more recently they stated that outcomes are no longer being tracked (Singer, 2018).

the communities most concerned' (Metcalf & Crawford, 2016, p. 2). While designers of online suicide prevention and mental health support are likely to be aware of the sensitivity and importance of this work, the sheer number of users they are aiming to support suggests that their sense of accountability for each one will be limited.

Additionally, in the case of suicide predictions based on data from electronic health records, data related to individuals who are considered to be high risk can be fed back to their primary care provider. Importantly, that provider is able to look at the output of the algorithm and place a categorisation of high risk into context. The patient's medical history, including past attempts, past and present diagnoses, and major life events can be taken into account along with the AI generated prediction. Given the inevitability of false positives resulting from the use of predictive analytic tools in health care, being able to place an assessment of someone's suicidality within the context of their care could be extremely important in terms of ruling out false positives, as would occur within screening for suicidal thoughts or intentions. The case of the chatbot offers a stark contrast, in that all that is generated is a predictive value for suicidality for each user, which can only be understood in the context of their activity on the site. Moreover, while health care providers have an ongoing relationship with patients and can test the efficacy of these interventions, most social media sites, with limited opportunities for follow up, have little ability to assess the effectiveness of these interventions. If a user fails to return to the site, could it be because they did in fact die by suicide, or because they resented the intervention brought about by the site? With the exception of sites like Facebook and Twitter, in which a memorial may be set up by friends and family, there is little way to tell. Unfortunately, these differences appear to be especially difficult to avoid in contexts that are not linked with mainstream health care. In such projects, while the initial stages of design, the development of the algorithm and predictability of the tools, may perform expertly, the final stages of the process, when it matters most, are often the least thought through.

5. Rethinking the design space in suicide prevention technologies

5.1 Expectations as grounds for ethical design

Understanding the many ways in which the expectations of users in online communities differ from those of patients in health care settings can shed light on why Koko's intervention, which was developed with purely beneficent aims, came across as so offencive to users. Attending to the intended audience of those seeking support in the context of an online community reveals that some users may have no desire to interact with more formal health care settings. This may be because they prefer the support of peers, they fear the potential repercussions of disclosing sensitive information to health care professionals, or because they have had negative experiences in such settings. In online communities, users may also have no idea that their data is being collected or for what purposes it might be used, and may or may not be aware of the limited protections that apply to their health data in online settings. Furthermore, they may not expect or desire that an intervention will take place if their data suggests they are in danger of self harm. The reception of such an unexpected, and possibly unwanted, intervention may be negative, especially if the intervention disrupts a safe space for users by involving a chatbot or police officer, with which the user has no previous relationship. These differences demonstrate why it may be especially egregious for machine learning products to collect data and intervene in some settings, while it may be ethically permissible in other settings. As Nissenbaum argues, 'the contexts in which activities are grounded shape expectations that, when unmet, cause anxiety, fright, and resistance' (Nissenbaum, 2011, p. 38). We must pay attention to the norms and expectations present in different help-seeking contexts in order to develop new suicide prevention technologies that respect the individuals whose data is being used.

Informed consent offers a useful lens in which to futher illuminate the ways in which expectations can ground ethical design of health technologies. While neither of the cases involved a process of informed consent, as one accessed electronic health records without obtaining consent and the other drew upon publicly available data on Reddit, there are important differences related to the grounds for refraining from informed consent in these cases. Consent has multiple functions: it informs the participant of what is to be expected, is an opportunity to ask questions or clarify one's expectations, and gives individuals a chance to consider the risks and benefits of their participation. This process allows individuals to determine if the proposed research or intervention is in their best interest, as defined by them, taking into account their own values and interests, and is not something that can be done by those who are unaware of their values and interests (e.g., how they weight privacy versus safety). While these functions are important aspects of consent, there are some settings in which it may be ethically permissible to forgo consent because the benefits outweigh the harms. One example is within emergency medicine, in cases in which a life can be saved, but consent for treatment is impossible because the patient is unconscious. Similarly, in the case involving a suicide prediction tool developed through electronic health records, a consent waiver is often given for research which only involves the use of patient records and no intervention, as long as the risks are considered to be minimal and the research is likely to have benefit.

While both cases of suicide prediction technologies may appear to involve potential benefits that outweigh the potential risks, the act of failing to inform a individual patient or user about the use of their data takes on a different hue in these two cases. As discussed above, the system of health care operates on a foundation of trust. Individuals are aware that their data is being collected and may be shared with others for the benefit of themselves or others. If the violation of trust or harm that resulted from the use of that data was so great as to outweigh the benefits (e.g., jeopardising the likelihood that future participants would enrol in research), it would not be justified or permitted. In contrast, in the case of the chatbot intervening on users in distress on Reddit, there was no knowledge or understanding on the part of users that their data was being collected and would be used in the development of such an intervention. While the intervention was developed in line with the site's terms and conditions, which each user agrees to before joining and contributing to Reddit, such policies notices are notoriously incomprehensible and ignored (Jensen & Potts, 2004; Milne, Culnan, & Greene, 2006). The violation of trust and community norms was clearly significant, as demonstrated by the outraged comments of users after the fact. Responding to these reactions, the chatbot designers acknowledged the importance of 'educating users and moderators about who they were and what their goal was' (Popper, 2017). This suggests that in spaces in which the collection and use of health data is not normalised and protected, the importance of informed consent is greater than in traditional clinical settings.

5.2 Takeaways

Machine learning systems depend on large amounts of data, and the vast amounts of online data now available open up seemingly limitless opportunities for machine learning applications. However, as we have demonstrated, the limits of appropriate machine learning systems in the highly sensitive space of suicide prevention may be grounded in user expectations. Thus understanding and identifying the norms of online communities and users may be especially important for designing machine learning systems that are both effective and acceptable for preventing suicide.

A key takeaway from this discussion is the importance of thinking about what online spaces mean to people and ensuring that introducing a machine learning intervention to a space is not too disruptive and does not introduce more problems than it solves. When designing for a particular context, it is worth considering in advance: what is the purpose of this space, and what do users expect to take place (and not take place) within it? Identifying the norms and culture of an online community can sometimes be easy; oftentimes, community values, risks of participation (e.g., trigger warnings) and ground rules are explicitly posted, clearly written, and visible in online emotional support communities. At other times, it may be necessary for researchers to spend time within online settings or to conduct preliminary research in order to get a sense for what might constitute an intrusion and what might align with user expectations. Of course, such an assessment may reveal that the expectations within a community or population are diverse and pluralistic, providing a challenge for those hoping to take them into account during design. In such cases, where expectations may be highly variable, we recommend taking into consideration the most uncomfortable or least knowledgeable user in one's design, in order to ensure that as many users as possible are aware of the data collection and use that is taking place and that as few as possible are likely to be distressed by the appearance of a sudden intervention.

Design methods, such as value-sensitive design for understanding community values, which seek to identify the norms present within communities of users before developing technologies for a particular context, and participatory design, which involves the engagement of users as designers, are promising when working in high-stakes settings such as suicide (Friedman, Kahn, Borning, & Huldtgren, 2013; Thieme et al., 2013). Such design approaches that emphasise shared community values and stakeholder design input reduce the likelihood that novel technologies will introduce unexpected and unwelcome agents into a community and cause harm. As the human rights organisation Article 19 has pointed out, 'the pervasive and invisible nature of AI systems, coupled with their ability to identify and track behaviour, can have a significant chilling effect on the freedom of expression. This can take place through self-censorship, altered behaviour in public spaces and private communications alike' (Article 19, 2018, p. 8). In order to prevent such effects, which can damage both relationships and research, many working within research ethics have recognised the importance of establishing trusting relationships with community gatekeepers early on, especially when one is working with vulnerable or sensitive populations (Duarte, Brendel, Degbelo, & Kray, 2018; Johansson, Gulliksen, & Lantz, 2015; Woelfer, 2014).

This approach to working closely with stakeholders of sensitive populations could be highly beneficial when designing machine learning systems for suicide prevention. After their initial unsuccessful attempt disrupted the community, the creators of Koko's chatbot realised that engaging with gatekeepers early on in the design and intervention process could vastly improve the chances of the chatbot's acceptance in the community: 'Helping [moderators] understand and embrace the service is a containable problem, one that we're already having good success with' (Popper, 2017). While the consent of gatekeepers is already required for interventions within r/SuicideWatch (the rules state 'Surveys and research participation requests must be pre-approved by the mod team'), this approval may not be sufficient to prevent harmful effects within the community, and further community agreement or permissions may need to be sought out (Reddit).

As discussed above, the importance of engaging users in informed consent is more important in settings in which suicide prevention technologies are not expected. The fact that users did not anticipate any intervention by Koko's chatbot appears to be a significant part of why they were distressed by it. It is important to warn users of potential interventions, especially when collecting highly sensitive health information or implementing invasive measures, such as calling law enforcement to their homes. In a discussion related to online methods for suicide-related research, Michaels et al. suggest that 'in the consent process, participants should be informed that their responses to questions about suicide will be reviewed for their safety and that if their responses indicate they may be at risk for suicide, the experimenter will contact them via phone ... Similarly, participants should be informed that if their risk for suicide is determined to be imminent and/or serious, the experimenter may (based on the specific research context) call a mobile crisis team or 911.' (Michaels, Chu, Silva, Schulman, & Joiner, 2015). It is reasonable to assume that reactions to the chatbot may have been less negative, or even positive, had it been designed to ask participants for consent to use their information and informed them of how it might act on that information. At a minimum, users ought to be warned that their data is being collected and that this collection may result in interventions directed at them, so they have the opportunity to opt out of participating in an online community if they so wish. Such warnings could be incorporated into existing warnings related to community norms to increase the likelihood that users will become aware of the possibility of unexpected and/or unwelcome interventions.

Because of the expectations associated with many online activities, it is especially important to inform users if one is creating personal health information from non-health data related to their online activities. Furthermore, until the gap is closed between laws and regulations that protect health data disclosed in HIPAA covered settings and non-HIPAA covered settings, it is crucial that those designing technologies for suicide prevention are aware of the risks posed to users when one derives highly sensitive personal health information about them from their public online behaviour. Creating a suicide risk score for an individual based on their tweets or chatroom comments is not merely an exercise in one's algorithmic abilities; if accessed by other parties, such a score could lead to shame and stigma, experiences of forced treatment, denials in health insurance, as well as housing and employment discrimination.

When possible, those developing suicide prevention technologies online should attempt to integrate such technologies with existing services, increasing the likelihood that familiar users or faces may be part of the intervention if and when it does occur. Avoiding unwelcome visits from law enforcement or 'disturbing' encounters with chatbots is of central importance during crucial moments of intervention in suicide prevention (Popper, 2017). The presence of clinicians or other trained supportive humans-in-the-loop may be crucial to the safety and efficacy of machine learning approaches on devices and within emotionally supportive communities.

For example, many socially and clinically integrated machine learning approaches for preventing suicide are exploring the design space beyond AI chatbots. Recent work on passive suicide screening of text messages on smartphones shows a promising just-in-time and opt in machine learning approach designed to be integrated with clinical care (Nobles, Glenn, Kowsari, Teachman, & Barnes, 2018). Users who opt in to use the tool are connected with their care team when significant risk is detected, which could be especially appropriate for patients recently released from inpatient settings who are high-risk. Similarly, Crisis Text Line uses machine learning to predict the suicide risk of users who seek help and prioritises which users should be connected immediately with their trained counsellors (Crisis Text Line; Dinakar, Chen, Lieberman, Picard, & Filbin, 2015). Other machine learning systems are not designed to predict risk, but rather to improve support. 7 Cups of Tea, an emotional support online community, uses machine learning to suggest active listening skills to trained counsellors during online chats (7 Cups). Scaffolding peers to provide effective support is also explored by O'Leary et al. in a prototype that gives users prompts during supportive chats for safe and effective encounters (O'Leary, Schueller, Wobbrock, & Pratt, 2018). Bernecker et al. developed a massively scalable online training course for peer therapists wherein machine learning could play a role in tailoring content for each individual's learning style (Bernecker, 2017).

These examples of machine learning systems for intervention demonstrate a spectrum of approaches, from AI chatbots to peer-based scaffolding to clinical integration. In this chapter, we chose to explore the extreme ends of this design space—lone chatbots and clinical decision support tools—as a way to draw out ethical implications of using personal information to predict suicide in contrasting settings. In exploring these extreme cases, we have demonstrated that the ethical permissibility of machine learning systems for suicide prevention is influenced by user expectations related to data collection, the intended audience of the data, and the accountability of humans-in-the-loop. These expectations could be used as design considerations for those reflecting on the ethics of machine learning systems in this space.

6. Conclusion

In this chapter, we have suggested that paying attention to the norms and expectations that exist within different settings can support the design of better methodologies and technologies related to suicide prevention tools that utilise machine learning techniques. Drawing on two case studies in which similar technologies were developed in different contexts, we highlighted characteristics that influence expectations of patients and users related to: (1) the intended audience one is sharing health information with, (2) data collection, protections, and use; and (3) the nature of the intervention. We have argued that differences in norms and expectations specific to a context ought to guide ethical considerations within research and technology design for suicide prevention and have offered some examples of how this might be done. We hope that this work will encourage technologists and clinicians alike to explore the ways in which particular settings that are drawn upon for machine learning technologies can impact the way users and patients experience tools for suicide prevention, and help contribute to the improvement of such tools.

Acknowledgements

Thank you to Abe Friesen, as well as the editors of this volume, for helpful feedback on earlier versions of this chapter. This work was funded in part by the National Institute for Health Research (NIHR) Oxford Biomedical Research Centre, grant BRC-1215-20008.

References

7 Cups. Retrieved from https://www.7cups.com/.

Article 19. (2018). *Privacy and freedom of expression in the age of artificial intelligence*. Retrieved from https://www.article19.org/wp-content/uploads/2018/04/Privacy-and-Freedom-of-Expression-In-the-Age-of-Artificial-Intelligence-1.pdf.

Barak-Corren, Y., Castro, V. M., Javitt, S., Hoffnagle, A. G., Dai, Y., Perlis, R. H., ... Reis, B. Y. (2016). Predicting suicidal behavior from longitudinal electronic health records. *American Journal of Psychiatry, 174*(2), 154–162.

Bernecker, S. L. (2017). *Testing a skills training course for use in a peer-delivered mental health intervention*.

Bernecker, S. L., Banschback, K., Santorelli, G. D., & Constantino, M. J. (2017). A web-disseminated self-help and peer support program could fill gaps in mental health care: Lessons from a consumer survey. *JMIR Mental Health, 4*(1).

Biddle, L., Cooper, J., Owen-Smith, A., Klineberg, E., Bennewith, O., Hawton, K., ... Gunnell, D. (2013). Qualitative interviewing with vulnerable populations: Individuals' experiences of participating in suicide and self-harm based research. *Journal of Affective Disorders, 145*(3), 356–362.

Carson, B. L., Farrelly, T., Frazer, R., & Borthwick, F. (2015). Mediating tragedy: Facebook, aboriginal peoples and suicide. *Australasian Journal of Information Systems, 19*.

Crisis Text Line. Retrieved from https://www.crisistextline.org/.

Dinakar, K., Chen, J., Lieberman, H., Picard, R., & Filbin, R. (2015). *Mixed-initiative real-time topic modeling & visualization for crisis counseling*. Paper presented at the Proceedings of the 20th international conference on intelligent user interfaces.

Duarte, A. M. B., Brendel, N., Degbelo, A., & Kray, C. (2018). Participatory design and participatory research: An HCI case study with young forced migrants. *ACM Transactions on Computer-Human Interaction (TOCHI), 25*(1), 3.

DuBrul, S. A. (2014). The icarus project: A counter narrative for psychic diversity. *Journal of Medical Humanities, 35*(3), 257–271. https://doi.org/10.1007/s10912-014-9293-5.

Eichenberg, C. (2008). Internet message boards for suicidal people: A typology of users. *CyberPsychology & Behavior, 11*(1), 107–113.

Farrelly, T. (2008). The Aboriginal suicide and self-harm help-seeking quandary. *Aboriginal and Islander Health Worker Journal, 32*(1), 11.

Friedman, B., Kahn, P. H., Borning, A., & Huldtgren, A. (2013). Value sensitive design and information systems. *Early Engagement and New Technologies: Opening Up the Laboratory*, 55–95 (Springer).

Jensen, C., & Potts, C. (2004). *Privacy policies as decision-making tools: An evaluation of online privacy notices* (Paper presented at the Proceedings of the SIGCHI conference on Human Factors in Computing Systems).

Johansson, S., Gulliksen, J., & Lantz, A. (2015). *User participation when users have mental and cognitive disabilities*. Paper presented at the Proceedings of the 17th International ACM SIGACCESS Conference on Computers & Accessibility.

Jordan, M. I., & Mitchell, T. M. (2015). Machine learning: Trends, perspectives, and prospects. *Science, 349*(6245), 255–260. https://doi.org/10.1126/science.aaa8415.

Lakeman, R., & FitzGerald, M. (2009). The ethics of suicide research: The views of ethics committee members. *Crisis: The Journal of Crisis Intervention and Suicide Prevention, 30*(1), 13.

Lamb, H. R., Weinberger, L. E., & DeCuir, W. J. (2002). The police and mental health. *Psychiatric Services, 53*(10), 1266–1271. https://doi.org/10.1176/appi.ps.53.10.1266.

Metcalf, J., & Crawford, K. (2016). Where are human subjects in big data research? The emerging ethics divide. *Big Data & Society, 3*(1). https://doi.org/10.1177/2053951716650211, 2053951716650211.

Michaels, M. S., Chu, C., Silva, C., Schulman, B. E., & Joiner, T. (2015). Considerations regarding online methods for suicide-related research and suicide risk assessment. *Suicide and Life-Threatening Behavior, 45*(1), 10–17.

Milne, G. R., Culnan, M. J., & Greene, H. (2006). A longitudinal assessment of online privacy notice readability. *Journal of Public Policy & Marketing, 25*(2), 238–249.

Naslund, J. A., Aschbrenner, K. A., Marsch, L. A., & Bartels, S. J. (2016). The future of mental health care: Peer-to-peer support and social media. *Epidemiology and Psychiatric Sciences, 25*(2), 113–122. https://doi.org/10.1017/S2045796015001067.

Nissenbaum, H. (2009). *Privacy in context: Technology, policy, and the integrity of social life*. Stanford University Press.

Nissenbaum, H. (2011). A contextual approach to privacy online. *Daedalus, 140*(4), 32–48.

Nobles, A. L., Glenn, J. J., Kowsari, K., Teachman, B. A., & Barnes, L. E. (2018). *Identification of imminent suicide risk among young adults using text messages*. Paper presented at the Proceedings of the 2018 CHI Conference on Human Factors in Computing Systems.

O'Connor, E., Gaynes, B., Burda, B., Williams, C., & Whitlock, E. (2013). *Screening for suicide risk in primary care: A systematic evidence review for the US preventive service task force agency for healthcare research and quality*. Report No: 13–05188-EF-1.

O'Leary, K., Bhattacharya, A., Munson, S. A., Wobbrock, J. O., & Pratt, W. (2017). *Design opportunities for mental health peer support technologies*. Paper presented at the CSCW.

O'Leary, K., Schueller, S. M., Wobbrock, J. O., & Pratt, W. (2018). *"Suddenly, we got to become therapists for each other": Designing peer support chats for mental health*. Paper presented at the Proceedings of the 2018 CHI Conference on Human Factors in Computing Systems.

Paolacci, G., & Chandler, J. (2014). Inside the Turk: Understanding mechanical turk as a participant pool. *Current Directions in Psychological Science, 23*(3), 184–188. https://doi.org/10.1177/0963721414531598.

Popper, B. (March 2, 2017). The empathy layer. *The Verge*. Retrieved from https://www.theverge.com/2017/3/2/14764620/koko-social-network-mental-health-depression-app-kik.

Price, M. (2005). The challenge of training police officers. *Journal of the American Academy of Psychiatry and the Law, 33*(1), 50–54.

Reddit. SuicideWatch. Retrieved from https://www.reddit.com/r/SuicideWatch/.

Rhodes, R. (2001). Understanding the trusted doctor and constructing a theory of bioethics. *Theoretical Medicine and Bioethics, 22*(6), 493–504. https://doi.org/10.1023/A:1014430208720.

Ribeiro, J., Franklin, J., Fox, K. R., Bentley, K., Kleiman, E. M., Chang, B., & Nock, M. K. (2016). Self-injurious thoughts and behaviors as risk factors for future suicide ideation, attempts, and death: A meta-analysis of longitudinal studies. *Psychological Medicine, 46*(2), 225–236.

Richardson, H.,S., & Belsky, L. (2012). The ancillary-care responsibilities of medical researchers: An ethical framework for thinking about the clinical care that researchers owe their subjects. *Hastings Center Report, 34*(1), 25–33. https://doi.org/10.2307/3528248.

Robinson, J., Rodrigues, M., Fisher, S., Bailey, E., & Herrman, H. (2015). Social media and suicide prevention: Findings from a stakeholder survey. *Shanghai Arch Psychiatry, 27*(1), 27–35. https://doi.org/10.11919/j.issn.1002-0829.214133.

Rosen, G. (2017). *Getting our community help in real time*. Retrieved from https://newsroom.fb.com/news/2017/11/getting-our-community-help-in-real-time/.

Sharac, J., Mccrone, P., Clement, S., & Thornicroft, G. (2010). The economic impact of mental health stigma and discrimination: A systematic review. *Epidemiology and Psychiatric Sciences, 19*(3), 223–232.

Singer, N. (2018, Dec. 31). In screening for suicide risk, Facebook takes on tricky public health role. *The New York Times*. Retrieved from https://www.nytimes.com/2018/12/31/technology/facebook-suicide-screening-algorithm.html.

Tastrom, K. (June 27, 2017). When calling the police results in harm, not help. *Slate*. Retrieved from http://www.slate.com/articles/health_and_science/medical_examiner/2017/06/charleena_lyles_death_reminds_me_that_for_many_calling_the_cops_can_be_risky.html.

Thieme, A., Wallace, J., Johnson, P., McCarthy, J., Lindley, S., Wright, P., … Meyer, T. D. (2013). *Design to promote mindfulness practice and sense of self for vulnerable women in secure hospital services*. Paper presented at the Proceedings of the SIGCHI Conference on Human Factors in Computing Systems.

United States. (2004). *The Health Insurance Portability and Accountability Act (HIPAA)*. Washington, D.C.: U.S. Dept. of Labor, Employee Benefits Security Administration.

U.S. Department of Health and Human Services. (2016). *Examining oversight of the privacy & security of health data collected by entities not regulated by HIPAA*. Retrieved from https://www.healthit.gov/sites/default/files/non-covered_entities_report_june_17_2016.pdf.

Wise, J. (2017). Mental health: Patients and service in crisis. *BMJ, 356*, j1141.

Woelfer, J. P. (2014). Engaging homeless young people in HCI research. *Interactions, 21*(1), 54–57.

Zhang, A. J., Albrecht, L., & Scott, S. D. (2018). Using twitter for data collection with health-care consumers: A scoping review. *International Journal of Qualitative Methods, 17*(1). https://doi.org/10.1177/1609406917750782, 1609406917750782.

Zuckerberg, M. (2018). *A blueprint for content governance and enforcement*. Retrieved from https://www.facebook.com/notes/mark-zuckerberg/a-blueprint-for-content-governance-and-enforcement/10156443129621634/.

CHAPTER NINE

The ethics of early identification and intervention in psychosis

Charlotte A.L. Marriott[a] and Matthew R. Broome[b,*]
[a]Worcestershire Health and Care NHS Trust, Worcester, England, United Kingdom
[b]Institute for Mental Health, University of Birmingham, Birmingham, England, United Kingdom
*Corresponding author: E-mail: M.R.Broome@bham.ac.uk

Contents

1. Introduction

In this chapter we will explore the ethics of early identification and intervention in psychosis. To start, we will consider what psychosis is, both in the context of first episode and clinical high-risk states, and what we mean by early. We will then look at the ethical issues related to this diagnosis, its identification and treatment, including those relating to stigma, the impact on families, use of medication and psychosocial interventions, the effectiveness of Early Intervention services, clinical research and challenges for the future. We will bear in mind fundamental principles of medical ethics — autonomy, justice, beneficence and non-maleficence — throughout.

Developments in Neuroethics and Bioethics, Volume 2
ISSN 2589-2959
https://doi.org/10.1016/bs.dnb.2019.05.003

1.1 What is psychosis?

Psychosis is a broad term that describes a collection of symptoms characterised by perceptual disturbance, including delusions (false beliefs that are held with a strong, unshakeable conviction), hallucinations (sensory perceptions in the absence of a real stimulus) and disorder of the form of thought (disorganised thinking and speech). Delusions and hallucinations are usually referred to as positive symptoms, where positive means in addition to the normal mental state and experience, but there are also negative symptoms of psychosis, including low mood, loss of motivation or apathy, loss of interest or anhedonia, social isolation and withdrawal from normal occupation, which may precede the development of positive symptoms. The presence of psychotic symptoms may be indicative of a severe and enduring mental illness, such as schizophrenia, schizoaffective disorder or bipolar affective disorder (where affective means mood); an organic disease of the brain such as dementia, epilepsy or tumour; part of another psychiatric disorder such as a severe depressive episode; or a one-off episode of illness related to stress or substance misuse, such as amphetamine, cocaine or cannabis (Henquet et al., 2005). Hence, it is worth bearing in mind that the term psychosis is used to pick out a collection of symptoms or experiences, rather than a precise diagnosis. Isolated psychotic symptoms can also be experienced by members of the general population who are not help-seeking and do not meet the criteria for diagnosis of a psychotic illness, with prevalence rates of 4–28.4%, median 5% (Eaton, Romanoski, Anthony, & Nestadt, 1991; Kendler, Gallagher, Abelson, & Kessler, 1996; van Os, Linscott, Myin-Germeys, Delespaul, & Krabbendam, 2009). This finding has led to the development of the continuum model of psychosis, where psychotic symptoms are in continuity with normal experiences but are quantitatively different (Johns & van Os, 2001).

Psychotic illnesses create personal suffering and also pose a significant public health problem, with estimated prevalence rates of non-organic (*functional*) psychosis of 4.5 per thousand in the general population of Great Britain (Brugha et al., 2005). Schizophrenia was the ninth leading cause of disability worldwide in 1990 and bipolar affective disorder the sixth (Murray & Lopez, 1996) with enormous implications for health services and socioeconomic costs, as well as family and carer burden. Schizophrenia contributes 13.4 (95% UI: 9.9–16.7) million years of life lived with disability to the global burden of disease (Charlson et al., 2018). In 2007 approximately 220,000 people in England and Wales had a diagnosis of schizophrenia, accounting for around 30% of total expenditure on adult mental health and social care services (Cooper et al., 2012).

The course of the illness is variable: it can be acute in presentation and resolve completely with treatment, without further relapse; be part of a chronic disease state with ongoing symptoms; or be of a relapsing and remitting nature. Figures from research suggest that, with treatment, 27% patients can achieve remission within 4 weeks and 45% within 5 years (Andreasen et al., 2005), but relapse is a common occurrence and 20–30% of patients may become treatment-resistant (Kane, Honigfeld, Singer, & Meltzer, 1988). To further confuse the picture, there are a number of terms for brief psychotic episodes, including; acute and transient psychotic disorder (ATPD), brief psychotic disorder (BPD), brief intermittent psychotic symptoms (BIPS) and brief limited psychotic symptoms (BLIPS), which do not exhibit prognostic differences in terms of risk of recurrence, although their long-term prognosis is better than that of first episode of schizophrenia in remission (Fusar-Poli et al., 2016).

Schizophrenia is termed a non-affective psychosis, as it is traditionally not associated with alterations in mood state, whereas bipolar affective disorder, schizoaffective disorder and severe depressive episodes with psychotic symptoms are termed affective psychoses. However, these distinctions, and the Kraepelinian dichotomy in the functional psychoses, are increasingly challenged (Craddock & Owen, 2005, 2010).

Psychotic symptoms most commonly present in adolescence and early adulthood, and if not treated effectively at the earliest opportunity can have devastating consequences for long-term functional outcomes. There is a second peak of occurrence in women in their 40s and people in their 60s, but older people have not traditionally been the targets of Early Intervention in Psychosis (EIP) services, set up to provide care for people in the 14–35 year age group, although this is now changing with the development of extension of the age range to 14–65 years (NICE, 2015). EIP services aim to identify and treat patients in their first episode of psychosis (FEP), regardless of precise diagnosis, utilising a descriptive phenomenological approach to psychopathology[1] (Marwaha, Thompson, Upthegrove, & Broome, 2016), or those identified as being at risk of developing psychosis, as we

[1] Karl Jaspers coined the term phenomenology to refer to a patient's ability to describe their subjective experience and the doctor's ability to recognise and understand it in relation to their knowledge of normal and abnormal mental states (Catone, Lindau, & Broome, 2014).

will discuss later. One of the aims of Early Intervention in Psychosis services is to reduce the duration of untreated psychosis (DUP).

1.2 The duration of untreated psychosis

The duration of untreated psychosis (DUP) refers to the time between the development of the first symptoms of psychosis and the initiation of treatment. An emerging psychotic illness may initially present with social withdrawal, isolation and loss of occupational or educational role, so the longer this state continues the more harm is done to a person's functioning and quality of life (Marshall et al., 2005). Long DUP may be predicted by poor insight, social isolation and good coping skills, which allow masking of symptoms and lead to a delay in help-seeking (Drake, Haley, Akhtar, & Lewis, 2000). Some recent research from Denmark has shown that social variables can have an impact on DUP, with a long DUP (more than 12 months) being associated with older age at onset, female gender, cannabis misuse and living in peripheral municipalities (Hastrup, Haahr, Jansen, & Simonsen, 2018). A recent study of a large cohort of first episode psychosis (FEP) patients in England showed that socioeconomic deprivation and unemployment had a significant impact on long DUP, but that being in contact with mental health services prior to the onset of psychosis significantly reduced the DUP by up to 53 days (Reichert & Jacobs, 2018).

It is, however, unclear as to why a long DUP should predict poor outcome. There exists a theory that untreated psychosis is inherently neurotoxic (Wyatt, 1991) because of postulated effects of dopaminergic hyperactivity, prolonged hypothalamic-pituitary-adrenal activation, and persistent activity of catecholamines (Anderson, Voineskos, Mulsant, George, & McKenzie, 2014). Other research evidence throws the neurotoxicity hypothesis into doubt. They demonstrate that there is little correlation between DUP and neurocognitive functioning or brain morphology (Anderson et al., 2014). Rather, morphological changes such as grey matter volume loss actually predate the onset of psychosis (This is the neurodevelopmental hypothesis (Owen, O'Donovan, Thapar, & Craddock, 2011; Weinberger, 2017)). They may be adversely affected by subsequent treatment with antipsychotic medications (Ho, Andreasen, Ziebell, Pierson, & Magnotta, 2011; Lieberman et al., 2005), cannabis use (Bangalore et al., 2008) and other lifestyle factors. The fact that treatment with antipsychotic medications may result in brain volume loss leads to an important ethical dilemma – we know that treating a first episode of psychosis as early as possible improves symptoms and functional outcomes, so to do so would be dictated by the principle of beneficence. But these gains must be weighed up against the possible harms of antipsychotic use. The principle of non-maleficence requires these must always be prescribed at the lowest therapeutic dose.

Whether psychosis is neurotoxic, neurodevelopmental or a bit of both, there is good evidence that reducing the DUP improves outcomes for patients at 6 and 12 months. This includes reducing total symptoms (both depression and anxiety symptoms and positive and negative symptoms of psychosis); being much more likely to achieve remission of psychosis; and improving overall functioning and social functioning, as would be dictated by the principle of beneficence (Marshall et al., 2005). The Treatment and Intervention in Psychosis early detection Study (TIPS) long-term follow-up (Hegelstad et al., 2012) found that a significantly higher percentage of early detection patients had recovered at 10-year follow-up than standard detection patients, which are attributed to higher employment rates in the early detection group.

1.3 What is early intervention?

Early Intervention in Psychosis (EIP) services were set up in order to identify patients suffering from their first episode of psychosis (FEP) and start treatment at the earliest opportunity, in order to reduce the DUP and thus improve long-term outcomes (Marwaha et al., 2016). As such, it can be classed as secondary prevention – preventing further harms resulting from a disorder which has already developed. Intervening at this early stage seems, on the face of it, to pose few ethical challenges, as patients are already suffering from psychotic symptoms, which can be very distressing and severely affect quality of life, as discussed above. Indeed, not intervening might seem to be ethically unacceptable. However, there are many ethical concerns including use of diagnostic terminology, the associated stigma, risks of medical treatment versus no treatment, use of psychosocial and psychotherapeutic approaches, involvement of the family, issues of capacity and consent and involvement in research studies. We will look at each of these ethical issues in more detail in the following sections.

Writing about the vulnerability to psychosis, McGlashan and Johannessen (1996) highlight the importance of altering the course of the disease, in the premorbid and onset phases, when the neurobiological deficit processes are at their most active but neuroplasticity can also be retained or reversed. Birchwood, Todd, and Jackson (1998) describe a critical period of early psychosis in which biopsychosocial influences are developing and are subject to change and long-term outcome can be predicted. Hence EIP services also aim to identify individuals at risk of developing psychosis; those in the At-Risk Mental State (ARMS), who may, or may not, transition into full psychosis. Until recently identification of the At-Risk Mental State was a research strand of EIP, rather than part of regular practice. The Comprehensive Assessment of At-Risk Mental States (CAARMS) was developed to provide a valid and reliable method for identifying prodromal states (Yung et al., 2005). CAARMS enables an assessment of psychopathology thought to indicate imminent development of psychosis, and can determine if an individual meets criteria for being at ultra-high risk (UHR) for psychosis. However, there are notable difficulties in accurately identifying the ultra-high risk state, which can present with very non-specific, subtle symptoms (Dimitrakopoulos, Kollias, Stefanis, & Kontaxakis, 2015; Yung et al., 1996). As such, there is a high risk of identifying false-positive cases (when the mental state abnormality indicates a different, non-psychotic, psychiatric disorder) and false-false positive cases, when the subject is at risk of psychosis but the symptoms are masked by good coping strategies or good social support, for example (Yung et al., 1996). Strauss (1969) and van Os, Hanssen, Bijl, and Ravelli (2000) postulated that rather than being an all-or-nothing phenomenon, discrete from normal experience, psychosis may actually be on a continuum of severity, and many subsequent studies of voice hearing in healthy individuals have supported this view, as described by Johns and van Os (2001). There is, therefore, a risk of stigmatising healthy individuals by labelling them as at risk of psychosis, treating them in mental health services, and even offering pharmacological and psychosocial interventions that might not be necessary and could contravene the principle of non-maleficence by doing more harm than good. In addition, over the years the transition rates from UHR states to full psychosis have been in decline (Yung et al., 2007), leading to questions about the validity of an UHR construct, especially as the majority of individuals identified as UHR will never develop psychosis. Is the decline due to better identification and swift, effective treatment, or is it due to more identification of false-positive cases, with the associated stigma and possible exposure to medication and other treatment modalities that may harm someone who may never develop a psychotic illness?

Cornblatt, Lencz, and Kane (2001) argue that interest in prevention has outpaced the necessary scientific and ethical underpinnings for clinical trials involving the schizophrenia prodrome, which is essentially a retrospective construct, since it cannot be conclusively identified as the prelude to psychosis until the psychosis becomes apparent. It is also worth noting that the At-Risk Mental State predicts a first episode of psychosis, but that a first episode of psychosis does not necessarily lead to a diagnosis of schizophrenia. There is a need for the defining signs and symptoms of prodromal states to be validated in prospective, longitudinal studies, but for the time being, such signs and symptoms can rouse clinical suspicion only, encouraging watchful waiting, rather than active treatment. Early social and role deficits appear to be stable risk factors for psychosis as such functional deficits are characteristic of schizophrenia and are, therefore, appropriate targets for early intervention, including psychoeducation and psychosocial therapies (Cornblatt et al., 2012).

2. The impact of early intervention on patients and their families

2.1 Stigma and diagnosis

Stigma is defined by Goffman (1963) as 'an attribute that is deeply discrediting', which leads a stigmatised person to be 'reduced ... from a whole and usual person to a tainted or discounted one'. Elliott, Ziegler, Altman, and Scott (1982) describe stigma as 'a form of deviance that leads others to judge individuals as illegitimate for participation in an interaction, because they are incompetent, unpredictable, inconsistent, or a threat to the interaction'. Stigma, prejudice and discrimination are inter-related and influenced by people's knowledge, attitudes and behaviour, which can be deeply entrenched in society in what Miranda Fricker terms 'the collective social imagination' (Fricker, 2007). People with schizophrenia are often perceived as being hard to talk to, unpredictable and dangerous to others, and this prejudice can contribute to social isolation, distress and employment difficulties (Crisp, Gelder, Rix, Meltzer, & Rowlands, 2000), in addition to those difficulties caused by the psychosis itself.

Stigma has a marked effect on the quality of life of patients with psychosis, and mental health services must therefore approach diagnosis with caution, particularly as the patient population may be very young and at a crucial phase of social and occupational development. The diagnosis of schizophrenia can lead to impoverishment, social marginalisation and poor quality of life (Thornicroft, Brohan, Rose, Sartorius, & Leese, 2009), affecting patients' ability to make and keep friends and relationships, and find and keep a job. A review of schizophrenia and employment identified low rates of employment (between 10 and 20% in Europe), but higher rates amongst patients with first episode of psychosis (Marwaha & Johnson, 2004). Stigma and discrimination are listed as significant factors affecting employment rates, alongside a fear of losing benefits and a lack of professional help. Employment, however, is correlated with good outcomes, including social functioning, symptom reduction, quality of life and self-esteem.

The approach that views mental illness as an illness like any other (for example, equating a mental illness to a physical illness, such as diabetes mellitus, with a clear biological basis) may be unhelpful. This approach seems to absolve the mentally ill person of any responsibility for their behaviour and, therefore, may increase, rather than reduce, the general public's prejudice, in terms of people's likelihood to perceive schizophrenic patients as untreatable, dangerous and unpredictable, with associated fear and desire for social distance (Read, Haslam, Sayce, & Davies, 2006). Patients can also self-stigmatise, internalising society's negative perceptions about them, affecting their self-esteem, self-efficacy, insight and adherence to treatment (Fung, Tsang, & Corrigan, 2008). This anticipated discrimination has a significant impact on quality of life, which may be more powerful than that involved in actually experiencing negative discrimination. A study by Thornicroft et al. (2009) showed that anticipated discrimination affects 64% patients with schizophrenia when applying for work (only 29% actually experienced discrimination), and 55% when looking for a close relationship (only 27% actually experienced discrimination.) In addition, 72% patients described feeling the need to conceal their diagnosis. Professionals must bear these issues in mind when making a diagnosis of first episode psychosis or schizophrenia, as the potential benefits to patients of having a diagnosis that can guide appropriate treatment must be weighed against the potential harms from the associated stigma.

Stigma also has a bearing on family and carer involvement and includes both general and associative stigma. Associative stigma is the stigma of being associated with another stigmatised individual, and is determined by cultural, psychosocial and personal factors from the patient as well as the family members. A study showed that the prevalence of associative stigma might be as high as 86%, with family members reporting feelings of depression, guilt and shame, whilst trying to hide their relative's illness, or to explain what it is really like (Catthoor et al., 2015). Catthoor et al. (2015) found that patients' disruptive behaviour is the most accurate predictor of high associative stigma. Another study showed that relatives of people with any mental illness frequently think that the patients would be better off dead (18%)

and may experience suicidal thoughts themselves (10%) (Ostman & Kjellin, 2002). Family stigma is found to be low in the early stages of a psychotic illness (Wong et al., 2009), but ethnic differences exist, with some groups reporting a sense of shame and a need to conceal the patient's illness, demonstrating the need for cultural sensitivity when working with patients and their families.

Due to the issues of stigma associated with the diagnosis of schizophrenia, EIP services tend to use the terms acute and transient psychosis, psychosis not otherwise specified (NOS), first episode psychosis or drug-induced psychosis, terms that communicate diagnostic uncertainty, as these diagnoses are subject to change over time (Hobbs, Stanbrook, & Chakraborty, 2017). This diagnostic uncertainty can create therapeutic challenges, for example with regards to psychoeducation, discussions about prognosis and relapse prevention, and whether or not to continue with antipsychotic medication and for how long. This is particularly pertinent for the community teams that take over a patient's care once they have completed the usual three years in the EIP service and may not be sure of the direction of care. It is important to note that only a proportion of patients at ultra-high risk (UHR) of psychosis will transition to psychosis, as discussed later, and only 75% of those who transition to FEP will progress to a schizophrenia diagnosis (Yung et al., 2003). As discussed above, there is a risk of identifying false positive and false-false positive cases, so caution in applying diagnostic labels is indicated ethically, to reduce the risk of stigmatising individuals, with associated impact on quality of life. There is also the risk of treating false positive patients with antipsychotic medication, with the inherent side-effect profile causing significant harm (contrary to the principle of non-maleficence), including abnormal movements, weight gain, metabolic syndrome, hyperprolactinaemia, and cardiac problems, to name but a few.

Epistemic injustice, as described by Fricker (2007) and quoted by Crichton, Carel, and Kidd (2017) is a term that describes the harm done to a person in their capacity as a knower, that is, someone who gives knowledge (testifies) or makes sense of their experiences (interprets). Negative stereotypes can give rise to epistemic injustice, when a person is not taken seriously, believed or treated as credible, because of the negative attitudes towards them, which may be unconscious. This affects many groups of people including, for example, women (sexism) and people of different ethnic origin (racism), as well as the physically unwell and mentally unwell. Patients with psychotic symptoms, for example, may be reluctant to disclose their experiences for fear that they will be stigmatised, or because they know that they may be diagnosed with a psychotic illness and possibly be detained in hospital and treated against their will. If a patient discloses psychotic symptoms the psychiatrists might assume that the symptoms are more severe than they really are because the patient is perceived as being unable to conceal them anymore. Fear of stigma and negative stereotypes among those at risk of developing a psychotic illness can make early intervention difficult, and those who have been diagnosed and feel stigmatised might disengage from services and experience more frequent relapses or more severe illness. Epistemic injustice can have wider implications for society, including reinforcing negative public perceptions of mental illnesses, contributing to further stigma, and explaining why funding for mental health services is disproportionate to requirements even while parity of esteem, which also rests on the principle of justice, has been a high-profile concept since 2011 (Millard & Wessely, 2014). In the phenomenon known as diagnostic overshadowing (a term coined by Reiss, Levitan, and Szyszko (1982)), psychiatric patients may also experience epistemic injustice when they present with symptoms of physical ill health which are attributed to their mental disorder and which, again contrary to the notion of parity of esteem resting on the principle of justice, may not be given sufficient credibility by physicians.

2.2 Impact on families

Relatives and friends of patients suffering from psychosis make up the support network and suffer along with the patient. The UK700 study showed that many relatives of patients with psychotic illnesses experience psychological distress but that patient-relative contact is beneficial for the patients, with shorter inpatient admissions and lower rates of parasuicide (Harvey et al., 2001). Research indicates that relatives want to be involved at an early stage, with a preference for early detection and treatment, so as to know how to behave towards the patient and how to cope with feelings of shame and guilt, preventing suffering for all involved (Lauber & Rossler, 2003). The BIOMED I study showed that relatives of patients with schizophrenia experience high levels of burden when they have poor coping resources and limited social support, so family interventions must have a social focus, to increase the support network and reduce stigma (Magliano et al., 1998). An important part of Early Intervention services' work is to include family and carers at an early stage in psychoeducation, treatment planning and therapeutic interventions, including behavioural family therapy (BFT).

Key statements related to families and carers from the Psychosis and Schizophrenia in Adults NICE Quality Standard (NICE, 2015) include, as quoted; *family members of adults with psychosis or schizophrenia are offered family intervention* and *carers of adults with psychosis or schizophrenia are offered carer-focused education and support programmes*, placing family and carer involvement squarely in national guidance. A systematic review of the literature regarding implementing family involvement in the treatment of patients with psychosis identified a number of facilitating and hindering factors (Eassom, Giacco, Dirik, & Priebe, 2014). For example, patients report concerns about placing their relatives in a position of power over them. When this is in ways that might jeopardize the patient's autonomy it here invokes potential issues over the ethical principle of autonomy. Both families and staff members report concerns about burdening the family and making the patient's symptoms worse. There can also be problems with engaging families, due to scepticism, lack of motivation, reluctance or refusal to engage, but this can be ameliorated by early intensive efforts to involve family members and presenting the approach enthusiastically. The goal is a collaborative approach, where teams, patients and their families work together in the interests of good patient care, the benefits of which include good engagement with treatment plans; the recognition of, and swift response to, early warning signs of relapse; and better outcomes, including fewer inpatient admissions, shorter inpatient stays and better quality of life reports by patients (in line with the ethical principle of beneficence). These benefits need to be balanced against the burden placed on families to fill the role of professionals, without remuneration, and the impact this can have on family relationships, challenging the principle of non-maleficence. In addition, it is important to ensure that families are not made to feel patronised or criticised, that the approach is flexible and culturally sensitive, and that the needs of the whole family are identified and addressed, with a recognition of family dynamics.

When considering the ethics of treating patients in Early Intervention services we need to be clear about what we are treating and whether the treatment constitutes primary or secondary prevention. Primary prevention aims to prevent disease before it occurs, in this case preventing transition from the At-Risk Mental State to psychosis by intervening in the prodrome. Secondary prevention aims to reduce the impact of disease once it has already occurred, in this case treating patients with first episode psychosis at the earliest stage, in order to improve long-term outcomes, including reducing relapse rates.

In clinical practice in the UK patients in EIP services are usually experiencing their first episode of psychosis, rather than being in the At-Risk Mental State or prodromal phase, hence EIP services are mostly concerned with secondary prevention. The ethics of treating these two groups are quite different. Once someone is suffering from psychotic symptoms amounting to a first episode with associated distress and functional impairment the need for treatment, with a biopsychosocial approach, is clearer, to follow the ethical principle of beneficence by alleviating symptoms, reducing suffering and improving functioning. The thorny ethical questions occur more starkly when considering the treatment of patients in prodromal, pre-psychotic states. As discussed above, the prodrome is a retrospective construct that can only be accurately identified as such once a person makes the transition to psychosis, which only a small proportion of patients do; 18% after 6 months, 22% after 1 year, 29% after 2 years and 36% after 3 years of follow-up (Fusar-Poli et al., 2012). To treat patients in a supposed prodrome first requires them to come under the care of an EIP service. Since only a very small proportion (4.1%) of patients in prodromal states receive this attention (Ajnakina et al., 2017), there is risk of stigmatising them from the outset, challenging the principle of non-maleficence. The ethical trade-offs are complex here. Whilst being monitored for possible transition to psychosis patients will be offered psychological interventions (for example psychoeducation, behavioural family therapy) and social ones (education and employment support, carer input, social activities). They may also be offered pharmacological treatments, depending on the clinical rationale, including antidepressants and/or antipsychotics, which can have adverse as well as favourable effects.

3. Are EIP services effective?

3.1 Evidence for intervening in the prodrome and early stages of psychosis

There have been a number of high-profile research studies on the impact of intervening in the prodromal phase or at-risk mental state. The Recognition and Prevention (RAP) program studied 62 adolescents in three clinical high-risk groups with non-specific prodromal symptoms including, (1) negative and non-specific symptoms such as social isolation, (2) emerging moderate attenuated positive symptoms and (3) severe attenuated, sub-psychotic positive symptoms. The study found that four neurodevelopmental risk factors (Cognitive deficits, Affective disturbances, Social Isolation and School failure – the CASIS cluster) were equally impaired across the three groups, supporting the concept of an underlying biological vulnerability to psychosis (the vulnerability core, which may be necessary for schizophrenia to develop). The study concluded that treating patients in these high-risk groups with antidepressants and/or antipsychotic medication can confer a neuroprotective effect; moderately symptomatic youngsters did well over the 1 year follow-up observation period, but 47% of severely symptomatic (although sub-psychotic) patients converted to psychosis within 1 year (Cornblatt et al., 2003).

The suggestion that antidepressants are beneficial in the treatment of psychosis is the focus of much current research into effective medical interventions, as depressive symptoms commonly occur in prodromal states, 76% (Yung et al., 2003); acute psychosis, 70%, (Birchwood, Iqbal, Chadwick, & Trower, 2000); and post-psychosis, 54% (Birchwood et al., 2000). Upthegrove (2009) writes that depression can be accurately distinguished from negative symptoms of psychosis, with the aid of structured assessments, and should be a focus of pharmacological and psychological treatments in order to achieve a full functional recovery.

The Lambeth Early Onset (LEO) study showed that patients with early psychosis treated in an intensive, case-managed service with community based interventions and low dose antipsychotic medication had lower rates of relapse and dropout, improved measures of social and vocational functioning, satisfaction, quality of life and medication adherence at 18 month follow-up, compared to treatment as usual (Craig et al., 2004). The Danish OPUS study demonstrated similar benefits at 2 year follow-up, including better symptom control and less substance misuse, but over 10 years most of the positive short-term effects had diminished, with no differences between OPUS and Treatment as Usual (TAU) regarding income, work-related outcomes or marital status (Secher et al., 2015).

A Cochrane review evaluated the effects of early detection, phase-specific treatments and specialised early intervention teams in the treatment of prodromal states and first episode psychosis (Marshall & Rathbone, 2011). The review concluded that there is emerging but inconclusive evidence that some interventions are helpful in the prodrome, with evidence in favour of specialist teams vs. standard care, but the personal and social, and so ethical consequences of treating people who may never become unwell need further research. Phase-specific treatments for patients with FEP may help with employment, and there are inconclusive benefits of family therapy in FEP. Interestingly, the authors conclude that there is no evidence to support the benefits of early detection of patients in their first episode of psychosis, and that much of the supporting evidence for EIP services is circumstantial (based on the correlation between duration of untreated psychosis and outcome), rather than definitive (from robust clinical trials). A more recent network meta-analysis showed no evidence of significantly better efficacy of any one intervention over the others for the prevention of psychosis (including combinations of medications, psychological interventions, and needs-based interventions) at 6 and 12 months, concluding that further experimental research is needed (Davies et al., 2018).

3.2 Are the gains of EIP maintained?

There is a concern that although patients under the care of EIP services for 3 years do really well with the intense input, after their 3 years are up their care is transferred to generic services, such as Community Mental Health Teams (treatment as usual, TAU) and the gains may not be maintained. Unfortunately, both the OPUS and LEO studies demonstrated that the early gains identified were not maintained at 5 year follow-up (Bertelsen et al., 2008; Gafoor et al., 2010). The long-term follow-up of the PACE clinic cohort of ultra-high risk (UHR) patients identified that UHR patients remain at risk of transition to psychosis over 10 years, but the highest risk is in the first 2 years, and individuals with a long duration of symptoms and poor early functioning may need closer monitoring. During the 15 year follow-up period, 114 of the 416 participants were known to have developed a psychotic illness, with an overall transition rate of 34.9% over 10 years, so ongoing monitoring of this cohort is indicated in TAU (Francey et al., 2013). Marshall and Rathbone (2011) conducted a systematic review of the literature to evaluate the short and longer-term (5 years) effects of early detection, phase-specific treatments and specialised EIP teams in the treatment of people with prodromal symptoms or first episode psychosis. Their findings led them to question whether the gains of EIP services are maintained once patients return to treatment as usual, and they suggest that further longitudinal prospective studies are required. Patients likely to relapse do so in the first 12 months following discharge from EIP to TAU, predicted by numbers of relapses prior to discharge (Kam, Singh, & Upthegrove, 2015), but potentially mitigated by improved emphasis on relapse prevention and early post-discharge monitoring, with targeted interventions to address substance misuse and functional recovery. There is also the problem of patients transitioning between different mental health teams, introducing what can be a time of uncertainty and discontinuity.

3.3 EIP and mental health policy in England

The research in favour of Early Intervention in Psychosis services has led these specialised teams to become embedded in mental health care. In response to the NHS England Mental Health Taskforce, an Early Intervention in Psychosis Access and Waiting Time Standard (NICE, 2016) was developed in 2016 in England in order to improve access to EIP services for all people (aged 14–65) experiencing their first episode of psychosis, including that associated with trauma or substance misuse, and those in the At-Risk Mental State (ARMS). The standard requires that >50% patients experiencing FEP commence a National Institute for Health and Care Excellence (NICE) recommended package of care within 2 weeks of referral (NICE, 2016). The aim of the standard is to improve the quality of care received by patients with first episode psychosis and improve their ability to recover, and has an ambition towards continuous improvement with year-on-year gains.

Recent work has been investigating and debating the cost-effectiveness of EIP services. One study showed that a comprehensive package of care may cost more than treatment as usual but can improve quality of life more significantly, with a favourable cost-benefit ratio (Rosenheck et al., 2016). A longitudinal retrospective observational study in England showed that patients in EIP services were more likely to have favourable outcomes (including employment, accommodation in mainstream housing and emotional wellbeing scores), and costs to the National Health Service (NHS) were lower due to fewer inpatient admissions, lower acute hospital outpatient costs, and lower accident and emergency costs, compared to higher community mental health costs (Tsiachristas, Thomas, Leal, & Lennox, 2016). The debate continues as to whether separate EIP services are necessary or whether an integrated approach would be ultimately more cost-effective, as resources would not be stripped from generic services to fund specialist teams (Castle & Singh, 2015). This is particularly important to remember in a resource limited healthcare system like the NHS.

4. Looking to the future

Research evidence is emerging in favour of the benefits of Early Intervention in Psychosis services, in parallel with the promotion of early intervention for other neuropsychiatric disorders such as those involving dementia (Robinson, Tang, & Taylor, 2015) and Early Intervention in Eating Disorders services (Lang et al., 2016; Treasure & Russell, 2011).

As described, there is a need for ongoing longitudinal, prospective studies of the long-term impact of EIP, including relapse rates, functional outcomes, socioeconomic costs, and more. There is a need for further research into effective treatment strategies, including possible pharmacological treatment of prodromal states, and the impact of psychosocial therapies. The long-term socioeconomic benefits of EIP services will need to be robustly demonstrated in order for these specialist and well-resourced services to continue to be funded in times of reduced financial resource, with a clear advantage over treatment as usual being of paramount importance. There is an imperative for EIP services to be expanded to serve patients aged 14–65, as per the Access and Waiting Time Standard (NICE, 2016), which has funding implications and risks stripping further resource from already hard-pressed generic community mental health teams.

5. Conclusions

We have discussed a number of ethical issues involved in early intervention in psychosis, including accurate identification of patients at risk of psychosis; thorny issues around definitive diagnosis and the impact that has in terms of stigma and family dynamics; and the imperative to intervene early when to do so will improve functional outcomes. The difficult question is what that intervention should be, with tricky risk/benefit decisions to be made, and the knowledge that the early gains may not be long-term, especially once treatment as usual resumes in community teams. Professionals must always have in mind the ethical principles of autonomy, justice, beneficence and non-maleficence when making these difficult decisions with patients.

Conflict of interest

Matthew Broome and Charlotte Marriott do not report any conflicts of interest.

References

Ajnakina, O., Morgan, C., Gayer-Anderson, C., Oduola, S., Bourque, F., Bramley, S., … David, A. S. (2017). Only a small proportion of patients with first episode psychosis come via prodromal services: A retrospective survey of a large UK mental health programme. *BMC Psychiatry, 17*(1), 308. https://doi.org/10.1186/s12888-017-1468-y.

Anderson, K. K., Voineskos, A., Mulsant, B. H., George, T. P., & McKenzie, K. J. (2014). The role of untreated psychosis in neurodegeneration: A review of hypothesized mechanisms of neurotoxicity in first-episode psychosis. *Canadian Journal of Psychiatry, 59*, 513–517. https://doi.org/10.1177/070674371405901003.

Andreasen, N. C., Carpenter, W. T., Kane, J. M., Lasser, R. A., Marder, S. R., & Weinberger, D. R. (2005). Remission in schizophrenia: Proposed criteria and rationale for consensus. *American Journal of Psychiatry, 162*(3), 441–449. https://doi.org/10.1176/appi.ajp.162.3.441.

Bangalore, S. S., Prasad, K. M. R., Montrose, D. M., Goradia, D. D., Diwadkar, V. A., & Keshavan, M. S. (2008). Cannabis use and brain structural alterations in first episode schizophrenia — A region of interest, voxel based morphometric study. *Schizophrenia Research, 99*(1–3), 1–6. https://doi.org/10.1016/j.schres.2007.11.029.

Bertelsen, M., Jeppesen, P., Petersen, L., Thorup, A., Øhlenschlæger, J., le Quach, P., ... Nordentoft, M. (2008). Five-year follow-up of a randomized multicenter trial of intensive early intervention vs standard treatment for patients with a first episode of psychotic illness. *Archives of General Psychiatry, 65*(7), 762. https://doi.org/10.1001/archpsyc.65.7.762.

Birchwood, M., Iqbal, Z., Chadwick, P., & Trower, P. (2000). Cognitive approach to depression and suicidal thinking in psychosis. 1. Ontogeny of post-psychotic depression. *The British Journal of Psychiatry: The Journal of Mental Science, 177*, 516–521. Retrieved from http://www.ncbi.nlm.nih.gov/pubmed/11102326.

Birchwood, M., Todd, P., & Jackson, C. (1998). Early intervention in psychosis. The critical period hypothesis. *The British Journal of Psychiatry. Supplement, 172*(33), 53–59. Retrieved from http://www.ncbi.nlm.nih.gov/pubmed/9764127.

Brugha, T., Singleton, N., Meltzer, H., Bebbington, P., Farrell, M., Jenkins, R., ... Lewis, G. (2005). Psychosis in the community and in prisons: A report from the British National Survey of psychiatric morbidity. *American Journal of Psychiatry, 162*(4), 774–780. https://doi.org/10.1176/appi.ajp.162.4.774.

Castle, D. J., & Singh, S. P. (2015). Early intervention in psychosis: Still the 'best buy'? *British Journal of Psychiatry, 207*(04), 288–292. https://doi.org/10.1192/bjp.bp.114.160630.

Catone, G., Lindau, J. F., & Broome, M. R. (2014). Phenomenological psychopathology and the neurosciences. *Journal of Psychopathology, 20*(4), 358–365.

Catthoor, K., Schrijvers, D., Hutsebaut, J., Feenstra, D., Persoons, P., De Hert, M., ... Sabbe, B. (2015). Associative stigma in family members of psychotic patients in Flanders: An exploratory study. *World Journal of Psychiatry, 5*(1), 118–125. https://doi.org/10.5498/wjp.v5.i1.118.

Charlson, F. J., Ferrari, A. J., Santomauro, D. F., Diminic, S., Stockings, E., Scott, J. G., ... Whiteford, H. A. (2018). Global epidemiology and burden of schizophrenia: Findings from the global burden of disease study 2016. *Schizophrenia Bulletin, 44*, 1195–1203. https://doi.org/10.1093/schbul/sby058.

Cooper, S., Crawford, M., Etherington, A., Jayakumar, S., Lemmey, S., Marsh, R., ... Zalewska, K. (2012). *Report of the National Audit of Schizophrenia* (pp. 1–160). Royal College of Psychiatrists. Retrieved from http://www.rcpsych.ac.uk/pdf/NAS National report FINAL.pdf.

Cornblatt, B. A., Carrion, R. E., Addington, J., Seidman, L., Walker, E. F., Cannon, T. D., ... Lencz, T. (2012). Risk factors for psychosis: Impaired social and role functioning. *Schizophrenia Bulletin, 38*(6), 1247–1257. https://doi.org/10.1093/schbul/sbr136.

Cornblatt, B. A., Lencz, T., & Kane, J. M. (2001). Treatment of the schizophrenia prodrome: Is it presently ethical? *Schizophrenia Research, 51*(1), 31–38. https://doi.org/10.1016/S0920-9964(01)00236-5.

Cornblatt, B. A., Lencz, T., Smith, C. W., Correll, C. U., Auther, A. M., & Nakayama, E. (2003). The schizophrenia prodrome revisited: A neurodevelopmental perspective. *Schizophrenia Bulletin, 29*, 633–651. https://doi.org/10.1093/oxfordjournals.schbul.a007036.

Craddock, N., & Owen, M. J. (2005). The beginning of the end for the Kraepelinian dichotomy. *British Journal of Psychiatry, 186*(05), 364–366. https://doi.org/10.1192/bjp.186.5.364.

Craddock, N., & Owen, M. J. (2010). The Kraepelinian dichotomy – going, going ... but still not gone. *British Journal of Psychiatry, 196*(02), 92–95. https://doi.org/10.1192/bjp.bp.109.073429.

Craig, T. K. J., Garety, P., Power, P., Rahaman, N., Colbert, S., Fornells-Ambrojo, M., & Dunn, G. (2004). The Lambeth Early Onset (LEO) team: Randomised controlled trial of the effectiveness of specialised care for early psychosis. *BMJ, 329*(7474), 1067. https://doi.org/10.1136/bmj.38246.594873.7C.

Crichton, P., Carel, H., & Kidd, I. J. (2017). Epistemic injustice in psychiatry. *BJPsych Bulletin, 41*(2), 65–70. https://doi.org/10.1192/pb.bp.115.050682.

Crisp, A. H., Gelder, M. G., Rix, S., Meltzer, H. I., & Rowlands, O. J. (2000). Stigmatisation of people with mental illnesses. *The British Journal of Psychiatry: The Journal of Mental Science, 177*, 4–7. Retrieved from http://www.ncbi.nlm.nih.gov/pubmed/10945080.

Davies, C., Cipriani, A., Ioannidis, J. P. A., Radua, J., Stahl, D., Provenzani, U., ... Fusar-Poli, P. (2018). Lack of evidence to favor specific preventive interventions in psychosis: A network meta-analysis. *World Psychiatry: Official Journal of the World Psychiatric Association (WPA), 17*(2), 196–209. https://doi.org/10.1002/wps.20526.

Dimitrakopoulos, S., Kollias, C., Stefanis, N. C., & Kontaxakis, V. (2015). Early psychotic experiences: Interventions, problems and perspectives. *Psychiatrike = Psychiatriki, 26*(1), 45–54. Retrieved from http://www.ncbi.nlm.nih.gov/pubmed/25880383.

Drake, R. J., Haley, C. J., Akhtar, S., & Lewis, S. W. (2000). Causes and consequences of duration of untreated psychosis in schizophrenia. *British Journal of Psychiatry, 177*(06), 511–515. https://doi.org/10.1192/bjp.177.6.511.

Eassom, E., Giacco, D., Dirik, A., & Priebe, S. (2014). Implementing family involvement in the treatment of patients with psychosis: A systematic review of facilitating and hindering factors. *BMJ Open, 4*(10), e006108. https://doi.org/10.1136/bmjopen-2014-006108.

Eaton, W. W., Romanoski, A., Anthony, J. C., & Nestadt, G. (1991). Screening for psychosis in the general population with a self-report interview. *The Journal of Nervous and Mental Disease, 179*(11), 689–693. Retrieved from http://www.ncbi.nlm.nih.gov/pubmed/1940893.

Elliott, G. C., Ziegler, H. L., Altman, B. M., & Scott, D. R. (1982). Understanding stigma: Dimensions of deviance and coping. *Deviant Behavior, 3*(3), 275–300. https://doi.org/10.1080/01639625.1982.9967590.

Francey, S. M., Yuen, H. P., Amminger, G. P., Lin, A., Thompson, A., Wood, S. J., ... Simmons, M. (2013). Long-term follow-up of a group at ultra high risk ("Prodromal") for psychosis. *JAMA Psychiatry, 70*(8), 793. https://doi.org/10.1001/jamapsychiatry.2013.1270.

Fricker, M. (2007). *Epistemic injustice. Power and the ethics of knowing*. Oxford University Press.

Fung, K. M. T., Tsang, H. W. H., & Corrigan, P. W. (2008). Self-stigma of people with schizophrenia as predictor of their adherence to psychosocial treatment. *Psychiatric Rehabilitation Journal, 32*(2), 95–104. https://doi.org/10.2975/32.2.2008.95.104.

Fusar-Poli, P., Bonoldi, I., Yung, A. R., Borgwardt, S., Kempton, M. J., Valmaggia, L., ... McGuire, P. (2012). Predicting psychosis: Meta-analysis of transition outcomes in individuals at high clinical risk. *Archives of General Psychiatry, 69*(3), 220–229. https://doi.org/10.1001/archgenpsychiatry.2011.1472.

Fusar-Poli, P., Cappucciati, M., Bonoldi, I., Hui, L. M. C., Rutigliano, G., Stahl, D. R., ... McGuire, P. K. (2016). Prognosis of brief psychotic episodes. *JAMA Psychiatry, 73*(3), 211. https://doi.org/10.1001/jamapsychiatry.2015.2313.

Gafoor, R., Nitsch, D., McCrone, P., Craig, T. K. J., Garety, P. A., Power, P., & McGuire, P. (2010). Effect of early intervention on 5-year outcome in non-affective psychosis. *British Journal of Psychiatry, 196*(05), 372–376. https://doi.org/10.1192/bjp.bp.109.066050.

Goffman, E. (1963). *Stigma: Notes on the management of spoiled identity*. Simon and Schuster.

Harvey, K., Burns, T., Sedgwick, P., Higgitt, A., Creed, F., & Fahy, T. (2001). Relatives of patients with severe psychotic disorders: Factors that influence contact frequency. *British Journal of Psychiatry, 178*(03), 248–254. https://doi.org/10.1192/bjp.178.3.248.

Hastrup, L. H., Haahr, U. H., Jansen, J. E., & Simonsen, E. (2018). Determinants of duration of untreated psychosis among first-episode psychosis patients in Denmark: A nationwide register-based study. *Schizophrenia Research, 192*, 154–158. https://doi.org/10.1016/j.schres.2017.05.026.

Hegelstad, W. V., Larsen, T. K., Auestad, B., Evensen, J., Haahr, U., Joa, I., … McGlashan, T. (2012). Long-term follow-up of the TIPS early detection in psychosis study: Effects on 10-year outcome. *American Journal of Psychiatry, 169*(4), 374–380. https://doi.org/10.1176/appi.ajp.2011.11030459.

Henquet, C., Krabbendam, L., Spauwen, J., Kaplan, C., Lieb, R., Wittchen, H.-U., & van Os, J. (2005). Prospective cohort study of cannabis use, predisposition for psychosis, and psychotic symptoms in young people. *BMJ, 330*(7481), 11. https://doi.org/10.1136/bmj.38267.664086.63.

Ho, B.-C., Andreasen, N. C., Ziebell, S., Pierson, R., & Magnotta, V. (2011). Long-term antipsychotic treatment and brain volumes: A longitudinal study of first-episode schizophrenia. *Archives of General Psychiatry, 68*(2), 128–137. https://doi.org/10.1001/archgenpsychiatry.2010.199.

Hobbs, M., Stanbrook, R., & Chakraborty, N. (2017). Frequency of change of diagnoses in first-episode psychosis. *Progress in Neurology and Psychiatry, 21*(01), 6–8. Retrieved from https://www.progressnp.com/article/frequency-change-diagnoses-first-episode-psychosis/.

Johns, L. C., & van Os, J. (2001). The continuity of psychotic experiences in the general population. *Clinical Psychology Review, 21*(8), 1125–1141. Retrieved from http://www.ncbi.nlm.nih.gov/pubmed/11702510.

Kam, S. M., Singh, S. P., & Upthegrove, R. (2015). What needs to follow early intervention? Predictors of relapse and functional recovery following first-episode psychosis. *Early Intervention in Psychiatry, 9*(4), 279–283. https://doi.org/10.1111/eip.12099.

Kane, J., Honigfeld, G., Singer, J., & Meltzer, H. (1988). Clozapine for the treatment-resistant schizophrenic. A double-blind comparison with chlorpromazine. *Archives of General Psychiatry, 45*(9), 789–796. Retrieved from http://www.ncbi.nlm.nih.gov/pubmed/3046553.

Kendler, K. S., Gallagher, T. J., Abelson, J. M., & Kessler, R. C. (1996). Lifetime prevalence, demographic risk factors, and diagnostic validity of nonaffective psychosis as assessed in a US community sample. The National Comorbidity Survey. *Archives of General Psychiatry, 53*(11), 1022–1031. Retrieved from http://www.ncbi.nlm.nih.gov/pubmed/8911225.

Lang, K., Glennon, D., Mountford, V., McClelland, J., Koskina, A., Brown, A., & Schmidt, U. (2016). Early intervention for eating disorders. In *Encyclopedia of feeding and eating disorders* (pp. 1–6). Singapore: Springer Singapore. https://doi.org/10.1007/978-981-287-087-2_185-1.

Lauber, C., & Rossler, W. (2003). Relatives and their attitude to early detection of schizophrenic psychosis. *Psychiatric Bulletin, 27*(4), 134–136. https://doi.org/10.1192/pb.27.4.134.

Lieberman, J. A., Tollefson, G. D., Charles, C., Zipursky, R., Sharma, T., Kahn, R. S., … HGDH Study Group. (2005). Antipsychotic drug effects on brain morphology in first-episode psychosis. *Archives of General Psychiatry, 62*(4), 361. https://doi.org/10.1001/archpsyc.62.4.361.

Magliano, L., Fadden, G., Madianos, M., Caldas de Almeida, J. M., Held, T., Guarneri, M., … Maj, M. (1998). Burden on the families of patients with schizophrenia: Results of the

BIOMED I study. *Social Psychiatry and Psychiatric Epidemiology, 33*(9), 405–412. https://doi.org/10.1007/s001270050073.

Marshall, M., Lewis, S., Lockwood, A., Drake, R., Jones, P., & Croudace, T. (2005). Association between duration of untreated psychosis and outcome in cohorts of first-episode patients: A systematic review. *Archives of General Psychiatry, 62*(9), 975–983. https://doi.org/10.1001/archpsyc.62.9.975.

Marshall, M., & Rathbone, J. (2011). Early intervention for psychosis. *Schizophrenia Bulletin, 37*(6), 1111–1114. https://doi.org/10.1093/schbul/sbr110.

Marwaha, S., & Johnson, S. (2004). Schizophrenia and employment. *Social Psychiatry and Psychiatric Epidemiology, 39*(5), 337–349. https://doi.org/10.1007/s00127-004-0762-4.

Marwaha, S., Thompson, A., Upthegrove, R., & Broome, M. R. (2016). Fifteen years on – early intervention for a new generation. *British Journal of Psychiatry, 209*(03), 186–188. https://doi.org/10.1192/bjp.bp.115.170035.

McGlashan, T. H., & Johannessen, J. O. (1996). Early detection and intervention with schizophrenia: Rationale. *Schizophrenia Bulletin, 22*(2), 201–217. Retrieved from https://academic.oup.com/schizophreniabulletin/article-abstract/22/2/201/1928232.

Millard, C., & Wessely, S. (2014). Parity of esteem between mental and physical health. *BMJ, 349*, g6821. https://doi.org/10.1136/bmj.g6821.

Murray, C. J. L., Lopez, A. D., Harvard School of Public Health, World Health Organization, & World Bank. (1996). *The global burden of disease : A comprehensive assessment of mortality and disability from diseases, injuries, and risk factors in 1990 and projected to 2020.* Harvard School of Public Health on behalf of the World Health Organization and the World Bank. Retrieved from http://www.hup.harvard.edu/catalog.php?isbn=9780674354487.

NICE. (2015). *Psychosis and schizophrenia in adults. Guidance and guidelines.* NICE. Retrieved from https://www.nice.org.uk/guidance/qs80.

NICE. (2016). *Implementing the early intervention in psychosis access and waiting time standard: Guidance.*

van Os, J., Hanssen, M., Bijl, R. V., & Ravelli, A. (2000). Strauss (1969) revisited: A psychosis continuum in the general population? *Schizophrenia Research, 45*(1–2), 11–20. Retrieved from http://www.ncbi.nlm.nih.gov/pubmed/10978868.

van Os, J., Linscott, R. J., Myin-Germeys, I., Delespaul, P., & Krabbendam, L. (2009). A systematic review and meta-analysis of the psychosis continuum: Evidence for a psychosis proneness–persistence–impairment model of psychotic disorder. *Psychological Medicine, 39*(02), 179. https://doi.org/10.1017/S0033291708003814.

Ostman, M., & Kjellin, L. (2002). Stigma by association: Psychological factors in relatives of people with mental illness. *The British Journal of Psychiatry: The Journal of Mental Science, 181*, 494–498. Retrieved from http://www.ncbi.nlm.nih.gov/pubmed/12456519.

Owen, M. J., O'Donovan, M. C., Thapar, A., & Craddock, N. (2011). Neurodevelopmental hypothesis of schizophrenia. *The British Journal of Psychiatry: The Journal of Mental Science, 198*(3), 173–175. https://doi.org/10.1192/bjp.bp.110.084384.

Read, J., Haslam, N., Sayce, L., & Davies, E. (2006). Prejudice and schizophrenia: A review of the "mental illness is an illness like any other" approach. *Acta Psychiatrica Scandinavica, 114*(5), 303–318. https://doi.org/10.1111/j.1600-0447.2006.00824.x.

Reichert, A., & Jacobs, R. (2018). Socioeconomic inequalities in duration of untreated psychosis: Evidence from administrative data in England. *Psychological Medicine, 48*(5), 822–833. https://doi.org/10.1017/S0033291717002197.

Reiss, S., Levitan, G., & Szyszko, J. (1982). Emotional disturbance and mental retardation: Diagnostic overshadowing. *American Journal of Mental Deficiency, 86*(6), 567–574. Retrieved from http://www.ncbi.nlm.nih.gov/pubmed/7102729.

Robinson, L., Tang, E., & Taylor, J. P. (2015). Dementia: Timely diagnosis and early intervention. *BMJ, 350*, h3029. https://doi.org/10.1136/bmj.h3029.

Rosenheck, R., Leslie, D., Sint, K., Lin, H., Robinson, D. G., Schooler, N. R., ... Kane, J. M. (2016). Cost-effectiveness of comprehensive, integrated care for first episode psychosis in the NIMH RAISE early treatment program. *Schizophrenia Bulletin, 42*(4), 896–906. https://doi.org/10.1093/schbul/sbv224.

Secher, R. G., Hjorthoj, C. R., Austin, S. F., Thorup, A., Jeppesen, P., Mors, O., & Nordentoft, M. (2015). Ten-year follow-up of the OPUS specialized early intervention trial for patients with a first episode of psychosis. *Schizophrenia Bulletin, 41*(3), 617–626. https://doi.org/10.1093/schbul/sbu155.

Strauss, J. S. (1969). Hallucinations and delusions as points on continua function: Rating scale evidence. *Archives of General Psychiatry, 21*(5), 581–586. https://doi.org/10.1001/archpsyc.1969.01740230069010.

Thornicroft, G., Brohan, E., Rose, D., Sartorius, N., & Leese, M. (2009). Global pattern of experienced and anticipated discrimination against people with schizophrenia: A cross-sectional survey. *The Lancet, 373*(9661), 408–415. https://doi.org/10.1016/S0140-6736(08)61817-6.

Treasure, J., & Russell, G. (2011). The case for early intervention in anorexia nervosa: Theoretical exploration of maintaining factors. *British Journal of Psychiatry, 199*(01), 5–7. https://doi.org/10.1192/bjp.bp.110.087585.

Tsiachristas, A., Thomas, T., Leal, J., & Lennox, B. R. (2016). Economic impact of early intervention in psychosis services: Results from a longitudinal retrospective controlled study in England. *BMJ Open, 6*(10), e012611. https://doi.org/10.1136/bmjopen-2016-012611.

Upthegrove, R. (2009). Depression in schizophrenia and early psychosis: Implications for assessment and treatment. *Advances in Psychiatric Treatment, 15*(5), 372–379. https://doi.org/10.1192/apt.bp.108.005629.

Weinberger, D. R. (2017). The neurodevelopmental origins of schizophrenia in the penumbra of genomic medicine. *World Psychiatry: Official Journal of the World Psychiatric Association (WPA), 16*(3), 225–226. https://doi.org/10.1002/wps.20474.

Wong, C., Davidson, L., Anglin, D., Link, B., Gerson, R., Malaspina, D., ... Corcoran, C. (2009). Stigma in families of individuals in early stages of psychotic illness: Family stigma and early psychosis. *Early Intervention in Psychiatry, 3*(2), 108–115. https://doi.org/10.1111/j.1751-7893.2009.00116.x.

Wyatt, R. J. (1991). Neuroleptics and the natural course of schizophrenia. *Schizophrenia Bulletin, 17*(2), 325–351. Retrieved from http://www.ncbi.nlm.nih.gov/pubmed/1679255.

Yung, A. R., McGorry, P. D., McFarlane, C. A., Jackson, H. J., Patton, G. C., & Rakkar, A. (1996). Monitoring and care of young people at incipient risk of psychosis. *Schizophrenia Bulletin, 22*(2), 283–303. Retrieved from http://www.ncbi.nlm.nih.gov/pubmed/8782287.

Yung, A. R., Phillips, L. J., Yuen, H. P., Francey, S. M., McFarlane, C. A., Hallgren, M., & McGorry, P. D. (2003). Psychosis prediction: 12-month follow up of a high-risk ("prodromal") group. *Schizophrenia Research, 60*(1), 21–32. https://doi.org/10.1016/S0920-9964(02)00167-6.

Yung, A. R., Yuen, H. P., Berger, G., Francey, S., Hung, T.-C., Nelson, B., ... McGorry, P. (2007). Declining transition rate in ultra high risk (prodromal) services: Dilution or reduction of risk? *Schizophrenia Bulletin, 33*(3), 673–681. https://doi.org/10.1093/schbul/sbm015.

Yung, A. R., Yung, A. R., Pan Yuen, H., Mcgorry, P. D., Phillips, L. J., Kelly, D., ... Buckby, J. (2005). Mapping the onset of psychosis: The Comprehensive Assessment of At-Risk Mental States. *Australian & New Zealand Journal of Psychiatry, 39*(11–12), 964–971. https://doi.org/10.1080/j.1440-1614.2005.01714.x.

CHAPTER TEN

Preparing for the unthinkable? The prevention of posttraumatic stress disorder and the limits of positive psychology

Ulrich Koch*
Clinical Research and Leadership, School of Medicine and Health Sciences, The George Washington University, Washington DC, United States
*Corresponding author: E-mail: koch@email.gwu.edu

Contents

'Suicide, drug abuse, posttraumatic stress disorder, panic, depression, divorce,' began General Casey [...]. 'What does positive psychology say about all that, Dr. Seligman?'

Seligman (2018), p. 311.

1. Introduction

This chapter revisits the controversies surrounding the Comprehensive Soldier Fitness (CSF) initiative, a program developed in 2008 by a team of psychologists led by Martin E.P. Seligman to prevent mental health problems among U.S. soldiers after their deployment to Afghanistan and Iraq. In the following sections, several aspects of these controversies are placed in their broader historical and theoretical contexts, including the ethical objections the initiative has raised, key concepts that underlie its

Developments in Neuroethics and Bioethics, Volume 2
ISSN 2589-2959
https://doi.org/10.1016/bs.dnb.2019.04.008

rationale, as well as Seligman's efforts to promote it through his popular writings. Although Seligman describes the program as based on objective empirical findings, I argue that it nevertheless carries with it normative implications. Not only does it perpetuate problematic ideas about what it means to fall mentally ill, it also relies on implicit assumptions about human development that are equally questionable. Positive psychologists such as Seligman, I contend, endorse a vision of psychological growth that imagines it as unfolding along a linear trajectory that is both continuous and predictable. Ironically, the plausibility of this vision is challenged by the very experience of trauma, the effects of which the CSF program is supposed to inoculate against.

I begin my discussion by providing some historical context for the emergence of the contemporary notion of posttraumatic stress and psychiatric prevention programs in the U.S. military.

2. The trauma of war and the prevention of psychiatric casualties

Ever since the American Civil War, military physicians have observed soldiers returning from the battlefields – often quite literally – shaken and distraught yet outwardly physically unharmed. It was the massive breakdown of soldiers in the trenches of World War I, however, that prompted authorities in war-waging nations to devise means not only to treat but also to prevent what by then were mostly considered disorders of psychological origin. Unleashing destructive forces on an unprecedented industrial scale, the Great War led to an epidemic of neuropsychiatric casualties that was difficult to ignore.

At the outbreak of the Second World War, the most promising strategy of preventing such breakdowns was the screening and rigorous selection of recruits. During the previous conflict, military psychiatrists had already experimented with what came to be considered an effective method of *secondary* prevention, namely treating psychiatric casualties with little delay, close to the front lines. The aim of these methods was to prevent the condition from becoming chronic and to return the soldier to the battlefield. In 1940, the U.S. government's Selective Service System implemented a thorough screening program to reduce the number of those who would have to be temporarily removed from the battlefield. The program was set up to identify inductees who supposedly showed signs of maladjustment to the vicissitudes of civilian life and were therefore presumed to succumb to the much harsher conditions in the military. In some instances, up to a quarter of those examined were deemed unfit to serve (Pols, 2007). But – to the surprise of many psychiatrists – selection alone proved insufficient to prevent psychiatric casualties during World War II, as in subsequent wars.

As the debates over America's participation in the war in Vietnam exposed political and cultural fault lines in the 1960s and 1970s, combat-induced psychiatric disorders became an issue not just of medical but also of great public concern. The introduction of the diagnosis posttraumatic stress disorder (PTSD) in 1980 reflected a sea change in the public conversation about the psychological hazards of war and of experiences of violence more broadly. Although, the rate of American soldier breakdowns had been lower in Vietnam than in the two previous wars (Dean, 1997). In the post-Vietnam era, psychiatric symptoms, now directly related to combat experiences, often surfaced only *after* soldiers had returned from a war zone. According to the causal mechanism that was postulated for PTSD such difficulties stemmed from a harmful experience, a trauma, coming back to haunt the victims in the form of involuntary memories (Young, 1997; 2008).

The broad application of the PTSD diagnosis had implications for how psychiatric casualties were treated as well as for how the military approached the prevention of combat-related disorders. Previous strategies included efforts to improve or stabilise troop morale, to educate soldiers about the mental challenges they may face during deployment, and, in the immediate post-World War II era especially, to sensitise the public to the difficulties returning soldiers were likely to experience as they readjusted to civilian life (Pols & Oak, 2007). The traumatic origin of many mental health problems, however, turned the attention of experts, as well as the public's concern, away from the social conditions that shape how individuals and societies respond to psychological trauma and towards the harmful events and the stress reactions they provoked.

Framing posttraumatic symptoms as a universal response to extreme stress accelerated and solidified what one could call the decontextualization of traumatic experience. In addition, many clinicians, who fought for the introduction of the PTSD diagnosis, on moral and political as much as on scientific grounds (Scott, 1990), were warry of attempts to complicate the matter. For too long, the medical establishment had lacked a diagnostic vocabulary that acknowledged the existence of posttraumatic suffering, as psychiatrists had mostly attributed the onset and persistence of combat-related disorders to the victim's predisposition, his or her personal history prior to the event, or the presumably hazardous incentives of the pension system (see e.g., Hacking, 1998; Leese, 2002; Leys, 2010; Micale, Lerner, & Rosenberg, 2001; Micale, 2009; Young, 1997). The focus on the event and its stress-inducing effects, in other words, quite effectively undermined a morally questionable line of clinical reasoning that, in hindsight, amounted to victim-blaming.

By 2008, when the U.S. Armed Forces solicited the help of the University of Pennsylvania psychologist Martin Seligman to develop a prevention program to enhance soldiers' psychological fitness, the pendulum had begun to swing back, and the question was addressed yet again why some individuals are more likely to succumb to trauma. Recent epidemiological studies suggested that not all who experience trauma develop symptoms of PTSD (cf. Jones & Wessely, 2007); and, in the 1990s, psychologists began researching risk and protective factors to help explain why. Together with his collaborators, Seligman adapted and tested a series of interventions drawn from the Penn Resiliency Program, which was developed to prevent depression in educational settings (Gillham, Jaycox, Reivich, Seligman, & Silver, 1990), to build resilience to trauma and other adverse situations among military personnel. Previously, the Army had tested the efficacy of other interventions designed to prevent PTSD, such as the Battlemind Training, that were applied post-deployment to ease the re-adjustment to the civilian environment (Adler, Bliese, McGurk, Hoge, & Castro, 2011). The initiative, christened Comprehensive Soldier Fitness (CSF), soon drew criticism from other psychologists who raised a number of ethical concerns.

Before engaging in this debate, however, I need to situate two concepts central to this initiative, *resilience* and *posttraumatic growth*, within the wider project of positive psychology – a scientific movement launched by Seligman and other social scientists, mostly psychologist, in the late 1990s.

3. Positive psychology, resilience and posttraumatic growth

At its heart, at least according to its most influential and outspoken ambassadors, positive psychology is about rejecting what was wrong with academic psychology at the end of the 20th century: the supposedly negative outlook that stemmed from the discipline's focus on disease and dysfunction, curtailing psychology's potential. Rather than simply "repairing damage" and "operating within a disease model of human functioning," Seligman and the creativity researcher Mihaly Csikszentmihalyi argued in a programmatic paper that psychological knowledge should be employed to improve also the functioning and wellbeing of "normal people" (Seligman & Csikszentmihalyi, 2000, pp. 5–8). Why simply contend with alleviating suffering, with bringing people back to normal, if you could strive for their improvement?

With ample financial support from a number of private funding bodies, positive psychologists set out to study positive human emotions, positive personality traits, and – in practice, to a noticeably lesser extent – positive institutions, hoping that their interventions would lead to positive developments for individuals and society as a whole (Seligman, 2018). Further attesting to the movement's success, educational programs in applied positive psychology began springing up at a number of universities around the globe, following a model pioneered at the University of Pennsylvania.

This self-styled movement of academic renewal was in many ways not as ground-breaking as its proponents claimed. Critics could point to humanistic traditions in psychology and their long-standing concern with what positive psychologists began describing as personal growth and flourishing. The impetus to research the conditions underlying mental wellbeing was also not entirely new. Already in 1948, the constitution of the World Health Organisation had defined health not merely as the absence of ill health, but as "a state of complete physical, mental and social well-being" (WHO, 1948). Positive psychology's concern with aggregate measures of happiness, treated as indices of social progress in post-industrial societies, was predated by the study of *quality of life*, promoted by the economist John Kenneth Galbraith in the 1960s (Galbraith, 1964). And in the late 1970s, after analysing health data of Holocaust survivors (Antonovsky, Maoz, Dowty, & Wijsenbeek, 1971), the Israeli-American medical sociologist Aaron Antonovsky (1979a) urged health researchers to study what he called *salutogenesis* (as opposed to pathogenesis), the social conditions and personality traits associated with faster recovery and a lower likelihood of falling ill in the first place; the factors that make and keep people healthy, in other words.

Similarly, both the concept of *resilience* as well as the notion of *posttraumatic growth* have their predecessors. While there exist no uniform definitions for both constructs (Maercker & Zoellner, 2004; Meredith et al., 2011), psychological resilience is often described as "the ability of an individual to 'bounce back' after experiencing stress," to restore a homeostatic state of physical and psychological equilibrium (Meadows, Miller, & Robson, 2015, p. 9). Posttraumatic growth, which is believed to sometimes occur in highly resilient individuals in the wake of extreme adversity, is "not simply a return to a baseline," as Richard Tedeschi and Richard Calhoun (both leading authorities on the matter) write, "— it is an experience of improvement [in the wake of trauma] that for some persons is deeply profound" (Tedeschi & Calhoun, 2004, p. 4).

Considerable conceptual overlap exists between resilience and earlier concepts such as *hardiness*, a personality trait, first observed among managers, that is thought to protect from the onslaught of daily stressors (Kobasa, 1979), or the supportive web of personal beliefs Antonovsky summed up as *sense of coherence* (Antonovsky, 1979b). The historical precursors of posttraumatic growth, on the other hand, more often trace back to discourses situated outside the confines of academic psychology. Throughout the last century, alpinists, psychotherapists — such as Victor Frankl, himself a Holocaust survivor — as well as physicians who shared their interest in near-death experiences have attempted to draw out the transformative potential of harrowing accidents, of coming to terms with a terminal diagnosis and other catastrophic experiences that today are readily classified as traumatic (Martinović, 2017). Moreover, the medical doctrine that an upsetting, frightful event may also have curative effects has a relatively long and visible tradition that reaches through the 19th and back into the late 18th centuries — a history which includes the infamous shock therapeutics of asylum medicine (Koch, 2014).

Positive psychology, demarcated from now obscure medical traditions, humanistic psychology and various other intellectual enterprises, successfully asserted itself as a discipline firmly grounded in contemporary psychological research. Unlike the recommendations found in the popular self-help literature, with which positive psychology shares numerous themes, Seligman (1999) contends that the interventions of positive psychology are evidence-based and informed by current scientific models. More recently, however, the validity of several empirical claims put forth by positive psychologists have been called into question; and their theoretical efforts have received at times damning critiques (e.g., Brown, Sokal, & Friedman, 2013; Lazarus, 2003; Brown, Lomas, & Eiroa-Orosa, 2017; see also Seligman, 2018, Chapter 22). Also, given the current replication crisis in the social sciences that has sowed doubts about the reliability of experimental findings in psychological research, some members of the movement have urged their colleagues to become more self-critical and circumspect with regards to their empirical claims (see Robbins & Friedman, 2018).

Seligman's collaboration with the US Army dates from a time before many of these critiques surfaced. It is the CSF initiative and the ethical concerns it has raised to which I will now turn. In this context, the question of the program's scientific validity is surely relevant, but a methodological critique of positive psychology is not what is undertaken here (for a critical review of the available evidence on the program's effectiveness see Steenkamp, Nash, & Litz, 2013). As we will see, in the case of traumatic suffering it proves difficult to disentangle moral concerns from scientific questions about human responses to extreme adversity. As such they pose problems that cannot be resolved by an improvement of the program's methods.

First, it is important to note that CSF was designed to promote wellbeing and prevent all types of mental health problems members of the military (and their families) might face, not just PTSD. The idea of posttraumatic growth, however, plays a central role in the program's rationale.

4. Growth and comprehensive fitness

In his autobiography, Seligman begins the origin story of his collaboration with the U.S. Armed Forces by recounting his response to a question about what positive psychology had to offer in support of the army's fight against mental illness (which served me as the epigraph for this chapter): "The human reaction to awful events like combat", he began his reply, "is bell-shaped." Individuals who fall into the lower end of the distribution succumb to trauma, with the known effects of PTSD, depression, drug abuse and suicide — "different names for the ways in which people's lives are ruined." The majority of individuals, however, are resilient, they may struggle initially but after a few months "they are back where they were before." At the higher, right end of the distribution are those who grow from their experiences, who "one year later are stronger — by our physical and psychological measures — than they were to begin with." Therefore, Seligman concluded his pitch, the army should strive to move "the whole curve right-ward, toward posttraumatic growth" (Seligman, 2018, pp. 311–312).

Seligman's answer, his choice of metaphors is telling. The analogy between physical strength and resilience, which can be trained like a muscle, is a powerful one. The prevention program developed by Seligman's team, together with Brigadier General Rhonda Cornum, a physician with a specialisation in urology, was conceived as a *training* program to enhance mental fitness. It consisted of three components. After an initial assessment phase to establish a baseline, participating soldiers completed online modules to improve their fitness in four domains (emotions, relationships, family, and spirituality). Unlike the additional psychoeducational unit on posttraumatic growth, these modules were not mandatory. More relevant and arguably more effective in fostering resilience was the week-long face-to-face Master Resilience Training drill sergeants received, who were then instructed to pass on what they had learned to their subordinates.

Why was the module on posttraumatic growth deemed important? In an interview with the *Harvard Business Review*, Seligman (2011c) mentions that, in his experience, a vast majority of soldiers were aware of the possible negative effects of trauma, but only a small minority had heard of posttraumatic growth. Soldiers' focus on the negative, on PTSD, and being unaware of the fact that only a minority respond to trauma with a chronic condition, he reckoned, may become an impediment to recovery. Beliefs about PTSD turn out to be self-fulfilling, he argued, a downward spiral wherein negative expectations and negative reactions mutually reinforce each other (Seligman, 2011c).

Building on Tedeschi's work, the educational module on posttraumatic growth attempted to fill this knowledge gap. Soldiers learn that initial strong emotional reactions to devastating experiences, along with shattered beliefs about oneself, others and the world, are indeed normal. They also learn techniques and attitudes that may lead to posttraumatic growth in the wake of trauma: techniques for controlling involuntary remembering, disclosing oneself to others, "creating a narrative in which trauma is seen as a fork in the road" that comes with new possibilities, and articulating a set of "life principles" (Seligman, 2011a, p. 4).

5. Preventing posttraumatic stress versus preventing trauma

Critics have argued that Seligman and his team naively subscribed to the operational imperatives of the U.S. Army, never questioning that the institution's needs might diverge from the needs of the soldiers undergoing the interventions. The clinical psychologist Roy Eidelson and his co-authors, for instance, have pointed out that the articles featured in a special issue of the *American Psychologist* devoted to the program fail to discuss any of its potential adverse effects (see, e.g., Cornum, Matthews, & Seligman, 2011). This is especially troubling, they contend, since prior to implementing the initiative, no pilot study was conducted, no informed consent obtained from the participants, nor was there an assessment of the program's ethical ramifications by an institutional review board (Eidelson, Pilisuk, & Soldz, 2011). (The army's, as well as Seligman's [2018], defence against the latter charge has been that the intervention, which reached more than one million soldiers, did not constitute a research project in the proper sense and that, therefore, other ethical standards should apply.) Might those soldiers who view combat "as a growth opportunity", Eidelson and his colleagues ask, pose a potential harm to their comrades, to civilians, or even to themselves, as they more readily engage in reckless behaviour? Will they be more inclined to participate in atrocities and other transgressions that they will later come to regret, leaving them "morally injured" (Eidelson et al., 2011, p. 4)?[1] Another issue they raise — one that perhaps more clearly illustrates the moral unease felt by these critics — points directly to the limits of positive psychology's individualist outlook: They charge that the special issue contributors, who were all involved in the program's execution, systematically disregard signs of other, systemic sources of traumatic stress unrelated to combat but instead to the structural conditions of the military, such as the experience of sexual assault among servicewomen. This "calls less for building a [...] resilient outlook", they note, "than for recognition of how the commonplace victimisation of women in war should be vociferously prevented" (Eidelson et al., 2011, p. 4).

For Seligman and his collaborators, preventing PTSD or other trauma-related disorders is not the same as — and does not amount to — preventing trauma. And they've devoted themselves to the former. It is positive psychology's commitment to preventing posttraumatic symptoms (and enabling posttraumatic growth) while showing little interest in what causes them, I believe, that provokes moral concern.

To psychological researchers, differentiating between an event and the person experiencing it is sound scientific practice. In the case of trauma, however, this seemingly straightforward task proves stubbornly difficult

[1] With this remark, the authors reference more recent clinical theorising that stresses the moral dimension of psychologically harmful war experiences. Soldiers may be left morally injured, clinical researchers such as Brett Litz and Shira Maguen argue, after engaging in "unnecessary acts of violence" or failing to prevent them (Litz et al., 2009). The idea of moral injury, first popularised by the psychiatrist Jonathan Shay (1995), coincides with a shift in focus away from the fear-inducing aspects of trauma to the moral, more existential questions that often arise from combat experiences. Accordingly, those who call for its study see moral injury as distinct from PTSD, which, in their view, has become too inclusive as a diagnostic category (see also MacLeish, 2018).

(as the enduring debates about the DSM's so-called "stressor criterion" show; see, e.g., McNally, 2003; Roberts et al., 2012). Trauma is a *relational* concept that cannot be delimited without taking into account how the event affects the traumatised. The distinction between objective event and subjective experience is, moreover, difficult to uphold since we are not *morally indifferent* to traumatic events. Trauma leads to human suffering, often involves actors who possess agency and is, therefore, usually avoidable (Hampe, 2007).

What's more, during most of the 20th century, the recognition that posttraumatic symptoms of various kinds were indeed *caused* by trauma, and not by a pre-existing condition or congenital weakness, hinged upon the conviction that those who experienced such events were so-called normal individuals and that it was instead the horrific, abnormal events that victims had to endure that caused posttraumatic ailments (Koch, 2014). In other words: normative, historically variable assumptions about what constitutes an abnormal experience inadvertently shape what is identified as a traumatic stressor. Combat once belonged to the class of unpleasant experiences a normal (male) individual was expected to endure. This is no longer the case. In 1987, the revised third edition of the Diagnostic and Statistical Manual of Mental Disorders, the DSM-III-R, defined a "traumatic stressor" as an "event outside the range of normal human experience" (APA, 1987, p. 247).

Because PTSD may constitute a normal reaction to an abnormal experience, acknowledging trauma as such often includes a call to action to prevent it. (The very act of remembering trauma, in all its individual and collective, public and private forms, has recently become a moral as much as a therapeutic act.) Diverting attention away from the events to their psychological effects threatens to obfuscate the moral urgency that trauma has come to represent. Implementing a strategy to prevent *symptoms* without signalling a commitment to also prevent what *causes* them becomes ethically questionable, especially if the causes are morally objectionable human actions.

A hypothetical example can illustrate this point. Only recently have discriminatory police practices that single out African-American men entered public awareness in the USA. Suffering violence at the hands of the police will undoubtedly lead to symptoms of posttraumatic stress. However, the idea to prevent such symptoms by offering resilience training for members of a corresponding subpopulation, say, African-American males between the ages of 6 and 10, would, rightly so, raise a moral outcry.

The implementation of such measures – even if they had been found to be efficacious – would indicate a kind of moral defeat. It would signal to future victims of police brutality that any attempts to tackle systemic racism are either futile or not a worthy moral cause. The responsibility for dealing with the fallout of a societal problem, moreover, would be burdened onto those who already bear the brunt of it.

For a society at war, however, the detached verdict that trauma is inevitable may be easier to accept. Given the politically charged climate at the time CSF was implemented, it comes as no surprise to Seligman (2018) that most of his critics also opposed the wars in Afghanistan and Iraq. Their critiques, so he alleged, were politically motivated. But again, the case of trauma illustrates that the boundary between science and morality is often difficult to draw and even more so to uphold. There is no neutral ground here. Treating the trauma of war as inevitable reflects a moral attitude (whether and how one can mount an ethical defence of war is another matter, of course). In this case, positive psychology's focus on the positive in situations where others see little to be positive about, does not keep it from affirming that there are impactful negative experiences.

Trauma may present a limit case for positive psychology because of the inescapability of moral questions in relation to human suffering caused by violent experiences. Yet it also raises a broader question, namely whether the discipline can live up to its claim that a focus on the positive side of human existence and human growth can avoid engaging with its troubling, dark sides. My answer to this question hinges on how positive psychologists understand personal growth and how it in turn relates to mental suffering (see also Lazarus, 2003).

6. Growth as self-optimisation

Critics of positive psychology, especially those steeped in psychology's humanistic tradition, have scolded its proponents for a lack of theoretical clarity and historical depth when it comes to explicating notions like "flourishing" and "growth", so central to its mission (cf. MacDonald, 2018). Before pointing out the conceptual flaws, it is instructive to first ask how positive psychologists determine growth. Seligman's pitch to General Casey contains a simple, yet by no means trivial answer to this question: Growth is measured.

As mentioned, Seligman likes to invoke the metaphor of strength. In one of his articles published in the popular press, he cites Friedrich Nietzsche: "That which does not kill us makes us stronger" (Seligman, 2011a, p. 101). But he also likes to reference the more prosaic topic of psychometrics (Seligman, 2018, p. 312). Posttraumatic growth becomes tangible as a measurable improvement. The implication being that growth is the linear expansion of a set of personal attributes. Moreover, by invoking the image of a normal distribution, Seligman suggests that there is a continuity between dysfunctional, normal, and optimal long-term reactions to adversity. Just like bodily strength, resilience is malleable and can be augmented, which, in Seligman's telling, lays the foundation for potential posttraumatic growth: Ordinary resilience and extraordinary posttraumatic growth can be placed on different points of the same, continuous dimension.

Leaving aside questions concerning the validity of the bell-shaped, normal distribution of human responses to adversity, the very concept of posttraumatic growth raises the question whether the idea of such a continuity is at all plausible. Surely, one can *anticipate* the possibility of having a profound personal crisis in the wake of a traumatic experience, but how can a soldier, or anyone else, *prepare* for having his or her fundamental beliefs shattered? Indeed, if posttraumatic growth does entail a "deeply profound" transformation, a "paradigm shift", as Tedeschi and Calhoun (2004) suggest, it would imply that for the one who has yet to undergo this transformation its end-state must remain opaque. It is hard to see what help a narrative blueprint laying out the possibility of growth would provide to those who later struggle to find that new "fork in the road", as Seligman likes to describe it.

Also, from an empirical vantage point, things become more ambiguous, not less. For instance, researchers in the field have discussed the different outcomes resilience can have and have asked whether it may consist not in returning to a previous norm but in establishing new ones, a "new normal" (Bonanno, 2005). In the *American Psychologist* special issue devoted to CSF, Tedeschi and the Harvard psychologist Richard McNally (2011) mention that resilience should not be confounded with posttraumatic growth (Tedeschi & McNally, 2011). In fact, empirical findings suggest that *less* resilient individuals are *more* likely to experience posttraumatic growth (Levine, Laufer, Stein, Hamama-Raz, & Solomon, 2009). Tedeschi and McNally believe this to be the case because more resilient individuals struggle less and, therefore, also have fewer opportunities to grow.[2] Moreover, given the inconsistent track record of previous PTSD prevention programs, they caution against implementing the posttraumatic growth module before testing its efficacy in a randomised controlled trial (Tedeschi & McNally, 2011).

Seligman does not waste much ink writing about the possibility that there are events one cannot prepare for. Particularly in his books and articles addressing a general readership, he consistently focuses on the kind of adversities one should be able to inoculate oneself against. One of the take-away messages from his first trade book *Helplessness: on depression, development, and death* (Seligman, 1975) was that because helplessness is *learned* it could also be unlearned, or at least prevented by creating environments that fostered a *belief* in control. Urged by his publisher to tend to the, as it were, flipside of what he had described in his first book (Seligman, 2018), he then went on to write the immensely successful *Learned Optimism* (2006), which clearly reflects a shift in focus yet is an extension of the same line of reasoning that highlights the role of beliefs and an optimistic mindset in overcoming adversity. Prior to positive psychology's annexation of the topic, personal transformations in the wake of near-death experiences were thought of in almost transcendental terms. Seligman, in his popular accounts at least, narrows it down to a linear increase in psychological fortitude.

However, his suggestion to move "the curve rightward towards posttraumatic growth" conflicts with one of the fundamental principles with which he advocated for positive psychology (Seligman, 2018, p. 312). Because the absence of mental illness is not the same as wellbeing and human flourishing, he had deemed a research program necessary that was devoted to studying such positive qualities and developing techniques to augment them. "Good and bad, positive and negative" are not the kind of opposites that "lie on the same dimension", Seligman explains. Rather, "each side *forms its own distinct world* with properties not deducible from the absence of the properties of the other world" (Seligman, 2018, p. 269). Therefore, he argues, the positive side of human existence forms a distinct area of study. However, to propose that resilience protects against mental disorders while its lack leads to mental dysfunction amounts to just that: placing the positive

[2] Of course, this leaves out other possible interpretations of this finding that would challenge the preventive strategy also endorsed by these authors. For instance, it is plausible that certain active, problem-focused coping strategies, which indicate high resilience, are simply less effective in the aftermath of trauma.

(resilience) and the negative (mental dysfunction) on opposite ends of a dimension. Only if a decrease in mental dysfunction can be deduced from an augmentation of resilience does it make sense to declare the latter an effective method to prevent the former. Either resilience constitutes "its own distinct world" and does not necessarily inoculate against mental illness, although the program might nevertheless lead to more posttraumatic growth, or an increase in resilience protects from falling mentally ill.

Seligman, at times, seems to endorse the first option. In his popular writings the focus is on posttraumatic growth. "The army is not a hospital", he stated back in 2008 during his initial pitch to General Casey, and although the military should continue to spend money to treat those whose "lives are ruined", such efforts should be supplemented by measures to increase resilience (2018, pp. 312–313). In an article penned for the *Harvard Business Review*, he explores how the program could be applied to civilian life, particularly the business world. "The mandatory module, on posttraumatic growth, is" touted as "highly relevant for business executives". Throughout the article he threads the case vignette of the Penn MBA graduate "Douglas", who quickly bounces back after having lost his job, and within a year has "grown because of the experience". During times of "failure and stagnation", Seligman suggests, resilience training can help to "create an army of Douglases who can turn their most difficult experiences into catalysts for improved performance" (Seligman, 2011a, p. 102).

When advocating for resilience and posttraumatic growth, then, the prevention of mental illness is not one of his main talking points. When it comes to promoting the effectiveness of the CSF interventions, however, he could point out its proven effectiveness in preventing psychiatric problems, including an observed decrease in substance use disorders and a lower incidence rate of PTSD, depression, and panic disorder, among soldiers who were in some way reached by the Master Resilience Training (Seligman, 2018, p. 323). However, one should note that conclusive randomised control studies about resilience programs in the military are scarce and the effectiveness of resilience training is still being debated (Meadows et al., 2015).

The program's efficacy, or, rather, the lack of sufficient proof thereof, is one matter that may raise doubts about the utility of resilience training (Bonanno, Westphal, & Mancini, 2011). Another matter concerns the ways in which the language of resilience and posttraumatic growth may shape laypersons' beliefs about mental illness. Clearly, the expert knowledge conveyed in the module has normative implications. Participants and those who read about the program in the popular press learn, albeit in very broad terms, how someone *should* respond to trauma. Seligman himself pointed out that seemingly neutral descriptions can become self-fulfilling or, to put it differently, prescriptive (see also Hacking, 1995). This holds true also for the descriptions given by positive psychologists about the beneficial long-term reactions to adversity. Posttraumatic stress is anticipated as a *personal* struggle, and posttraumatic growth as a *personal* task – trauma becomes an opportunity to grow. Given the nature of the program, participants also learn that a lack of resilience or ineffective coping leads to mental illness.

The anthropologist Allan Young notes that, in the 1970s, the interest in human resilience evolved from the study of individuals who withstood adversity and that resilience "was initially represented as being an unusual and even extraordinary quality" (Young, 2008, p. 39). Today, Seligman is not alone in thinking that resilience is rather common. Indeed, some degree of resilience is believed to be the norm. Seligman empirically grounds this assumption in experimental data accumulated while studying learned helplessness in laboratory animals (and sometimes in humans) during the 1960s and 1970s (cf. Seligman, 1975, 2011a). Although, it must be noted that, at the time, these experiments were not designed to elucidate resilience or related constructs. Metaphorically speaking, resilience, previously a scarce resource, is now considered a widely available one that we are encouraged to grow and capitalise on.

The simple traumatogenic model of mental illness Seligman likes to invoke proposes that mental health problems such as depression and anxiety are the outcome of an interaction between an adverse event and a person's resilience (cf. Seligman, 2011a, 2011c, 2018). Not only are traumatic events completely decontextualized in this rendering, but also questions about the *external* conditions that sustain posttraumatic symptoms – or that might positively influence recovery – are left out. The question, in other words, of why some fall into the lower end of the distribution is only inadequately addressed, if at all. A lack of resilience, presented as a quasi-natural resource but in actuality a set of social and psychological skills, is the only viable candidate for an explanation. What happens to someone *after* an adverse event, for instance, seems to matter very little.

While most can now consider themselves resilient, our reactions to adversity have at the same time become optimizable, which requires training and self-monitoring. This leaves one with the question whether such preventive measures that encourage us to grow from traumatic experiences or warn us of the dangers of showing lasting signs of its impact resurrect the stigmatising attitudes of the past. Little more than four decades ago, having a "mental breakdown" in response to trauma was widely considered a moral failure or a sign of inferiority. The current program's posttraumatic growth training module teaches the participants that breaking down in the face of adversity is "normal". Yet, this initial struggle can be overcome, according to Seligman's upbeat message, by drawing on one's resilience. The emphasis on posttraumatic growth runs the risk of replacing the old stigma, as a marker of deficiency and spoiled identity, with a less severe, free-floating stigma that attaches itself only to those who withdraw from the struggle to personally grow from the experience. To escape the blemish of this kind of stigma, one would have to overcome the temporary state of victimhood one has been thrust into.[3]

7. Conclusion: growth as progress

PTSD and other trauma-related disorders represent an anomaly within current classification systems insofar as their aetiology, by definition, is known. This poses, I have argued, an ethical challenge for interventions designed to prevent PTSD. Because traumatic events are today believed to be at the root of posttraumatic suffering, the decision to prevent symptoms of the latter, instead of the conditions that lead to the former, becomes morally charged and, at times, ethically questionable. Yet, as I have also attempted to show, the ethical concerns CSF raises extend beyond questions about the program's applicability in a given context. Rather, the interventions themselves reflect a narrow, reductive perspective on personal growth and human development. Interventions such as undergoing the educational module on post-traumatic growth are reflexively shaped both by positive psychology's stated mission and the methodological premises underlying the quantitative study of human behaviour and experience. Particularly in Seligman's popular treatments of the topic, it becomes apparent that CSF training does not only aim at increasing knowledge about mental processes or providing trainees with skills to help them overcome adversity. The

[3] It could be argued that CSF does not adequately address the stigma of mental illness, which carries numerous costs not just for individuals who suffer from mental disorders but also for healthcare systems and national economies by significantly increasing their disease burden. Considering the attitudinal and institutional barriers soldiers face when seeking treatment for mental health problems (Vogt, 2011), extolling the virtues of posttraumatic growth could add to those barriers rather than diminish them.

dissemination of knowledge also serves a *motivational* end; the training ultimately aims at changing participants' outlook on adversity, including their beliefs about their innate ability to overcome hardships. But this inadvertently blurs the line between educational training and guided self-help intervention. Seligman effectively appeals to the participants' faith in science to *persuade* them to place their faith in themselves.

Underlying positive psychology's mission, after all, is the idea that perfecting the human condition can be achieved by scientific means. Rather than following the advice of self-help gurus or engaging in the unexamined practices handed down to us by tradition, academic psychology is seen as the proprietor of targeted, more effective, more *rational* techniques. It is questionable, however, whether scientific knowledge holds the key to many of the pursuits positive psychologists have made their own. Questions about the good life once fell under the purview of philosophy, and the majority of humans continue to turn to religion, not only for answers but also to partake in practices and share in belief systems that imbue their lives with meaning. In our present age, scientific insights, of course, cannot and should not be excluded when tackling questions about human fulfilment. But although scientific findings may prompt a reflection on values, the sciences themselves do not possess the means to *evaluate* ends. Scientific knowledge seeks to explain and predict; it does not provide us with the kind of knowledge that helps adjudicate purposes and deliberate questions of meaning (see Smith, 2005). It is not clear how the *telos* of human striving can be inferred from scientific insights.

The purported ingredients of a good life identified by positive psychologists, for instance, are unsurprising and often stand on questionable theoretical grounds. Seligman's proposal that well-being or flourishing is comprised of (at least)[4] five interrelated elements: positive emotions, engagement, relationships, meaning, and accomplishment (PERMA), is not entirely implausible, but it also appears somewhat arbitrary and, surely, debatable (see Seligman, 2011b).[5] An "element" is defined as "what free, nonsuffering people choose to pursue for its own sake" (Seligman, 2018, p. 260). Unlike

[4] This is not yet a definitive list and may be revised soon (Seligman, 2018).

[5] Seligman (2018) provides the readers of his autobiography a glimpse into the process that led to the list: "I had posited three elements of wellbeing. [...] But in teaching MAPP [Master's in Applied Positive Psychology] and thinking about what questionnaires to put on the [Authentic Happiness] website, I now became convinced that there were five elements. [...] I revised the theory after considering how commonly people pursue achievement doggedly for its own sake: even if it brings no happiness, no flow, no relationships, no meaning" (pp. 260–1).

the ones of the periodic table, however, there exists no systematic relationship between these elements, nor do they all refer to the same *type* of entity. We don't pursue meaning in the same sense that we pursue positive emotions or *flow* (engagement). Most entries on Seligman's list – relationships, meaning, and accomplishment – have less to do with what motivates people and more so with what many of them value. (When we fall in love or commit to a friendship, we do it because of another person and not because we pursue relationships for their own sake.) And, of course, everyone can think of elements they would like to see on this list because they value them highly. Does knowledge fall under accomplishment or is it irrelevant as a source of well-being? What about self-determination?

Seligman emphasises that his theory of well-being is empirically grounded and therefore merely descriptive. He does not wish to tell people how they should live their lives. Whether and how much someone draws on these sources to flourish is said to vary (Seligman, 2011b). Clearly, members of individualist societies value their independence and choices in pursuing the good life. The success of positive psychology would be more difficult to comprehend would it present to its users a list of routines that are part of a greater whole, a web of beliefs and practices that explicitly endorse a certain way of life or present a distinct worldview. Instead, positive psychologists prescribe isolated exercises, such as the daily routine of counting three blessings, that are eerily similar to some religious practices or those described in self-help books but do not form part of a cosmology. Such activities nevertheless reflect and, ultimately, promote a particular way of viewing the world. The values that positive psychology draws on are not random and disconnected but are linked to social norms, which are historically contingent. The elements that make up PERMA, for instance, are easily recognised as inherent to the different life domains (work, play, family life, spirituality) through the prism of which many individuals in post-industrialist societies view their recurring activities. The larger point being that norms can't be sidestepped when conceiving the life that is *worth* living. A set of postulated intrinsic motivations, empirically validated by self-report measures, is no substitute for contemplating what constitutes the good life.

Just as positive psychology cannot avoid giving certain values preference over others, it cannot escape the pull of anthropological assumptions. By promoting the gospel of continuous self-improvement through a particular kind of science that relies on the quantification of what it sets out to improve, positive psychologists unwittingly provide personal growth with a trajectory while leaving its endpoint unexamined. Seligman's narrow conceptualisation of posttraumatic growth illustrates this. For the fortunate few, trauma turns out to be an opportunity to expand along the lines drawn out for them by experts. The proverbial "new fork in the road" is really a short cut, or better: an acceleration on the straight and narrow path of incremental self-improvement.

In his autobiography, Seligman mocks unnamed "professors of cultural anthropology" who question the "undeniable reality of human progress". "[Y]ou have to be blinded by ideology", he writes, "not to see that almost everything we care about (except mental health) is better now: less pollution, more democracy, more human rights, less violence, longer life expectancy, lower child mortality, more music, more education, tastier food, healthier bodies and more poetry" (Seligman, 2018, p. 277).

The argument is empirically dubious and historically naïve.[6] For one, it conveniently fails to mention the cataclysms that have led to doubts about the inevitability of progress. Two World Wars, the Holocaust, the threat of nuclear annihilation and environmental collapse have put proponents of the idea that humans will steadily increase their control over their environment and curtail their own destructive tendencies on the defencive. Even during times of relative social stability, it has become clear that technological progress comes with risks, unintended consequences that can be harmful. In fact, without the efforts of environmental activists during the 1970s and 1980s, many of whom had their doubts about progress (Torgerson, 1999), pollution would not have decreased in subsequent decades. In short: at the beginning of the 21st century, it has become difficult to convincingly argue for the old, all-encompassing notion of inevitable progress, a *philosophy of history* that purports to know that where the world is headed is surely a better place. This does not mean, however, that one would be hard-pressed to find a cultural anthropologist who believes that progress occurs with regard to specific aspects of societal life; that we can, indeed, make progress when it comes to solving specific social problems, such as poverty or illiteracy. It is the idea of the inevitability of progress and that it equally applies to all facets of human life — an assumption that is as comforting as it is unverifiable — that has come under scrutiny in the 20th century, not the fact that progress does sometimes happen.

[6] A similar argument was recently expounded by the psychologist Steven Pinker (2018) in his controversial book *Enlightenment Now*.

Personal growth mustn't be understood as progress. More poems do not inevitably amount to better poetry. The philosopher Ian Hacking once remarked that "[u]nlike the natural sciences [philosophy] is not in the progress business", hence its "perennial themes" (Hacking, 1990, p. 347). I assume, many poets would say the same about poetry. It represents a rather narrow vision of personal growth when it is conceived as a linear progression towards an abstract goal like higher PERMA-scores that in itself remains vacuous. However, there are still numerous other practices and belief systems on offer that supply methods for sense-making that do not follow the linear trajectory of continuous improvement. One can strive to live a good life without compulsively trying to get better *at* life. In the vision forwarded by positive psychology, which expands a mythical notion of progress to personal development, less attention is given to the types of transformations that may lead beyond what can be anticipated. But more of the same, even if it is better, is no substitute for hope.

Acknowledgements

I would like to thank the editors of this volume, as well as Daniela Drago and John M. Young for their helpful comments on earlier versions of this chapter.

References

Adler, A. B., Bliese, P. D., McGurk, D., Hoge, C. W., & Castro, C. A. (2011). Battlemind debriefing and battlemind training as early interventions with soldiers returning from Iraq: Randomization by platoon. *Sport, Exercise, and Performance Psychology, 1*(S), 66–83.

Antonovsky, A. (1979a). *Health, stress, and coping*. San Francisco, CA: Jossey-Bass.

Antonovsky, A. (1979b). *Sense of coherence scale*. San Francisco, CA: Jossey-Bass.

Antonovsky, A., Maoz, B., Dowty, N., & Wijsenbeek, H. (1971). Twenty-five years later: A limited study of the sequelae of the concentration camp experience. *Social Psychiatry, 6*(4), 186–193.

APA. (1987). *Diagnostic and statistical manual of mental disorders, DSM-III-R* (3rd, revised ed.). Washington, DC: American Psychiatric Association.

Bonanno, G. A. (2005). Resilience in the face of potential trauma. *Current Directions in Psychological Science, 14*(3), 135–138.

Bonanno, G. A., Westphal, M., & Mancini, A. D. (2011). Resilience to loss and potential trauma. *Annual review of clinical psychology, 7*, 511–535.

Brown, N. J., Lomas, T., & Eiroa-Orosa, F. J. (Eds.). (2017). *The Routledge international handbook of critical positive psychology*. New York, NY: Routledge.

Brown, N. J., Sokal, A. D., & Friedman, H. L. (2013). The complex dynamics of wishful thinking: The critical positivity ratio. *American Psychologist, 86*(9), 801–813.

Cornum, R., Matthews, M. D., & Seligman, M. E. (2011). Comprehensive soldier fitness: Building resilience in a challenging institutional context. *American Psychologist, 66*(1), 4–9.

Dean, E. T. (1997). *Shook over hell: Post- traumatic stress, Vietnam, and the Civil war*. Cambridge, MA: Harvard University Press.
Eidelson, R., Pilisuk, M., & Soldz, S. (2011). The dark side of comprehensive soldier fitness. *American Psychologist, 66*(7), 643–644. discussion 646–647.
Galbraith, J. K. (1964). Economics and the quality of life. *Science, 145*(3628), 117–123.
Gillham, J. E., Jaycox, L. H., Reivich, K. J., Seligman, M. E. P., & Silver, T. (1990). *The Penn Optimism Program*. Unpublished manual, University of Pennsylvania.
Hacking, I. (1990). Two kinds of "new historicism" for philosophers. *New Literary History, 21*(1), 343–364.
Hacking, I. (1995). The looping effects of human kinds. In D. Sperber, D. Premack, & A. J. Premack (Eds.), *Causal cognition: A multidisciplinary debate* (pp. 351–394). New York, NY: Clarendon Press.
Hacking, I. (1998). *Rewriting the soul: Multiple personality and the sciences of memory*. Princeton University Press.
Hampe, M. (2007). Achilles' brain: Philosophical notes on trauma. *History of the Human Sciences, 20*(3), 85–103.
Jones, E., & Wessely, S. (2007). A paradigm shift in the conceptualization of psychological trauma in the 20th century. *Journal of Anxiety Disorders, 21*(2), 164–175.
Kobasa, S. C. (1979). Stressful life events, personality, and health: An inquiry into hardiness. *Journal of Personality and Social Psychology, 37*(1), 1–11.
Koch, U. (2014). *Schockeffekte. Eine historische Epistemologie des Traumas*. Zürich/Berlin: Diaphanes.
Lazarus, R. S. (2003). Does the positive psychology movement have legs? *Psychological Inquiry, 14*(2), 93–109.
Leese, P. (2002). *Shell shock: Traumatic neurosis and the British soldiers of the first world war*. London, UK: Palgrave McMillan.
Levine, S. Z., Laufer, A., Stein, E., Hamama-Raz, Y., & Solomon, Z. (2009). Examining the relationship between resilience and posttraumatic growth. *Journal of Traumatic Stress, 22*(4), 282–286.
Leys, R. (2010). *Trauma: A genealogy*. University of Chicago Press.
Litz, B. T., Stein, N., Delaney, E., Lebowitz, L., Nash, W. P., Silva, C., & Maguen, S. (2009). Moral injury and moral repair in war veterans: A preliminary model and intervention strategy. *Clinical Psychology Review, 29*(8), 695–706.
MacLeish, K. (2018). On 'moral Injury': Psychic fringes and war violence. *History of the Human Sciences, 31*(2), 146–182.
Maercker, A., & Zoellner, T. (2004). The Janus face of self-perceived growth: Toward a two-component model of posttraumatic growth. *Psychological Inquiry, 15*(1), 41–48.
Martinović, J. (2017). *Mort Imminente. Genѐse d'un phénomѐne scientifique et culturel*. Geneva: Métis Presses.
MacDonald, D. A. (2018). Taking a closer look at well-being as a scientific construct: Delineating its conceptual boundaries in relation to spirituality and existential functioning. In N. J. Brown, T. Lomas, & F. J. Eiroa-Orosa (Eds.), *The Routledge international handbook of critical positive psychology* (pp. 26–52). New York, NY: Routledge.
McNally, R. J. (2003). Progress and controversy in the study of posttraumatic stress disorder. *Annual Review of Psychology, 54*(1), 229–252.
Meadows, S. O., Miller, L. L., & Robson, S. (2015). Understanding resilience. In S. O. Meadows, L. L. Miller, & S. Robson (Eds.), *Airman and family resilience. Lessons from the literature*. Santa Monica, CA: Rand Corporation.
Meredith, L. S., Sherbourne, C. D., Gaillot, S. J., Hansell, L., Ritschard, H. V., Parker, A. M., & Wrenn, G. (2011). *Promoting psychological resilience in the U.S. Military*. Santa Monica, CA: Rand Corporation.

Micale, M. S. (2009). *Hysterical men: The hidden history of male nervous illness*. Harvard University Press.

Micale, M. S., Lerner, P., & Rosenberg, C. (2001). *Traumatic pasts: History, psychiatry, and trauma in the modern age, 1870–1930*. Cambridge University Press.

Pinker, S. (2018). *Enlightenment now: The case for reason, science, humanism, and progress*. New York, NY: Viking.

Pols, H. (2007). War neurosis, adjustment problems in veterans, and an ill nation: The disciplinary project of American psychiatry during and after world war II. *Osiris, 22*(1), 72–92.

Pols, H., & Oak, S. (2007). War & military mental health. *American Journal of Public Health, 97*(12), 2132–2142.

Robbins, B. D., & Friedman, H. L. (2018). The unavoidable role of values in positive psychology: Reflections in light of psychology's replicability crisis. In N. J. Brown, T. Lomas, & F. J. Eiroa-Orosa (Eds.), *The Routledge international handbook of critical positive psychology* (pp. 15–25). New York, NY: Routledge.

Roberts, A. L., Dohrenwend, B. P., Aiello, A., Wright, R. J., Maercker, A., Galea, S., & Koenen, K. C. (2012). The stressor criterion for posttraumatic stress disorder: Does it matter? *The Journal of Clinical Psychiatry, 73*(2), 264–270.

Scott, W. J. (1990). Ptsd in DSM-III: A case in the politics of diagnosis and disease. *Social Problems, 37*(3), 294–310.

Seligman, M. E. P. (1975). *Helplessness: On depression, development, and death*. New York, NY: WH Freeman.

Seligman, M. E. P. (1999). Positive social science. *Journal of Positive Behavior Interventions, 1*(3), 181–182.

Seligman, M. E. P. (2006). *Learned optimism: How to change your mind and your life*. New York, NY: Vintage.

Seligman, M. E. P. (2011a). Building resilience. *Harvard Business Review, 89*(4), 100–106.

Seligman, M. E. P. (2011b). *Flourishing. A new understanding of happiness and well-being—and how to achieve them*. Boston: Nicholas Brealey.

Seligman, M. E. P. (2011c). *Interview with Martin Seligman*. Retrieved from: https://hbr.org/2011/04/building-resilience.

Seligman, M. E. P. (2018). *The hope circuit: A psychologist's journey from helplessness to optimism*. New York, NY: Public Affairs.

Seligman, M. E. P., & Csikszentmihalyi, M. (2000). Positive psychology: An introduction. *American Psychologist, 55*(1), 5–14.

Shay, J. (1995). *Achilles in Vietnam. Combat trauma and the undoing of character*. New York, NY: Simon & Shuster.

Smith, R. (2005). Does reflexivity separate the human sciences from the natural sciences? *History of the Human Sciences, 18*(4), 1–25.

Steenkamp, M. M., Nash, W. P., & Litz, B. T. (2013). Post-traumatic stress disorder: Review of the comprehensive soldier fitness program. *American Journal of Preventive Medicine, 44*(5), 507–512.

Tedeschi, R. G., & Calhoun, L. G. (2004). Posttraumatic growth: Conceptual foundations and empirical evidence. *Psychological Inquiry, 15*(1), 1–18.

Tedeschi, R. G., & McNally, R. J. (2011). Can we facilitate posttraumatic growth in combat veterans? *American Psychologist, 66*(1), 19.

Torgerson, D. (1999). *The promise of green politics: Environmentalism and the public sphere*. Duke University Press.

Vogt, D. (2011). Mental health-related beliefs as a barrier to service use for military personnel and veterans: A review. *Psychiatric Services, 62*(2), 135–142.

WHO. (1948). *Constitution of the world health organization*. Geneva: WHO.

Young, A. (1997). *The harmony of illusions: Inventing post-traumatic stress disorder*. Princeton University Press.

Young, A. (2008). Posttraumatic stress disorder of the virtual kind: Trauma and resilience in post-9/11 America. In A. Sarat, N. Davidovitch, & M. Alberstein (Eds.), *Trauma & memory. Reading, healing, and making law* (pp. 21–48). Stanford University Press.

CHAPTER ELEVEN

Going too far? How the public health anti-obesity drives could cause harm by promoting eating disorders

Jacinta O.A. Tan[a,*], Suzana Corciova[b] and Dasha Nicholls[c]

[a]Specialist Child and Adolescent Mental Health Service, Aneurin Bevan University Health Board, Gwent, Wales, United Kingdom
[b]Higher Trainee in Child and Adolescent Psychiatry, London, United Kingdom
[c]Imperial College, London, United Kingdom
*Corresponding author: E-mail: jacinta.tan@wales.nhs.uk

Contents

Developments in Neuroethics and Bioethics, Volume 2
ISSN 2589-2959
https://doi.org/10.1016/bs.dnb.2019.04.009

1. Introduction

"Primum non nocere" (above all, do no harm) is a guiding principle of medicine, in spite of having contested and obscure origins (Gillon, 1985). Public health campaigns amongst the general population that aim to prevent disease or health conditions can raise ethical issues in several different ways. In this chapter, we will explore the public health campaigns and programmes in the United Kingdom aimed at reducing obesity and overweight, and examine the ethical implications and in particular the potential impact on one particular group of mental illnesses, eating disorders.

1.1 Obesity – an epidemic?

"Obesity Britain: Two-thirds of us are overweight and only two - Turkey and Malta - out of 53 European countries are fatter, UN study finds" shouts the Daily Mail (Spencer, 2018). "Obesity epidemic risks eyesight of young Scots women" says BBC Scotland (Alderson, 2018). "More children are obese than EVER before: One in three 10 and 11 year olds are now overweight, shocking figures reveal" the Daily Mail also says (Matthews, 2016). Lest anyone should think this is a British problem alone, "Australia's 'XXXXXXXXXL' schoolkids: Obese students wearing specially-designed uniforms have to visit the physio to physically survive the 6-h school day" is a news headline hot off the press as we write (Chapman, 2019).

Obesity is defined by a metric, Body Mass Index (weight in kilograms divided by height in metres squared), a metric that many consider flawed in its ability to determine health status (Tomiyama, Hunger, Nguyen-Cuu, & Wells, 2016). The threshold criterion for obesity in adults is usually taken as BMI of 30 and above, the norm being 18.5 to 25 and overweight from BMI 25–30. For children, overweight is defined as being over one standard deviation body mass index for age and sex (or the 85th centile of weight for height), and obesity is defined as two standard deviations body mass index for age and sex (or the 95th centile of weight for height) respectively (WHO, 2019). There has been growing consensus that obesity is a major health issue not just in the United Kingdom and other industrialised countries, but globally. The World Health Organisation says in its report 'Taking Action on Childhood Obesity': "Childhood obesity is one of the most serious global public health challenges of the 21st century, affecting every country in the world. In just 40 years the number of school-age children and adolescents with obesity has risen more than

10-fold, from 11 million to 124 million (2016 estimates). In addition, an estimated 216 million were classified as overweight but not obese in 2016." (WHO, 2018b).

There is general scientific acceptance of a correlation between obesity and poor physical health. Obesity is associated with higher rates of diabetes, heart problems and some forms of cancer, as well as earlier mortality. However, there is a U-shaped relationship between weight and health status; it is in normal weight ranges that health is optimal, with both extremes of underweight and overweight being related to increasing ill-health and mortality (Aune et al., 2016). Nonetheless, it is estimated that using BMI alone misclassifies almost a quarter of the US population in relation to their cardiovascular health status both among the overweight and obese and among those at a healthy weight (Tomiyama et al., 2016). Furthermore, there is discrepancy or bias in application of BMI categories at the overweight compared to the underweight end of the BMI spectrum if risk of death is the primary concern, in that mortality at BMI 18.5 (lower end of the normal range) is comparable to mortality at BMI 35 (severe obesity) (Aune et al., 2016). It is therefore perhaps unsurprising that public perception of health risks associated with overweight and obesity differ from medical perspectives. This is particularly the case for parent's perception of their children; in a UK sample of children measured during a routine measurement programme, most parents (79%) of children classified as overweight or obese did not perceive their child's weight to be a health risk (Cinelli & O'Dea, 2016). When invited to derive cut offs for obesity and underweight, parents were more likely to classify their children as underweight when they are at the 0.8th centile or below, and overweight at the 99.7th centile or above, values that differ greatly from the BMI centile cut-offs for underweight (2nd centile) and overweight (85th centile) used in the clinical and scientific community (Black et al., 2015).

Arguably therefore, not only is the moral panic potentially harmful in its own right, it may be flawed in its basic premise regarding health risks and potential benefits. This is because overweight and obesity may not be as clearly or directly correlated to medical harms as might initially appear. Furthermore, trying to use moral panic to induce high levels of concern and spur action amongst parents may be less effective than expected. What is undeniable is that obesity rates are rising, a trend that all wish to see change. It is difficult to determine exactly how much eating and weight anxiety and weight stigma, factors potentially modifiable through policy change, perpetuate this trend. Below we will outline some of the mechanisms through which they may operate.

2. The public health campaigns and strategies for obesity in the United Kingdom

Obesity and overweight are seen as preventable and modifiable conditions (WHO, 2018b). The World Health Organisation advocates a global strategy for diet, physical activity and health (WHO, 2003). This strategy, which has been signed by member nations, largely focuses on policy-level changes that aim to alter the environment to encourage more physical activity, healthier diet and consequently reduction of overweight and improvement of health status.

2.1 Governmental public health strategies in the United Kingdom

An example of national level strategies is the current public consultation by the Welsh Government on its strategy: Healthy Weight, Healthy Wales – our ambitions to prevent and reduce obesity in Wales (Welsh Government, 2019). In it, the Welsh Government suggests a national, whole system approach to obesity in the population which involves creating leadership, creating healthy environments, creating healthy settings, and enabling healthy people. Examples of creating healthy environments are helping businesses to reformulate and offer healthier food choices and limiting television advertising of less healthy foods in public places, as well as limiting bulk discounts of food which encourage overstocking and overeating; creating more opportunities for active commuting and physical activity for all. Examples of creating healthy settings are mandating healthier food choices and increased physical activities in nurseries, schools, colleges and places of employment and strengthening school healthy weight programmes. Examples of enabling healthy people are improving opportunities through changes in routine healthcare for people to have help to achieve normal weights, as well as improving the National Health Service clinical obesity treatment pathways.

In 2007, the UK Government commissioned the Foresight programme to identify possible ways of tackling obesity in society over the next 40 years (Kopelman, Jebb, & Butland, 2007). Policies included include pricing and fiscal measures, school-based and worksite interventions, interventions in the primary care setting (e.g. prescribing physical activity), reformulation of products, transport policies (e.g. subsidies for active commuting instead of cars), child health measurement, nutritional information in schools, changes in portion sizes, food labelling in stores, informative logos conveying positive or negative characteristics of the product, traffic light schemes and nutrient lists on packaged foods. Over 77 policies have been implemented for children under five years old alone (Viner, personal communication).

Two examples of policy changes included in the Foresight programme that have created some controversy include food labelling and introduction of the National Child Measurement Programme. Food labelling is increasingly considered a crucial component in strategies to tackle unhealthy diet, as a possible tool to empower and facilitate healthier food choices. Different food labelling schemes have been devised: traffic light (green, amber or red according to levels of fat, saturated fat, salt and sugar per 100g), Guideline Daily Amount (GADA) and other types (e.g. Front of pack logos). Cecchini and Warin (2016) have shown in a review that traffic light labelling is the most effective scheme and increases the number of people selecting a healthier option by 29.36% (Cecchini & Warin, 2016). The review was restricted to peer-reviewed studies that were designed to evaluate the impact of food labelling in terms of likelihood of selection of either a healthier option or calorie choice/intake. Studies were randomised and included a control population with either a before/after or a case/control design. The other schemes had less effectiveness: GADA 14.69% and other types 11% respectively.

Greater consumption of foods prepared away from home, which tend to have a higher calorie content, is an important contributor of excessive weight gain in adults and children. Consumers often underestimate the caloric content of food in restaurants as well, and rarely consult nutrition information even if provided. Pulos and Leng assessed whether labelling restaurant menus with information of the nutrient content would cause consumers to alter ordering patterns (Pulos & Leng, 2010). The study showed a consequent reduction in the calories, fat and sodium content in the food ordered. On average the reduction of intake was 15 fewer calories on average, and the data indicated that only 20% of costumers choose a certain food on the basis that it had a lower calorie content. Recent evidence suggests that children and adolescents use menu labelling when making food purchasing decisions (Lillico, Hanning, Findlay, & Hammond, 2015; Sacco, Lillico, Chen, & Hobin, 2017; Seward, Block, & Chatterjee, 2018). Findings indicate that many consumers do not report seeing calorie labelling information and very few reported using labelling to purchase fewer calories. However, studies performed in student cafeterias settings have shown the opposite (Cinelli & O'Dea, 2016; Elbel et al., 2013; Elbel, Kersh, Brescoll, & Dixon, 2009; Kiszko, Martinez, Abrams, & Elbel, 2014). More recent work supports the idea that labelling improves the accuracy in estimating the number of purchased calories (Sacco et al., 2017; Seward et al., 2018). In summary, the evidence of the effect of menu labelling on food choices is mixed. Some data show that it might promote lower calorie purchases, while others indicate that it might have little impact on fast-food purchases.

The National Child Measurement Programme (NCMP), a nationally mandated public health programme led by Public Health England, was introduced in 2006 in England. A similar pilot programme exists in Wales, but not in Scotland and Northern Ireland. The NCMP provides the data for the child excess weight indicators in the Public Health Outcomes Framework, and is part of the government's approach to tackling child obesity. Height and weight are measured by school nurses in children starting primary school/reception year (4/5 years old) and at the end of year 6 (10/11 years). The measures are then translated into Body Mass Index (BMI) scores based on centiles. Its function was to ensure appropriate targeting of weight management resources. Standard cut offs are applied and the following categories were identified: obese, overweight, healthy weight and underweight. Initially the NCMP was intended as surveillance, but following the introduction of routine parental feedback, it became more of a screening programme with the intention to increase awareness of unhealthy body weight and prompt parental action.

Positive effects of child measurements have been shown as improving the percentage of parents able to recognise their child as being overweight following weight feedback (21.9%–37.7% increase) (Falconer et al., 2014). More than a third of parents sought further weight information following receiving a letter that their child was overweight, though this did not translate into lifestyle changes. The proportion of parents that had children who were measured as overweight or obese that perceived their children weight as a health risk was still low at follow up, but higher than at the baseline. Intriguingly, ethnic minority groups have reported greater changes to lifestyle and plans for help-seeking (Falconer et al., 2014); the reasons for this finding are not clear.

2.2 Non-governmental public health campaigns in the United Kingdom

Most organisations do not have the luxury of sweeping legislation or national policy-level approaches to obesity centred around modification of the environment or population behaviours. As a result, many campaigns focus on members the public themselves, attempting to influence and motivate them to change their attitudes and behaviours around overweight and obesity. These vary in their tone and emphasis.

Change4Life is a United Kingdom National Health Service public health campaign which attempts to positively motivate people to become more active and eat healthier diets (NHS, 2018). It provides its rationale is "By the time we reach middle age, the majority of us could do with losing at least a bit of weight. Being overweight isn't just about the way we look. It can lead to more aches and pains, problems sleeping and people report a loss of energy and confidence. It also increases our chances of getting heart disease, type 2 diabetes and some cancers." Regarding children, its stated rationale is: "Nine out of 10 of our kids today could grow up with dangerous amounts of fat in their bodies. This can cause life-threatening diseases like cancer, type 2 diabetes and heart disease." Despite overweight being named in its rationale, most of the advice and imagery does not focus on overweight or obesity. Using bright cartoon figures – which are neither fat nor thin - it asks questions like "Are you ready for your fun family snack challenge?". It advises how to read food labels and reduce high sugar drinks. It also says, "When choosing snacks for kids, look for 100 calorie snacks, two a day max", which is a recent Public Health England recommendation (Public Health England, 2018).

Cancer Research UK, in contrast, decided to target the perception of obesity directly. It recently launched an advertisement campaign intended to persuade the public that obesity causes cancer and to motivate them to lose weight in consequence, with the slogan "OB_S_ _Y is a cause of cancer" (Cancer Research UK, 2018). This slogan has been widely publicised by the charity, appearing on public billboards across the UK. The slogan also appears on its website, above boxes to click through to answer questions it expects to arise: "Does obesity cause cancer?", "Am I overweight?", and "How can I lose weight and be healthier?". Under "Does obesity cause cancer?" the answer provided is: "Fat is active, telling other cells what to do. And if there is too much fat in the body, then the signals it sends around the body can cause damage." (Cancer Research UK, 2018).

3. Eating disorders

Eating disorders are a group of mental disorders associated with feeding and eating behaviours that are not explained by other health conditions (WHO, 2018a). Eating disorders have high morbidity and high mortality (Smink, van Hoeken, & Hoek, 2013; Steinhausen, 2002). At the same time, they can be difficult to diagnose and even harder to treat.

Eating disorders are characterised by disturbed eating behaviour associated with concern about weight and shape and/or distress about eating. Patients are often young and highly vulnerable yet ambivalent about recovery and may be difficult to engage with treatment (Tan, Hope, & Stewart, 2003; Tan, Stewart, Fitzpatrick, & Hope, 2007; Vitousek & Manke, 1994). Without specific knowledge of these conditions, patients with eating disorders can evade detection, delaying time to diagnosis and treatment and influencing long term outcome (Joint Commissioning Panel for Mental Health, 2013a). Healthcare costs for eating disorders in the NHS across the UK have been recently estimated as £3.9—4.6 bn with an overall economic cost likely to be more than £6.8-8 bn per year (PwC (Price Waterhouse Cooper LLP) & Beat, 2015). Over 1.6 million people in the UK are estimated to be affected by eating disorders and this is likely to be an underestimate as many people do not seek help (Joint Commissioning Panel for Mental Health, 2013b). In addition, it is reported that the number of people diagnosed with eating disorders has increased by 15% between 2000 and 2009 (Micali, Hagberg, & Treasure, 2013). It is not clear if this is due to increased help seeking, due to better diagnosis or due to other factors such as better education and awareness amongst healthcare professionals and the public. Given the difficulty of engaging patients with eating disorders in treatment, and the scarcity of resource provision for specialist eating disorder treatment, it is likely that specialist eating disorder treatment teams see a minority of the sufferers. It is also likely that a large burden of care would fall on families (Joint Commissioning Panel for Mental Health, 2013a; Perkins, Winn, Murray, Murphy, & Schmidt, 2004; Winn, Perkins, Murray, Murphy, & Schmidt, 2004).

Core features of both anorexia nervosa and bulimia nervosa are fear of being fat and primacy of weight and shape in self-evaluation. Anorexia nervosa is characterised by significant self-induced weight loss associated with abnormal self-evaluation and thoughts such as a fear of fatness, and overwhelming drive to be thin, and misperception of oneself as being too fat

(WHO, 2018a). Bulimia nervosa is characterised by similar thoughts with frequent binges (subjective loss of control and overeating) with compensatory mechanisms for overeating aimed at preventing weight gain, such as vomiting and using laxatives (WHO, 2018a). Both these eating disorders and their variants can lead to severe immediate and long-term health consequences including risk to life, through extreme binging, the compensatory behaviours, and malnutrition associated with severe food restriction or abnormal diets. Almost all organ systems in the body can be adversely affected and medical assessment and treatment is often required (Birmingham & Treasure, 2010). Standardised mortality ratios are significantly higher than for Type 1 diabetes, epilepsy or asthma (Arcelus, AJ, Wales, & Nielsen, 2011).

3.1 Recent changes to eating disorder diagnoses

There have been important, recent changes to the classification of eating disorders. Previously in the WHO's International Classification of Disorders version 10 (ICD-10) the three main categories of eating disorders were anorexia nervosa, bulimia nervosa and eating disorder, unspecified (with other minor categories such as overeating associated with other psychological disturbances and vomiting associated with other psychological disturbances) (World Health Organization, 1992). In 2018 the WHO released the ICD-11 which has the following categories (WHO, 2018a):

- Anorexia Nervosa (AN)
- Bulimia Nervosa (BN)
- Binge eating disorder (BED)
- Avoidant-restrictive food intake disorder
- Pica
- Rumination-regurgitation disorder
- Other specified feeding or eating disorders (OSFED)
- Feeding or eating disorders, unspecified

The most significant changes to this classification are the introduction of Binge eating disorder (BED) and Avoidant-restrictive food intake disorder (ARFID) along with broadening of criteria for existing eating disorders (Al-Adawi et al., 2013), which follows similar changes in the DSM-5 diagnostic classifications. Of these diagnoses, the introduction of BED has been the most controversial, even though binge eating disorder have been known for 50 years. Studies suggest that BED has a different gender and age pattern from AN and BN, being more equally distributed amongst men and women, and also occurring in older age groups. Most importantly, BED is not infrequently associated with obesity and is also much higher in prevalence in the population than AN or BN (Kessler et al., 2013; Stice, Marti, & Rohde, 2013; Striegel-Moorel & Franko, 2003). Whilst BED tends to be associated with higher weight, binge eating *behaviour* is found across the weight spectrum, including in AN.

Avoidant-restrictive food intake disorder, in contrast, tends to occur much more in children, although it can occur at any age. Although lacking the characteristic thoughts and fears and about weight and shape, people who suffer from ARFID severely restrict their intake and can lose a great deal of weight. It is more evenly balanced than AN in terms of gender and is often associated with high degrees of obsessionality and anxiety (Eddy et al., 2015; Nicely, Lane-Loney, Masciulli, Hollenbeak, & Ornstein, 2014). The lack of emphasis on the value of weight, shape, and lack of body image disturbance means that the epidemiology of ARFID is likely to be less susceptible to societal values and drivers regarding the values and stigmas relating to weight and obesity (Izquierdo et al., 2019).

In addition to these main diagnoses, the Other specified feeding or eating disorders (OSFEDs) category includes diagnoses important to this discussion. The first is atypical AN, defined by the presence of all the characteristics of AN but the person has not reached a significantly low body weight. Patients with AN have often been overweight or obese and have undergone drastic weight loss to the point that their restrictive eating behaviours, exercise and other compensatory strategies have become out of the control. The other group is those with purging disorder in which, as the name implies, purging but not bingeing behaviour is the cardinal feature. For both diagnoses occur in the normal weight spectrum and are associated with high levels of medical risk.

3.2 Implications of changes in eating disorder diagnoses

The almost certain consequence of the broadening of eating disorder diagnostic criteria and the addition of new diagnostic labels, is that the number and range of people who can be diagnosed as having eating disorders increases. Bould et al. found that around 40% of adolescent girls engaged in some form of disordered eating behaviour, of which 11.4% met criteria for an eating disorder diagnosis (Bould, De Stavola, Lewis, & Micali, 2018). Current services are largely configured to treat severe or medically complex or risky eating disorders, particularly AN. The addition of both ARFID and BED will alter the demographics of those who can seek treatment for their eating disorders. BED prevalence rates appear to be considerably higher than the prevalence rates for AN and BN (Smink, Van Hoeken, Oldehinkel, & Hoek, 2014; Stice et al., 2013). Importantly, the context of increasing obesity and related increases in fear of fatness, weight and food preoccupation, and unhealthy weight loss strategies is likely to impact the number and nature of those presenting clinically, in particular with atypical AN. Da Luz et al. found more than sevenfold increases in the prevalence of obesity with comorbid binge eating from 1995 to 2015, and more than 11-fold increases in obesity with comorbid very strict dieting/fasting (Da Luz et al., 2017). Together these factors are likely to increase demand for treatment in services, and also change the nature and scope of eating disorder treatment services. Importantly, it suggests that where rates of obesity increase, rates of eating disorders increase.

4. The relationship and overlap between obesity and eating disorders

4.1 Obesity and mental disorders

The workgroup for the Diagnostic and Statistical Manual of Mental Disorders (DSM—5) section on feeding and eating disorders were asked to consider whether obesity should be included in the manual (American Psychiatric Association, 2014). In their carefully considered review to inform this decision, Marcus and Wildes concluded that: "Obesity is a condition of heterogeneous aetiology that is harmful for most individuals. However, there is scant evidence that obesity, in general, is caused by mental dysfunction. Although recent work examining the neurocircuitry of energy balance has suggested that mental dysfunction may be involved in the aetiology of specific obesity phenotypes, findings are too preliminary to support classification of obesity as a mental disorder. Nevertheless, there is evidence that obesity is related to mental disorder and many of the medications used to treat psychiatric illness." (Marcus & Wildes, 2009). The article did however re-emphasise the extent to which obesity is related to mental disorder and confirmed the importance of monitoring adiposity routinely among patients with psychiatric illness. This thinking appears to have extended to the recent ICD-11 revision (WHO, 2018a).

Nonetheless, there are many commonalities of obesity with mental health disorders, and many common risk factors. A full commentary on common risk factors for obesity and mental health is beyond the scope of this chapter, but include heritable factors, prenatal and perinatal risk factors, self-regulation and other cognitive factors such as risk taking and decision making, and socio-economic factors such as deprivation and sub-standard housing. Obesity is associated with increased rates of mood disorders, anxiety disorders, eating disorders, and personality disorders, especially in women and those with BMI >35, and weight status negatively impacts illness course and response to treatment in some of these groups. The association of attention deficit and hyperactivity disorder (ADHD) with obesity has also been shown in two metaanalyses (Cortese et al., 2015; Nigg et al., 2016), increasing the risk of later obesity by 30%–40%, particularly when persistent through childhood (White, Nicholls, Christie, Cole, & Viner, 2012). The association with obesity only emerged after childhood. The same study found that conduct problems also increased risk of adult obesity by approximately 40%–60%, particularly when present early in childhood or persistently throughout childhood.

Psychological correlates of obesity also include weight stigma, which we will discuss further, and impaired quality of life. Longitudinal studies show that obesity predicts the onset of psychiatric symptoms and vice versa, and that mental ill health can lead to obesity, not least through some psychopharmacological interventions.(Gariepy, Nitka, & Schmitz, 2010; Russell-Mayhew, McVey, Bardick, & Ireland, 2012; Strine et al., 2008). The nature and causes of these associations may be complex. Many psychopharmacological treatments increase appetite and many also increase tiredness and somnolence, leading to increased intake and/or decreased activity. Mental disorders themselves may be associated with decreased motivation, decreased activity and reduced socialisation. All these make it harder to be active and may lead to increases in body weight. Furthermore, eating may also become a coping mechanism for the distress of mental disorders. Finally, but very importantly, experiences of weight bias — being discriminated against or facing prejudice due to one's weight — is a major factor in lower self esteem and reluctance to participate in activity amongst overweight and obese children (Russell-Mayhew et al., 2012).

4.2 Obesity and eating disorders

With respect to the relationship between eating disorders and obesity, this can be summarised by considering obesity as comorbid with eating disorders and as a risk factor for eating disorders, and eating disorder psychopathology as a risk factor for obesity.

Comorbidity of obesity is, as already outlined, most commonly with BED although it could occur in all feeding and eating disorder diagnoses apart from AN. Those self-reporting bulimic or compulsive eating at age 30 had a median BMI of 22.9 kg/m^2 (Interquartile range: 21.0–26.5), of whom 61% were normal weight, 2% were underweight, and 37% were overweight or obese. These proportions were comparable in those not reporting binge eating (Nicholls, Statham, Costa, Micali, & Viner, 2016). A metanalysis of 36 studies looking at rates of binge eating behaviour (eating a large quantity of food and a subjective sense of loss of control (LOC) eating behaviour) and LOC eating alone (food quantity not relevant) in children and young people found an overall prevalence of binge/LOC eating of 26.3%, with 22.2% for binge eating and 31.2% for LOC eating, respectively (He, Cai, & Fan, 2017). Prevalence of binge eating in obese adults is slightly lower, at around 17%–19.4% (Bertoli et al., 2016; Pacanowski, Senso, Oriogun, Crain, & Sherwood, 2014). Thus most obese people do not binge eat, and binge eating behaviour is not always associated with overweight or obesity. However, binge eating behaviour in childhood is a risk factor for later eating disorders and obesity (Field, Garland, & Williams, 2003; Stice, Agras, & Hammer, 1999) and LOC eating predicts weight gain during middle childhood for those at risk of adult obesity (Tanofsky-Kraff et al., 2009).

The other potential mechanism through which obesity can be a consequence of eating disorder psychopathology relates to self-perception and body confidence. In a landmark study, Loth et al. (Loth, Watts, Van Den Berg, & Neumark-Sztainer, 2015) found that among overweight adolescents (n = 496) from Project EAT, those with the lowest body satisfaction at baseline had a nearly three unit greater increase in BMI at follow-up, compared with overweight girls in the high body satisfaction quartile. No such difference was found in boys. The authors concluded that high levels of body satisfaction during adolescence was not harmful, and in fact may be beneficial for girls in terms of long-term weight management. These findings are contrary to the commonly held belief that overweight young people ought to be dissatisfied with their bodies to motivate weight loss.

4.3 Obesity, dieting and anti-obesity campaigns as risk factors for eating disorders

Obesity has been demonstrated as a risk factor across the eating disorders (Berkowitz et al., 2016; Fairburn et al., 1998; Striegel-Moore et al., 2005). Obesity operates as a risk factor through two main mechanisms: dieting and body image dissatisfaction, particularly in combination with negative affect. Dieting is typically the trigger for both AN and BN. In AN restricted eating behaviour becomes maintained by possible aberrant reward systems and harm avoidance once the starvation syndrome sets in. In BN, restricted eating (dieting) leads after a time to uncontrolled binges, followed by guilt and shame, provoking purging behaviour and starvation, thus setting up a harmful binge purge cycle. Patton et al. found that recurrent dieting increased the risk of eating disorders by 18 times, and that moderate dieting increased risk by 5 times (Patton, Selzer, Coffey, Carlin, & Wolfe, 1999). And of course dieting is commonest among those who are overweight or obese, especially if accompanied by poor body image. Stice has recently elucidated the temporal pathway through which the key risk determinants for eating disorders operate (Stice & Van Ryzin, 2019). He found that almost half of the 51 adolescent girls who developed an eating disorder during an eight-year prospective study first perceived pressure to be thin, before showing body dissatisfaction, before showing onset of dieting and/or negative affect, before showing onset of the eating disorder. A further 29% of the sample had one of these steps out of order or did not cross one step in this model. This detailed description of the sequence of events in the onset of eating disorders chimes with clinical practice.

Dieting is prevalent in the population at all ages (Jeffery et al., 1984; Neumark-Sztainer, Wall, Larson, Eisenberg, & Loth, 2011). 60% of 35- to 65-year-old women and 32% of 45- to 65-year-old men report dieting, and fear of weight gain is found in almost 75% of women between 16 and 25 and in 45% of 25- to 55-year-old men (Slof-Op‘t Landt et al., 2017). Dieting and fear of weight gain were associated with each other and with BMI and gender (Slof-Op‘t Landt et al., 2017). There is a long and unresolved debate regarding the relationship between dieting and obesity in terms of causation. There is disagreement over whether dieting is a cause of weight gain/regain (so called weight cycling) and what the natural history of weight gain would be in those at risk of obesity in the absence of dieting. We do not attempt to resolve that debate here, but suffice to say that there do appear to be a proportion of the population for whom dieting/restrictive eating acts as a trigger for the onset of eating disorders. This suggests that obesity policy makers need to be mindful even as they target the population as a whole, that restrictive eating poses a risk to some individuals.

4.4 The risk to children and adolescents

One possible group at risk are those who are not overweight but perceive themselves as overweight or who are anxious about the risk of becoming overweight despite having a relatively low risk. In its extreme form, this type of body image distortion is a core feature of AN, but such beliefs are prevalent to a lesser degree in the population, particularly among women and younger girls.

The other group at risk are those who are prone to anxiety, and those who internalise rules very strictly or are prone to obsessional focus on detail. In both instances, children and young people may be particularly vulnerable because they have yet to learn to balance risks. The utility of making children anxious about the outcomes of obesity is unlikely to be high (fear has low impact on behaviour change generally, and even less in children and adolescents). While some argue that there is no harm in inducing some body dissatisfaction in order to achieve behavioural change (Heinberg, Thompson, & Matzon, 2001), evidence suggests this is not true. A 5-year longitudinal study found that, "in general, lower body satisfaction does not serve as a motivator for engaging in healthy weight management behaviors, but rather predicts the use of behaviors that may place adolescents at risk for weight gain and poorer overall health. Interventions with adolescents should strive to enhance body satisfaction and avoid messages likely to lead to decreases in body satisfaction." (Neumark-Sztainer, Paxton, Hannan, Haines, & Story, 2006).

In public health messages targeted at children, there is a risk of harm in terms of children internalising messages in overgeneralised terms, because they may not yet be at the developmental stage needed to able to appreciate the nuances involved. These harms often centre around diet, but can also extend to exercise. A common example is that a message given in school about not eating too much sugar because it is bad for health, with the intention of reducing intake of fizzy drinks or sweets, can easily become internalised as rule forbidding sugar intake. This may result in some children refusing to eat normal things like a slice of birthday cake at someone's party, or any food that is perceived to be bad or harmful because it contains sugar or fat. The 10 year old son of one of the authors came home insisting he must eat a certain type of breakfast cereal and no other cereal, because they had compared food labels in school and this type had the lowest sugar content. This overgeneralisation of internalised messages would not be unusual for how eating disorders start in preadolescents. If fear is too great (e.g. fear of dying of cancer) it can be overwhelming for children and lead to withdrawal or refusal. In the authors' experience as child and adolescent psychiatrists, behaviour change in middle childhood (roughly 6–12 years) is best promoted by very immediate rewards and encouragement of desired behaviours, rather than in simplistic messages that are easily misinterpreted or taken to extremes. In adolescents, the experience of eating disorder clinicians is that eating disorders often begin as an innocent attempt to lose weight or get fit, by dieting and/or increasing levels of exercise, which subsequently spiral out of control and are no longer within the person's control.

The second issue is excessive focus on detail and black and white thinking, again with an emphasis on those who are more developmentally at risk. Well-established risk factors for eating disorders are perfectionism and a strong ability to focus on details (Hilbert et al., 2014). Specific groups such as those with autism spectrum disorders and prepubertal children have a degree of cognitive inflexibility and tend to adhere very strictly to rules. Advice to count calories or sugar cubes can result in obsessional preoccupation with numbers and counting. These features of emotional and cognitive processing become even more pronounced in the context of starvation. It is not unusual to find children who have internalised a rule that "fat is bad" in a literal way, and are impervious to the information that growing brains need fat to develop. Pinhas et al. reported four cases of childhood onset eating disorder whose illness had been triggered by messages taught in school by healthy eating initiatives, which led them to focus excessively on their own body shapes, exercise excessively and restrict food intake to the point of severe weight loss with medical symptoms (Pinhas et al., 2013).

In summary, tendencies to binge eating, dieting, body image dissatisfaction and concrete thinking styles all potentially mediate relationships between obesity, efforts to counteract obesity and eating disorders. Children and adolescents are particularly vulnerable to developing eating disorders in response to simplistic messages about obesity and nutrition, and also programmes such as child measurement in schools.

5. The ethical issues raised by the public health campaigns for obesity

There are several ethical issues that can be raised by public health campaigns for obesity. These include:

1. The efficacy of public health campaigns on individual behaviour and weight status;
2. Personal responsibility and behavioural choice in obesity;
3. Conflict between state policy and freedom of individual choice;
4. Potential for injustice of placing the health of the many (people who are obese) above the health of the few (people with eating disorders).
5. Potential for unintended harms to the intended audience, namely people who are overweight or obese;

6. Potential for unintended harms to the unintended audience, namely people who are not overweight or obese or vulnerable subgroups within the population;
7. Potential for shifts in societal norms and perceptions;

It is an axiom of modern medicine that interventions should be effective. This is particularly the case if an intervention has the potential to cause harm. The greater the potential harm, the more it needs to be justified by very clear, major benefits which more than counterbalance the harm.

5.1 Responsibility and stigmatisation

Obesity results from gene-environment interaction in vulnerable individuals (Kopelman et al., 2007). Genetic risk may operate through risk for obesity as a whole (metabolic risk, body composition) and it may be through risk for specific behavioural or psychological traits, such as appetite, food responsiveness or LOC eating. Yet even some of these highly heritable traits can be learnt or induced into operating in a so-called obesogenic environment. When it comes to obesity risk and our desire to eat, there is considerable individual variation, and those who have inherited a predisposition to strong appetite drive in response to food cues are particularly vulnerable in environments where fast food and takeaway shops have taken over from grocery or butchers' shops in the high street. Motivation is often cited as a key obstacle, but for some the biological drive is virtually insurmountable, however strong willed or motivated a person might be.

The general underlying assumption of public health discourse is that behavioural modification is the key to reducing obesity. Some national-level policy changes, such as those suggested in the Welsh Government consultation, appear to assume that modifying behaviour requires modification of the environment, for example by modifying the choices of food available, or providing opportunities for increased activity where there may have previously been none. For campaigns targeting the general public, however, there is an underlying assumption that individuals are always able to make a choice regarding behavioural modification. As a consequence, there may be a yet further assumption that people are able to take personal responsibility for (and show agency through) their own choices regarding their behaviours and thereby their own weight status.

Brownell et al. assert that the concept of personal responsibility has been central to social, legal and political approaches to obesity, evoking the language of blame, weakness and vice for people who are obese (Brownell, 1991; Brownell et al., 2010). They feel that this understates the influence of environmental determinants which may undermine personal responsibility and freedom in terms of weight status; in other words, overemphasis on personal responsibility fails to acknowledge the lack of ability to make choices regarding their weight status that people may experience. This becomes of even greater consideration when the targets for intervention are developmentally ill prepared for that responsibility. Brownell et al. suggest, in consequence, that policies that help modify the environment are the way forward, rather than the notion of personal responsibility (Brownell et al., 2010).

5.2 Restriction of freedoms, stigmatisation and human rights

The converse issue of lack of personal freedoms can also occur regarding behavioural change that requires the state's modification of the environment; namely that the will of the state may infringe on personal freedoms and choice. This is a battle that has already been fought in the arena of smoking, which until recently was a major contributor to preventable illness and death. Following the acceptance of scientific evidence that smoking is bad for the health of both the smoker and those around him and public health discourse in the media, legislation has increasingly restricted the freedom of people to smoke cigarettes wherever they choose, and taxes have increasingly made it less and less affordable (Asbridge, 2004). Although activist organisations such as Action on Smoking and Health (ASH) (http://ash.org.uk/home/) have sprung up attempt to defend the rights of smokers to smoke, there has been increasing restrictions on smoking and a decline of smoking, along with increasing stigmatisation of smoking (Garfinkel, 1997). Indeed, stigmatisation is an acknowledged public health tool unapologetically used against smoking (Bell, Salmon, Bowers, Bell, & McCullough, 2010).

The public health of smoking has become a focus on a more general discussion of the ethics of public health, and human rights (Mann, 2006). There are discussions about how stigmatisation of smokers may have the greatest negative impact on vulnerable groups who may have fewer rights or choices regarding their smoking behaviours, such as people with mental disorders (Warner, 2009).

The battleground for human rights against stigmatisation has now moved to weight. In an age when gender, sexuality and ethnic discrimination are unacceptable, weight bias is sometimes described as the last frontier of discrimination, with anti-obesity values being both widespread, unchallenged and socially acceptable (Puhl & Heuer, 2009, 2010). The Cancer Research UK OB_S_ _Y Causes Cancer campaign has received criticism that it stigmatises obese people (Elliott, 2018). Elliott is not wrong; furthermore healthcare professionals have both implicit and explicit pro-thin/anti-fat bias, and explicit anti-fat bias increasing from 2003 to 2013 (Tomiyama et al., 2015).

Ironically, far from helping reduce food intake, exposure to stigmatising material about obesity may cause people who self-identify as overweight to consume more (Major, Hunger, Bunyan, & Miller, 2014). Stigmatisation of body weight may also decrease motivation to seek healthcare (Drury & Louis, 2002). Puhl and Heuer assert, "Weight bias translates into inequalities in employment settings, health-care facilities, and educational institutions, often due to widespread negative stereotypes that overweight and obese persons are lazy, unmotivated, lacking in self-discipline, less competent, non-compliant, and sloppy. These stereotypes are prevalent and are rarely challenged in Western society, leaving overweight and obese persons vulnerable to social injustice, unfair treatment, and impaired quality of life as a result of substantial disadvantages and stigma." (Puhl & Heuer, 2009).

Stigmatisation itself is ethically questionable; it would be particularly unjust when people who are stigmatised for being overweight have little or no ability to alter their weight. Vulnerable groups it would be particularly unjust to stigmatise include those whose weight gain is a result of mental disorders or their medication for mental disorders; those who are overweight as a result of medical conditions or medications such as steroids and those whose weight gain is associated with an eating disorder such as binge eating disorder. Stigmatisation is also unjust when people are genetically predisposed to be overweight and/or to have greater difficulty in losing weight. Weight gain in people suffering from mental disorders is well established, and appears to be particularly associated with anxiety (Scott, McGee, Wells, & Browne, 2008). Obesity may occur for a range of reasons — decreased activity due to isolation and social anxiety; increased eating or drinking to self-soothe when experiencing difficult emotions; unbalanced diets due to difficulties accessing healthier food options for reasons such as low motivation, loss of enjoyment and food poverty; the side effect of psychiatric medications; and of course pleasurable, hedonic overeating. In binge eating disorders and other 'non-restricting' eating disorders such as bulimia nervosa, weight gain is caused by compulsive binges of large amounts of food, which the sufferers perceive they have little or no ability to control (WHO, 2018a). In the United Kingdom, as in many other developed countries, both childhood and adult obesity are inversely related to socioeconomic status (El-Sayed, Scarborough, & Galea, 2012; Kinra, Nelder, & Lewendon, 2000; Stamatakis, Primatesta, Chinn, Rona, & Falascheti, 2005) Therefore, the most disadvantaged people in developed countries are also the most likely to be stigmatised for their weight, and wrongly or unfairly assigned the burden of personal responsibility for their own plight.

5.3 The impact on parents and children when children are labelled as overweight

In addition to weight stigmatisation, there are other more general potential harms and moral dangers for the intended audience of public health campaigns and initiatives for overweight and obesity. Weight feedback has been criticised for overemphasising personal responsibility and not promoting an approach more focused on the environment and society. It has already been indicated that letters to parents about their children being overweight do not have any positive effect on lifestyle changes (Falconer et al., 2014), which calls into question the benefit of such interventions. There are, moreover, identified potential harms. Studies have shown that linking obesity with fatal illnesses in the feedback letter created panic and annoyance among parents, and they were unwilling to follow the recommendation of the letter to visit the GP to discuss the feedback about their child being overweight. Worse, attached information materials to the weight letter were in most cases dismissed and thrown away. Weight feedback may affect children's mental health, with both children and parents feeling stigmatised and affecting children's self-esteem on the long term. Parents may also reject the notion of their children may be overweight for fear that children will become over sensitised about bodies/weight (Falconer et al., 2014), and find the opinion that their child is overweight upsetting, disgusting and annoying (Nnyanzi, Summerbell, Ells, & Shucksmith, 2016). A group of parents in the Gainsbury study reported that family and peers collaborated in dismissing overweight feedback (Gainsbury & Dowling, 2018).

In summary, studies have shown that parents understand health risks related to childhood excess weight, but are unlikely to recognise that health risk related to their own children and do not recognise their children as being overweight when they are (Gainsbury & Dowling, 2018; Herrera et al., 2013; Nnyanzi et al., 2016). Thus feedback regarding overweight was not always accepted by parents, is considered as lacking credibility and is unlikely to lead to behaviour change. This raises serious questions about the value of identifying children and young people as overweight or obese in terms of parental response. But what about the value to young people themselves of being identified as overweight or obese?

Hunger et al. investigated the association between weight labelling and disordered eating in adolescent girls. The study found that girls that were labelled as too fat showed increased disordered eating cognitions and behaviours 5 years later. This was more pronounced when the labelling was coming from a family member (Hunger & Tomiyama, 2018). Importantly, this finding was independent of actual baseline BMI. In other words, self stigma on the basis of perceived weight was potentially as harmful as stigma from others. Nnyazi has looked at children's reaction (10—11 years old) following being informed of weight problems (overweight or obese). The children described feeling shocked, worried and anxious. They kept thinking about their obesity and these thoughts influenced them to change their peer relationships by moving towards children that were overweight or obese as well. They also felt over sensitised about their weight (Nnyanzi, 2016).

5.4 Public health initiatives and those not targeted

Public health initiatives can have deleterious effects on people for whom the messages are not intended (O'Dea, 2005). In particular, there are concerns that the anti-obesity message may cause or exacerbate incipient or established eating disorders, promote calorie-counting, unhealthy dieting behaviours or induce body dissatisfaction among susceptible young people (O'dea, 2004; Schwartz & Henderson, 2009). "Children obese at 11 doomed to early death" says the front page of the Daily Telegraph (Donnelly, 2018), visible for all to see in newstands at the entrance to the supermarket. Such headlines are explicitly designed to create moral panic and are unsophisticated in their target. In 2010 a UK minister went further, suggesting that calling obese people fat, so called plain speaking, was needed to "jolt people into losing weight", adding shame and stigma to the cultural and policy environment (Martin, 2010). We have argued that, not only are anxiety and shame ineffective methods for instilling behaviour change, they risk having adverse consequences in their own right, such as causing children and adolescents who are not even overweight to be overly concerned about their weight or to engage in rigid eating behaviours or dieting which can then lead to disordered eating.

If there are risks of an anti-obesity public health campaign for people who are not overweight or obese, can this harm to a relatively small number of people who have no potential to benefit from the campaign be justified by the benefit to a larger group of people? Public health is based on utilitarian principles, functioning on the basis that benefit to many may offset the inevitable small harms to a few. For example, the breast cancer mammogram screening programme is justified because the small increased risks due to radiation to women undergoing mammograms is offset by the great benefits of identifying and treatment breast cancer well before it is clinically detectable (for example, as a lump or abnormality of breast skin) or has spread. For the women who are exposed to the risk of a mammogram but do not have breast cancer, the benefit of being screened for a cancer they were also at risk for remains. The case of public health anti-obesity campaigns however is an issue of justice – would a public health campaign be ethically justified if it causes harm to people who are not even at risk of obesity and thus not the targets of the campaign, and might otherwise remain well? That would be more analogous to exposing people who do not have breasts to mammograms. Even men do have breasts, although much smaller ones; and they are not offered screening mammograms because they are at a much lower risk of breast cancer than women. Furthermore, it is not clear that the principle of 'Primum non nocere' would be able to allow even one individual to come to harm due to actions intended to help the many (others), without thought and readjustment.

6. Resolution

What then, is the way forward? Obesity is clearly a major public health burden, with a major cost in terms of health and wellbeing to individuals as well as to society. There is also clear evidence that reduction of obesity rates will improve the health of the population. A plethora of policies have been directed towards addressing the problem, ranging from the environmental and industry targeted through to the personal nutritional and behavioural targeted. Evidence for the effectiveness of many of these policies is limited to date. In the ethical balance of public health policy, demonstrating benefit is as important as avoiding harm. What is well evidenced is that weight stigmatisation or targeting individuals, particularly vulnerable individuals who may have less ability to exercise choice or alter behaviours, has potentially harmful effects. The harms of anti-obesity public health messages would be reduced if we stop promulgating negative, stigmatising public health messages about obesity, and challenge commonly-held implicit or explicit assumptions both within the medical and healthcare community and in society that people are able to exercise full personal responsibility for their weight status. Given the scale of obesity, it would be unthinkable and unfeasible to approach the problem other than through public health policy, so emphasis on the need to evaluate the harms as well as benefits in the impact of any policies, especially if the programmes entail explicit effort to induce anxiety. Engendering anxiety about common public health targets such as smoking, alcohol, wearing seat belts or the importance of vaccinations carries no such risks; engendering anxiety about eating or being fat does.

It is important to note that improving eating and activity levels is not about unidirectional change — for some people healthier eating and healthier activity levels may mean increased intake and reduced activity. For people who are overweight or obese, this change of emphasis and positive message with increased environmental support might be critical to helping reduce the sense of shame and increase their ability to alter their behaviour; while for those in the early stages of eating disorders or who are unhealthily low in weight or excessively exercising, it may help them to recover rather than validating and driving them further into their predicament. Neumark-Sztainer and Swartz and Henderson suggest that prevention of both obesity and eating disorders are compatible, through the use of more nuanced and less polemic positive messages about eating and exercising healthily and **in moderation** (Neumark-Sztainer, 2005; Schwartz & Henderson, 2009). Schwartz and Henderson explain: "Obesity prevention efforts exist because of the striking increase in childhood obesity; however, the root problem is poor diet and inactivity among American children. We suggest reframing the current public health efforts of obesity prevention into health promotion for all children. This would address one of the key obstacles to implementing new policies - the belief that children who are not overweight do not have to worry about improving eating or activity levels." Brownell et al. further suggest that public health efforts, which are currently disproportionately aimed at individual responsibility, should instead be made at policy level, aiming to modify the environment to support and promote healthier behaviours that can reduce obesity (Brownell et al., 2010). The Welsh Government's concept of promoting 'Healthy Environments', 'Healthy Settings' and then only 'Healthy People' within the context of the environments and settings is both ethically laudable and also most likely to be effective.

The contribution of ethics to the public health of obesity must be to challenge assumptions regarding personal responsibility currently held amongst many public health authorities; to point out the damage and injustices that deliberate stigmatisation as a public health strategy bring; and to protect those who are vulnerable to harm from misguided public health policies. The global, national and local public health strategies should focus on enabling and supporting healthy weight and improved health through education, better healthcare, increasing choices and environmental support.

References

Al-Adawi, S., Bax, B., Bryant-Waugh, R., Claudino, A. M., Hay, P., Monteleone, P., … Uher, R. (2013). Revision of ICD — status update on feeding and eating disorders. *Advances in Eating Disorders, 1*(1), 10—20. https://doi.org/10.1080/21662630.2013.742971.

Alderson, R. (7th December 2018). *Obesity epidemic risks eyesight of young Scots women. Online news article.* Scotland: BBC. https://www.bbc.co.uk/news/uk-scotland-edinburgh-east-fife-46488358.

American Psychiatric Association. (2014). *DSM-5.* Arlington, Virginia: American Psychiatric Association. Retrieved from: http://www.dsm5.org/Pages/Default.aspx.

Arcelus, J., AJ, M., Wales, J., & Nielsen, S. (2011). Mortality rates in patients with anorexia nervosa and other eating disorders: A meta-analysis of 36 studies. *Archives of General Psychiatry, 68*(7), 724—731. Retrieved from: https://doi.org/10.1001/archgenpsychiatry.2011.74.

Asbridge, M. (2004). Public place restrictions on smoking in Canada: Assessing the role of the state, media, science and public health advocacy. *Social Science & Medicine, 58*(1), 13—24.

Aune, D., Sen, A., Prasad, M., Norat, T., Janszky, I., Tonstad, S., … Vatten, L. J. (2016). BMI and all cause mortality: Systematic review and non-linear dose-response meta-analysis of 230 cohort studies with 3.74 million deaths among 30.3 million participants. *Bmj, 353*, i2156.

Bell, K., Salmon, A., Bowers, M., Bell, J., & McCullough, L. (2010). Smoking, stigma and tobacco 'denormalization': Further reflections on the use of stigma as a public health tool. A commentary on social science & medicine's stigma, prejudice, discrimination and health special issue (67: 3). *Social Science & Medicine, 70*(6), 795—799.

Berkowitz, S. A., Witt, A. A., Gillberg, C., Råstam, M., Wentz, E., & Lowe, M. R. (2016). Childhood body mass index in adolescent-onset anorexia nervosa. *International Journal of Eating Disorders, 49*(11), 1002—1009.

Bertoli, S., Leone, A., Ponissi, V., Bedogni, G., Beggio, V., Strepparava, M. G., & Battezzati, A. (2016). Prevalence of and risk factors for binge eating behaviour in 6930 adults starting a weight loss or maintenance programme. *Public Health Nutrition, 19*(1), 71—77.

Birmingham, C. L., & Treasure, J. L. (2010). *Medical management of eating disorders.* Cambridge University Press. https://doi.org/10.1016/j.nurpra.2009.02.007.

Black, J. A., Park, M., Gregson, J., Falconer, C. L., White, B., Kessel, A. S., … Kinra, S. (2015). Child obesity cut-offs as derived from parental perceptions: Cross-sectional questionnaire. *British Journal of General Practice, 65*(633), e234—e239.

Bould, H., De Stavola, B., Lewis, G., & Micali, N. (2018). Do disordered eating behaviours in girls vary by school characteristics? A UK cohort study. *European Child & Adolescent Psychiatry, 1—9.*

Brownell, K. D. (1991). Personal responsibility and control over our bodies: When expectation exceeds reality. *Health Psychology, 10*(5), 303.

Brownell, K. D., Kersh, R., Ludwig, D. S., Post, R. C., Puhl, R. M., Schwartz, M. B., & Willett, W. C. (2010). Personal responsibility and obesity: A constructive approach to a controversial issue. *Health Affairs, 29*(3), 379—387.

Cancer Research UK. (2018). *OB_S__TY is a cause of cancer.* Retrieved from: https://www.cancerresearchuk.org/about-cancer/causes-of-cancer/obesity-weight-and-cancer.

Cecchini, M., & Warin, L. (2016). Impact of food labelling systems on food choices and eating behaviours: A systematic review and meta-analysis of randomized studies. *Obesity Reviews, 17*(3), 201–210.

Chapman, A. (January 28, 2019). *Australia's "XXXXXXXXXL" schoolkids: Obese students wearing specially-designed uniforms have to visit the physio to physically survive the six-hour school day.* Mail Online. https://www.dailymail.co.uk/news/article-6638239/Kids-overweight-wear-specially-9XL-uniforms-school-physiotherapy-classes.html.

Cinelli, R. L., & O'Dea, J. A. (2016). Obesity prevention programs in children: Impact on weight, shape and food concern. *Current Obesity Reports, 5*(1), 88–96.

Cortese, S., Moreira-Maia, C. R., Fleur, D., Morcillo-Peñalver, C., Rohde, L. A., & Faraone, S. V. (2015). Association between ADHD and obesity: A systematic review and meta-analysis. *American Journal of Psychiatry, 173*(1), 34–43.

Da Luz, F. Q., Sainsbury, A., Mannan, H., Touyz, S., Mitchison, D., & Hay, P. (2017). Prevalence of obesity and comorbid eating disorder behaviors in South Australia from 1995 to 2015. *International Journal of Obesity, 41*(7), 1148.

Donnelly, L. (January 23, 2018). *Child obesity crisis: Millions doomed to early death as doctors call for urgent action over junk food. Telegraph.* https://www.telegraph.co.uk/news/2018/01/23/obese-children-doomed-early-death-doctors-warn/.

Drury, C. A. A., & Louis, M. (2002). Exploring the association between body weight, stigma of obesity, and health care avoidance. *Journal of the American Academy of Nurse Practitioners, 14*(12), 554–561.

Eddy, K. T., Thomas, J. J., Hastings, E., Edkins, K., Lamont, E., Nevins, C. M., … Becker, A. E. (2015). Prevalence of DSM-5 avoidant/restrictive food intake disorder in a pediatric gastroenterology healthcare network. *International Journal of Eating Disorders.* https://doi.org/10.1002/eat.22350.

El-Sayed, A. M., Scarborough, P., & Galea, S. (2012). Unevenly distributed: A systematic review of the health literature about socioeconomic inequalities in adult obesity in the United Kingdom. *BMC Public Health, 12*(1), 18.

Elbel, B., Kersh, R., Brescoll, V. L., & Dixon, L. B. (2009). Calorie labeling and food choices: A first look at the effects on low-income people in New York city. *Health Affairs, 28*(6), w1110–w1121.

Elbel, B., Mijanovich, T., Dixon, L. B., Abrams, C., Weitzman, B., Kersh, R., … Ogedegbe, G. (2013). Calorie labeling, fast food purchasing and restaurant visits. *Obesity, 21*(11), 2172–2179.

Elliott, A. (2018). *Cancer Research UK's 'obesity is a cause of cancer' campaign ignores the very real issue of medical fatphobia.* Retrieved from: https://metro.co.uk/2018/03/03/cancer-researchs-obesity-cause-cancer-campaign-ignores-real-issue-medical-fatphobia-7357941/.

Fairburn, C. G., Doll, H. A., Welch, S. L., Hay, P. J., Davies, B. A., & O'Connor, M. E. (1998). Risk factors for binge eating disorder: A community-based, case-control study. *Arch Gen Psychiatry, 55.* https://doi.org/10.1001/archpsyc.55.5.425.

Falconer, C. L., Park, M. H., Croker, H., Skow, Á., Black, J., Saxena, S., … Viner, R. M. (2014). The benefits and harms of providing parents with weight feedback as part of the national child measurement programme: A prospective cohort study. *BMC Public Health, 14*(1), 549.

Field, D., Garland, M., & Williams, K. (2003). Correlates of specific childhood feeding problems. *Journal of Paediatrics and Child Health, 39*(4), 299–304.

Gainsbury, A., & Dowling, S. (2018). 'A little bit offended and slightly patronised': Parents' experiences of national child measurement programme feedback. *Public Health Nutrition, 1–9.*

Garfinkel, L. (1997). Trends in cigarette smoking in the United States. *Preventive Medicine, 26*(4), 447–450.

Gariepy, G., Nitka, D., & Schmitz, N. (2010). The association between obesity and anxiety disorders in the population: A systematic review and meta-analysis. *International Journal of Obesity, 34*(3), 407.

Gillon, R. (1985). " Primum non nocere" and the principle of non-maleficence. *British Medical Journal (Clinical Research Ed.), 291*(6488), 130.

He, J., Cai, Z., & Fan, X. (2017). Prevalence of binge and loss of control eating among children and adolescents with overweight and obesity: An exploratory meta-analysis. *International Journal of Eating Disorders, 50*(2), 91–103.

Heinberg, L. J., Thompson, J. K., & Matzon, J. L. (2001). Body image dissatisfaction as a motivator for healthy lifestyle change: Is some distress beneficial? In R. H. Striegel-Moore, & L. Smolak (Eds.), *Eating disorders: Innovative directions in research and practice* (pp. 215–232). Washington, DC, USA: American Psychological Association. https://doi.org/10.1037/10403-011.

Herrera, J., Lockner, D., Kibbe, D., Marley, S. C., Trowbridge, F., & Bailey, A. (2013). Innovative tools help counselors discuss childhood obesity with parents. *Childhood Obesity, 9*(2), 144–149.

Hilbert, A., Pike, K. M., Goldschmidt, A. B., Wilfley, D. E., Fairburn, C. G., Dohm, F. A., ... Striegel, W. R. (2014). Risk factors across the eating disorders. *Psychiatry Research, 220*. https://doi.org/10.1016/j.psychres.2014.05.054.

Hunger, J. M., & Tomiyama, A. J. (2018). Weight Labeling and Disordered Eating Among Adolescent Girls: Longitudinal Evidence From the National Heart, Lung, and Blood Institute Growth and Health Study. *Journal of Adolescent Health, 63*(3), 360–362.

Izquierdo, A., Plessow, F., Becker, K. R., Mancuso, C. J., Slattery, M., Murray, H. B.,., & Thomas, J. J. (2019). Implicit attitudes toward dieting and thinness distinguish fat-phobic and non-fat-phobic anorexia nervosa from avoidant/restrictive food intake disorder in adolescents. Special Issue:Child and Adolescent Feeding and Eating Disorders and Avoidant/Restrictive Food Intake Disorder *International Journal of Eating Disorders, 52*(4), 419–427. April 2019.

Jeffery, R. W., Folsom, A. R., Luepker, R. V., Jacobs, D. R., Jr., Gillum, R. F., Taylor, H. L., & Blackburn, H. (1984). Prevalence of overweight and weight loss behavior in a metropolitan adult population: The Minnesota heart survey experience. *American Journal of Public Health, 74*(4), 349–352.

Joint Commissioning Panel for Mental Health. (2013a). *10 key messages on commissioning eating disorder services*. London. Retrieved from: https://www.jcpmh.info/wp-content/uploads/10keymsgs-eatingdisorders.pdf.

Joint Commissioning Panel for Mental Health. (2013b). *Guidance for commissioners of eating disorder services*. London. Retrieved from: https://www.jcpmh.info/wp-content/uploads/jcpmh-eatingdisorders-guide.pdf.

Kessler, R. C., Berglund, P. A., Chiu, W. T., Deitz, A. C., Hudson, J. I., Shahly, V., ... Benjet, C. (2013). The prevalence and correlates of binge eating disorder in the World Health Organization world mental health surveys. *Biological Psychiatry, 73*. https://doi.org/10.1016/j.biopsych.2012.11.020.

Kinra, S., Nelder, R. P., & Lewendon, G. J. (2000). Deprivation and childhood obesity: A cross sectional study of 20 973 children in Plymouth, United Kingdom. *Journal of Epidemiology & Community Health, 54*(6), 456–460.

Kiszko, K. M., Martinez, O. D., Abrams, C., & Elbel, B. (2014). The influence of calorie labeling on food orders and consumption: A review of the literature. *Journal of Community Health, 39*(6), 1248–1269.

Kopelman, P., Jebb, S., & Butland, B. (2007). Executive summary: Foresight "tackling obesities: Future choices" project. *Obesity Review, 8*(Suppl. 1). vi–ix.

Lillico, H. G., Hanning, R., Findlay, S., & Hammond, D. (2015). The effects of calorie labels on those at high-risk of eating pathologies: A pre-post intervention study in a university cafeteria. *Public Health, 129*(6), 732–739.

Loth, K. A., Watts, A. W., Van Den Berg, P., & Neumark-Sztainer, D. (2015). Does body satisfaction help or harm overweight teens? A 10-year longitudinal study of the relationship between body satisfaction and body mass index. *Journal of Adolescent Health, 57*(5), 559–561.

Major, B., Hunger, J. M., Bunyan, D. P., & Miller, C. T. (2014). The ironic effects of weight stigma. *Journal of Experimental Social Psychology, 51*, 74–80.

Mann, J. (2006). Health and human rights: If not now, when? *American Journal of Public Health, 2*(11), 113–1943.

Marcus, M. D., & Wildes, J. E. (2009). Obesity: Is it a mental disorder? *International Journal of Eating Disorders, 42*(8), 739–753.

Martin, D. (July 29, 2010). *Obese? Just call them fat: Plain-speaking doctors will jolt people into losing weight, says minister.* Daily Mail Online. https://www.dailymail.co.uk/news/article-1298394/Call-overweight-people-fat-instead-obese-says-health-minister.html.

Matthews, S. (November 3, 2016). *More children are obese than EVER before: One in three 10 and 11 year olds are now overweight, shocking figures reveal.* Mailonline. Published. https://www.dailymail.co.uk/health/article-3901430/The-childhood-obesity-crisis-laid-bare-One-THREE-10-11-year-olds-overweight-shocking-statistics-reveal.html.

Micali, N., Hagberg, K. W., & Treasure, J. L. (2013). Incidence of eating disorders in the UK: Findings from the UK general practice Research database. *The Lancet, 381*, S75. https://doi.org/10.1016/S0140-6736(13)60515-2.

Neumark-Sztainer, D. (2005). Can we simultaneously work toward the prevention of obesity and eating disorders in children and adolescents? *International Journal of Eating Disorders, 38*(3), 220–227.

Neumark-Sztainer, D., Paxton, S. J., Hannan, P. J., Haines, J., & Story, M. (2006). Does body satisfaction matter? Five-year longitudinal associations between body satisfaction and health behaviors in adolescent females and males. *Journal of Adolescent Health, 39*(2), 244–251.

Neumark-Sztainer, D., Wall, M., Larson, N. I., Eisenberg, M. E., & Loth, K. (2011). Dieting and disordered eating behaviors from adolescence to young adulthood: Findings from a 10-year longitudinal study. *Journal of the American Dietetic Association, 111*(7), 1004–1011.

NHS. (2018). *Change4Life.* Retrieved from: https://www.nhs.uk/change4life/about-change4life.

Nicely, T. A., Lane-Loney, S., Masciulli, E., Hollenbeak, C. S., & Ornstein, R. M. (2014). Prevalence and characteristics of avoidant/restrictive food intake disorder in a cohort of young patients in day treatment for eating disorders. *Journal of Eating Disorders.* https://doi.org/10.1186/s40337-014-0021-3.

Nicholls, D., Statham, R., Costa, S., Micali, N., & Viner, R. M. (2016). Childhood risk factors for lifetime bulimic or compulsive eating by age 30 years in a British national birth cohort. *Appetite, 105*, 266–273.

Nigg, J. T., Johnstone, J. M., Musser, E. D., Long, H. G., Willoughby, M. T., & Shannon, J. (2016). Attention-deficit/hyperactivity disorder (ADHD) and being overweight/obesity: New data and meta-analysis. *Clinical Psychology Review, 43*, 67–79.

Nnyanzi, L. A. (2016). Combating childhood obesity: Reactions of children aged 10–11 years towards the national child measurement programme. *Journal of Child Health Care, 20*(4), 464–472.

Nnyanzi, L. A., Summerbell, C. D., Ells, L., & Shucksmith, J. (2016). Parental response to a letter reporting child overweight measured as part of a routine national programme in England: Results from interviews with parents. *BMC Public Health, 16*(1), 846.

O'dea, J. A. (2004). Prevention of child obesity: 'First, do no harm'. *Health Education Research, 20*(2), 259–265.

O'Dea, J. A. (2005). Prevention of child obesity: "First, do no harm. *Health Education Research, 20*(2), 259–265. https://doi.org/10.1093/her/cyg116.

Pacanowski, C. R., Senso, M. M., Oriogun, K., Crain, A. L., & Sherwood, N. E. (2014). Binge eating behavior and weight loss maintenance over a 2-year period. *Journal of Obesity, 2014.*

Patton, G. C., Selzer, R., Coffey, C., Carlin, J. B., & Wolfe, R. (1999). Onset of adolescent eating disorders: Population based cohort study over 3 years. *BMJ, 318*(7186), 765–768.

Perkins, S., Winn, S., Murray, J., Murphy, R., & Schmidt, U. (2004). A qualitative study of the experience of caring for a person with bulimia nervosa. Part 1: The emotional impact of caring. *International Journal of Eating Disorders, 36*(3), 256–268. https://doi.org/10.1002/eat.20067.

Pinhas, L., McVey, G., Walker, K. S., Norris, M., Katzman, D., & Collier, S. (2013). Trading health for a healthy weight: The uncharted side of healthy weights initiatives. *Eating Disorders, 21*(2), 109–116.

Public Health England. (January 2, 2018). *PHE launches Change4Life campaign around children's snacking*. Retrieved from: https://www.gov.uk/government/news/phe-launches-change4life-campaign-around-childrens-snacking.

Puhl, R. M., & Heuer, C. A. (2009). The stigma of obesity: A review and update. *Obesity, 17*(5), 941–964.

Puhl, R. M., & Heuer, C. A. (2010). Obesity stigma: Important considerations for public health. *American Journal of Public Health, 100*(6), 1019–1028.

Pulos, E., & Leng, K. (2010). Evaluation of a voluntary menu-labeling program in full-service restaurants. *American Journal of Public Health, 100*(6), 1035–1039.

PwC (Price Waterhouse Cooper LLP), & Beat. (2015). *The costs of eating disorders: Social, health and economic impacts*. London. Retrieved from: https://www.b-eat.co.uk/assets/000/000/302/The_costs_of_eating_disorders_Final_original.pdf.

Russell-Mayhew, S., McVey, G., Bardick, A., & Ireland, A. (2012). Mental health, wellness, and childhood overweight/obesity. *Journal of Obesity, 2012.*

Sacco, J., Lillico, H. G., Chen, E., & Hobin, E. (2017). The influence of menu labelling on food choices among children and adolescents: A systematic review of the literature. *Perspectives in Public Health, 137*(3), 173–181.

Schwartz, M. B., & Henderson, K. E. (2009). Does obesity prevention cause eating disorders? *Journal of the American Academy of Child & Adolescent Psychiatry, 48*(8), 784–786.

Scott, K. M., McGee, M. A., Wells, J. E., & Browne, M. A. O. (2008). Obesity and mental disorders in the adult general population. *Journal of Psychosomatic Research, 64*(1), 97–105.

Seward, M. W., Block, J. P., & Chatterjee, A. (2018). Student experiences with traffic-light labels at college cafeterias: A mixed methods study. *Obesity Science & Practice, 4*(2), 159–177.

Slof-Op't Landt, M. C. T., van Furth, E. F., van Beijsterveldt, C. E. M., Bartels, M., Willemsen, G., de Geus, E. J., … Boomsma, D. I. (2017). Prevalence of dieting and fear of weight gain across ages: A community sample from adolescents to the elderly. *International Journal of Public Health, 62*(8), 911–919.

Smink, F. R. E., van Hoeken, D., & Hoek, H. W. (2013). Epidemiology, course, and outcome of eating disorders. *Current Opinion in Psychiatry, 26*(6), 543–548. https://doi.org/10.1097/YCO.0b013e328365a24f.

Smink, F. R. E., Van Hoeken, D., Oldehinkel, A. J., & Hoek, H. W. (2014). Prevalence and severity of DSM-5 eating disorders in a community cohort of adolescents. *International Journal of Eating Disorders, 47*(6), 610–619. https://doi.org/10.1002/eat.22316.

Spencer, B. (September 12, 2018). *Obesity Britain: Two thirds of us are overweight and only two - Turkey and Malta - out of 53 European countries are fatter, UN study finds*. The Daily Mail.

Published. https://www.dailymail.co.uk/news/article-6156965/Obesity-Britain-Two-thirds-overweight-Turkey-Malta-fatter-study-finds.html.

Stamatakis, E., Primatesta, P., Chinn, S., Rona, R., & Falascheti, E. (2005). Overweight and obesity trends from 1974 to 2003 in English children: What is the role of socioeconomic factors? *Archives of Disease in Childhood, 90*(10), 999–1004. https://doi.org/10.1136/adc.2004.068932.

Steinhausen, H. (2002). Reviews and overviews the outcome of anorexia nervosa in the 20th century. *Psychiatry: Interpersonal and Biological Processes*, (August), 1284–1293. https://doi.org/10.1176/appi.ajp.159.8.1284.

Stice, E., Agras, W. S., & Hammer, L. D. (1999). Risk factors for the emergence of childhood eating disturbances: A five-year prospective study. *International Journal of Eating Disorders, 25*(4), 375–387.

Stice, E., Marti, C. N., & Rohde, P. (2013). Prevalence, incidence, impairment, and course of the proposed DSM-5 eating disorder diagnoses in an 8-year prospective community study of young women. *Journal of Abnormal Psychology, 122*(2), 445–457. https://doi.org/10.1037/a0030679.

Stice, E., & Van Ryzin, M. J. (2019). A prospective test of the temporal sequencing of risk factor emergence in the dual pathway model of eating disorders. *Journal of abnormal psychology, 128*(2), 119.

Striegel-Moore, R. H., Fairburn, C. G., Wilfley, D. E., Pike, K. M., Dohm, F.-A., & Kraemer, H. C. (2005). Toward an understanding of risk factors for binge-eating disorder in black and white women: A community-based case-control study. *Psychological Medicine, 35*(6), 907–917.

Striegel-Moorel, R. H., & Franko, D. L. (2003). Epidemiology of binge eating disorder. *International Journal of Eating Disorders*. https://doi.org/10.1002/eat.10202.

Strine, T. W., Mokdad, A. H., Dube, S. R., Balluz, L. S., Gonzalez, O., Berry, J. T., … Kroenke, K. (2008). The association of depression and anxiety with obesity and unhealthy behaviors among community-dwelling US adults. *General Hospital Psychiatry, 30*(2), 127–137.

Tan, J. O. a, Hope, T., & Stewart, A. (2003). Anorexia nervosa and personal identity: The accounts of patients and their parents. *International Journal of Law and Psychiatry, 26*(5), 533–548. https://doi.org/10.1016/S0160-2527(03)00085-2.

Tanofsky-Kraff, M., Yanovski, S. Z., Schvey, N. A., Olsen, C. H., Gustafson, J., & Yanovski, J. A. (2009). A prospective study of loss of control eating for body weight gain in children at high risk for adult obesity. *International Journal of Eating Disorders, 42*(1), 26–30.

Tan, J., Stewart, A., Fitzpatrick, R., & Hope, R. A. (2007). Competence to make treatment decisions in anorexia nervosa: Thinking processes and values. *Philosophy, Psychiatry, & Psychology, 13*(4), 267–282. https://doi.org/10.1353/ppp.2007.0032.

Tomiyama, A. J., Finch, L. E., Belsky, A. C. I., Buss, J., Finley, C., Schwartz, M. B., & Daubenmier, J. (2015). Weight bias in 2001 versus 2013: Contradictory attitudes among obesity researchers and health professionals. *Obesity, 23*(1), 46–53.

Tomiyama, A. J., Hunger, J. M., Nguyen-Cuu, J., & Wells, C. (2016). Misclassification of cardiometabolic health when using body mass index categories in NHANES 2005–2012. *International Journal of Obesity, 40*(5), 883.

Vitousek, K., & Manke, F. (1994). Personality variables and disorders in anorexia nervosa and bulimia nervosa. *Journal of Abnormal Psychology, 103*(1), 137–147. https://doi.org/10.1037//0021-843X.103.1.137.

Warner, J. (2009). Smoking, stigma and human rights in mental health: Going up in smoke? *Social Policy and Society, 8*(2), 275–286.

Welsh Government. (2019). *Healthy Weight, Healthy Wales - we are seeking your views on proposals to prevent and reduce obesity in Wales*. Cardiff, Wales. Retrieved from: https://beta.gov.wales/healthy-weight-healthy-wales.

White, B., Nicholls, D., Christie, D., Cole, T. J., & Viner, R. M. (2012). Childhood psychological function and obesity risk across the lifecourse: Findings from the 1970 British cohort study. *International Journal of Obesity, 36*(4), 511.

WHO. (2003). *Obesity and overweight: Global strategy on diet, physical activity and health*. https://doi.org/10.1080/10810730903279694.

WHO. (2018a). *ICD-11 for mortality and morbidity statistics: Feeding or eating disorders*. Retrieved from: https://icd.who.int/browse11/l-m/en#/http%3A%2F%2Fid.who.int%2Ficd%2Fentity%2F1412387537.

WHO. (2018b). *Taking action on childhood obesity*. Retrieved from: http://apps.who.int/iris/bitstream/handle/10665/274792/WHO-NMH-PND-ECHO-18.1-eng.pdf.

WHO. (2019). *Childhood overweight and obesity*. https://www.who.int/growthref/en/.

Winn, S., Perkins, S., Murray, J., Murphy, R., & Schmidt, U. (2004). A qualitative study of the experience of caring for a person with bulimia nervosa. Part 2: Carers' needs and experiences of services and other support. *International Journal of Eating Disorders, 36*(3), 269–279. https://doi.org/10.1002/eat.20068.

World Health Organization. (1992). The ICD-10 classification of mental and behavioural disorders. *International Classification, 10*, 1–267. https://doi.org/10.1002/1520-6505(2000)9:5<201::AID-EVAN2>3.3.CO;2-P.

CHAPTER TWELVE

Prevention in prayer camps? The ethics of government engagement in Ghana's 2012 mental health legislation

Lauren A. Taylor*
Harvard Business School, Boston, MA, United States
*Corresponding author: E-mail: ltaylor@hbs.edu

Contents

1. Introduction

This chapter explores the ethical challenges faced by the Ghanaian government as it reconsidered its relationship to a sector of Pentecostal prayer camps offering services to people with serious mental illness leading up to the passage of new mental health legislation in 2012. It raises the question of under what conditions a government should recognise, regulate or partner with a group that has a track record of human rights abuses but

Developments in Neuroethics and Bioethics, Volume 2
ISSN 2589-2959
https://doi.org/10.1016/bs.dnb.2019.04.007

also serves a critical function in a health care landscape defined by pronounced resource scarcity.[1]

The chapter proceeds in several parts. First, I provide substantive background on the history of mental health policy in Ghana, the design of the formal mental health system and the historical role of the prayer camps. Second, I outline the ethical challenge faced by the government in relation to the prayer camps, including the most compelling arguments for and against engaging the prayer camps as legitimate actors in the mental health system. This debate was complicated by policymakers' desire to accomplish two prevention-related policy goals at once, (a) preventing torturous treatment of people with mental illness and (b) preventing worsening of symptom burdens among people with mental illness. Third, I frame the specifics of this ethical dilemma within a broader conversation about the ethics of engagement in the negotiations literature. Fourth, I describe the ways in which the ethical dilemmas are relevant more broadly for prevention efforts within global mental health. Finally, I provide a brief epilogue of how the Ghanaian government ultimately addressed the prayer camps in the Mental Health Act in 2012 and what has happened in Ghanaian mental health subsequently.

2. Substantive background

2.1 Significant treatment gap

Ghana lacks systematically-collected epidemiologic data on the prevalence of mental illness. As a result, it relies on WHO estimates suggesting

[1] A note on my relationship to this case is warranted so that readers may understand my own perspective and perhaps biases as it relates to the prayer camps. I have been studying Ghana's prayer camp sector in more and less formal ways now for the past 6 years. I first became aware of the challenges inherent in the prayer camps when working at a global health conference where a delegation of Ghanaian policymakers was discussing the topic and soon after traveled to Ghana as part of a policy consultation and support team from Yale University to advance this discussion within the Ministry of Health and mental health advocacy community ("GHLI Assists with Mental Health Care Reform in Ghana," 2011). Two years later, while enrolled in a Masters of Divinity program, I returned to the question of the prayer camps by writing my thesis on a unique collaboration between a team of psychiatrists and psychologists and the staff at one of the largest prayer camps in the Eastern Region (Ofori-Atta et al., 2018). This involved six trips to Ghana over two and a half years and more than a month of direct observation of the prayer camp as well as more than 50 interviews with policymakers, medical providers and prayer camp staff and leadership. Previously, I've written about my experiences with the prayer camp staff (Taylor, 2016a) and the ethics of a randomised control trial at the camp (Taylor & Sayeed, 2019).

that roughly 10% of any population suffers from neuropsychiatric conditions and 1% from severe mental illness (Roberts, Asare, Morgan, Adjase, & Osei, 2013). More recently published materials indicate potentially larger variation in these prevalence rates from country to country than the WHO estimates had recognised (Kessler et al., 2009). Ghanaian researchers have suggested that in a country of approximately 24 million people, 650,000 Ghanaians experience mental illness that is acute enough to warrant medical care ("Ghana Mental Health Situation Analysis," n.d.).

The mental health services available through the national health system, called the Ghana Health System (GHS), are inadequate to meet demand. Three in-patient psychiatric hospitals are currently operating and serve as the bedrock of the mental health system. Together, these facilities make available roughly 1500 beds to the most severely ill individuals in Ghana, and serve between 6300 and 8000 patients per year (Roberts et al., 2013). As of 2011, 7 of the 10 regional hospitals also had psychiatric wings that are designated for people with mental illness. None of the district hospitals are equipped in this way and the majority of primary care physicians prefer to refer people with mental illness to specialist, in-patient treatment rather than engage them on an out-patient basis due to persistent stigmas even within the provider community. Recent counts indicate between 15 and 18 licensed psychiatrists are working in either the public or private sector, but many have been placed in administrative (i.e., hospital CEO) or policy roles (Roberts et al., 2013; WHO, 2018). Psychiatric nurses, trained at the Bachelors level, therefore deliver the majority of care in hospital settings (Ofori-Atta, Read, & Lund, 2010). To the extent that community-based care currently exists, it is delivered by a combination of psychiatric nurses and community mental health officers. Given the limited scope of these resources, the WHO estimates there to be a 98% treatment gap among people with mental illness in Ghana ("Ghana Mental Health Situation Analysis," n.d.).

Apart from the resource scarcity, the GHS′ mental health system lacks a reputation as a provider of quality mental health care in many communities. The mental health care provided at GHS institutions does not meet their needs of the public on account of several shortfalls. One, the psychiatric hospitals are clustered in urban areas in the southern portion of the country (Roberts et al., 2013). This means that for many Ghanaians in the Northern Region, it may be a full days' worth of driving, or more, before they could arrive at a formal inpatient facility. Two, the psychiatric hospitals have recently developed a reputation for being expensive sources of care relative to other options (Ae-Ngibise et al., 2010). The question of expense is particularly concerning, as mental health services are specified as one of several that should be provided free of charge to all who have national health insurance. Still, informal and point-of-care payments are common, as are charges for psychotropic drugs to be purchased on the open market. Three, to seek care at a psychiatric facility remains incredibly stigmatising (Barke, Nyarko, & Klecha, 2011). Many Ghanaians will not walk by a psychiatric hospital for fear of being thought of as mentally ill by peers. As it does in many other countries, this label can prove durable and damaging not only to an individual but to an entire family. Four, the care provided at psychiatric hospitals is seen by many Ghanaians as being inherently misaligned with their understanding of mental illness. Much of Ghana's populace believe mental illness can be spiritual in nature, caused by evil spirits, a curse, or God's punishment. To the extent that psychiatric hospitals market themselves as working on a material or physical plane, they fail to connect with many Ghanaians deepest fears and concerns regarding mental illness (Read & Doku, 2012).

2.2 The challenge and role of the prayer camps

In part because of these failings on the part of the formal mental health system, many Ghanaians currently rely on faith-based prayer camps to meet their needs for mental health care. Several hundred, if not thousand, prayer camps are estimated to exist throughout Ghana, although no formal census exists of yet. Most camps identify as Christian but there are some Muslim camps ("The Most Religious Country In The World is …," 2015). Both Christian and Muslim camps exhibit a mix of these faith traditions and local, African cosmologies, making their theologies difficult to precisely define.

These prayer camps are one part of a large and vibrant traditional healing landscape in Ghana. One recent publication suggests that 70% of Ghanaians consult traditional healers for mental health problems (Barimah, 2013), and this percentage is likely higher in rural areas that are poorly served by the formal psychiatric system. Among Ghanaians who seek non-traditional mental health care, many who consider themselves Pentecostal Christian seek "healing and deliverance" in Christian prayer camps and healing centers. These camps employ prayers, fasting and at times sprinkling of holy water or other relic-based practices as a means of treating disease. Generally, prayer camps charge on a weekly or monthly basis for residential services, including those for mental health purposes. Local people may come and go from the premises for prayer services and vigils free of charge but are prompted to make one or more contributions in the course of a ceremony (Arias, Taylor, Ofori-Atta, & Bradley, 2016). Describing the function of the camps, anthropologist Ursuala Read wrote:

> *One pastor in Kintampo town is well-known in the area for his power in healing those who are mentally ill, and hundreds if not thousands of pastors offer similar services throughout the country. A shrine in a small rural community in Kintampo South district is also famed for healing madness and is visited by people from as far afield as the Ashanti region and sometimes beyond. Treatment at prayer camps and shrines often involves a lengthy stay of several months; sometimes up to a year or even more. Relatives are usually expected to stay with the patient at the prayer camps and shrines to provide day-to-day care. Most frequently this is the mother, but sometimes the father, sister or another relative takes this role*
>
> ***Ursula M. Read, Adiibokah, and Nyame (2009)***

Most prayer camps consider themselves to be places of spiritual refuge for people with a broad range of concerns, but several do have special facilities within them (sanatoriums) that cater specifically to people with mental illness. In their communications with the Ghanaian public and local communities, the camps position themselves as places in which one can seek deliverance, or divine intervention, from whatever it is that troubles them (Arias et al., 2016). The prospect that God's power may cure a mental illness further differentiates the prayer camps from psychiatric hospitals, where the most that many health care providers will promise is a long-term plan for medication management (Read, 2012). The prayer camps' cultural influence is such that many high-ranking members of Ghanaian society, including professional soccer players, musicians and politicians, have been known to visit the camps for church services and soul-searching retreat. A former President stayed at one prayer camp for several days in preparation for a 2008 national election (Daily Guide, 2008).

In 2011, Human Rights Watch published a critical report of the activities undertaken by staff treating people with mental illness at a subset of Ghanaian prayer camps. Titled "Like a Death Sentence," the report chronicled a wide range of abuses in the prayer camps that included chaining, corporeal punishment, and starvation (Ssengooba et al., 2012). Although this was not the first public report of these abuses, the Human Rights Watch report mobilised a powerful set of global actors to condemn the prayer camps. In 2013, the UN Office of the Human Rights Commissioner Special Rapporteur traveled to Ghana to inspect the prayer camps and make further recommendations about how care in these environments could be improved so as to ensure human rights are protected (Mendez, 2014).

In considering the history of abuse in the prayer camps, it is worth noting that few of the abuses are defended on religious grounds. Only what human rights advocates consider starvation has a clear theological root in several Biblical passages that reference fasting. The distinction between starvation and fasting rests on the degree to which people forego food voluntarily and remains a matter of debate between mental health advocates and prayer camp staff (Taylor, 2016a). The degree to which this practice is undertaken voluntarily by people in prayer camp sanatoria is difficult to ascertain given the influence that prayer camp staff have over people staying in the sanatorium. Prayer camps rarely defend chaining or corporeal punishment as theologically-grounded therapies for people with mental illness. Rather, the camps suggest these are means of last resort to control violent and aggressive individuals in their care (Mendez, 2014).

Historically, there has been little to no formal interaction between GHS and the prayer camps. The Mental Health Act passed in 1972, which was operative until the 2012 revision, made no mention of the informal sector and the government was loath to discuss the activities of the camp publicly (Ofori-Atta et al., 2010). The camps can register with the Traditional and Alternative Medicines (TAMs) office in the Ministry of Health, but they are not required to do so. If they do, they receive a certificate which both the camps and the bureaucrats in the TAM see as inconsequential. Ghana's Federal Drug Administration (FDA) is tasked with monitoring advertising claims related to health but has rarely engaged with the prayer camps. This is in part because the prayer camps do not advertise themselves for curing cancer, mental illness or any other specific disease. Rather, they suggest that deliverance from *any* problem is possible and the general nature of these claims allows them to circumvent the prohibitions on false advertising claims. As a result, the prayer camps have continued to thrive as a parallel system of mental health care with scant government oversight in spite of negative press from the global health and human rights community.

In practice, the relationship between the prayer camps and the formal mental health sector is complex and varies from provider to provider and camp to camp. Many people with mental illness will alternately seek in-patient care at psychiatric hospitals and prayer camps, bouncing from one site to the other when they or their family runs out of funds, patience or time in one location (Edwards, 2014). On an out-patient basis, a person with mental illness may be simultaneously being seen by both a Prophet or Prophetess at a prayer camp and a psychiatrist or psychiatric nurse at a medical facility. Some prayer camps will actively refer people with mental illness to local hospitals or psychiatric facilities to be seen by doctors, while others see no value in such a referral and prefer to keep patients on-site at the camp. Vice versa, there are reports of some physicians referring patients to seek spiritual care or support although many psychiatrists are outspoken critics of the prayer camps.

2.3 Mental health legislation

The Mental Health Act which had been issued in 1972 never took full effect on account of poor implementation planning (Doku, Wusu-Takyi, & Awakame, 2012; Ofori-Atta et al., 2010). In an effort to remedy the failures of said Act, several attempts at comprehensive reform were made and failed throughout the 1990s. Only in 2003 did the process which would ultimately result in the Mental Health Act of 2012 begin. The WHO lent technical advisory support beginning in 2004 (Awenva et al., 2010). The process that took place between 2004 and 2012 involved significant consultation with a variety of international and local stakeholders, including civil society/non-governmental organisations such as Basic Needs, Ghana Federation for the Disabled and Mind Freedom.

In an effort to support the reform of the mental health system, policymakers prioritised changes to governance structures, planning protocols and funding schemes for the provision of mental health in Ghana. For years, mental health advocates had felt that GHS administrators de-prioritised mental health in favor of physical health, most notably in terms of budget allocations. Bearing this experience in mind, their goal was to create a Mental Health Authority (MHA) to oversee the implementation of all policies related to mental health. This authority would report up to the Director of the Ghana Health Service and ultimately the Minister of Health (Walker & Osei, 2017) but could otherwise operate largely independently from the rest of GHS.

In addition to the new governance structures outlined in the law, policymakers placed significant emphasis on reforming the conditions inside the existing mental health infrastructure by establishing psychiatric wings at general district hospitals, mechanisms for people to report abuse, new protocols for involuntary treatment, clear prohibitions on prolonged chaining or seclusion and enhanced protections for vulnerable groups. These standards stood to both improve the quality of mental health services and encourage providers to treat patients in the least restrictive environment required by their condition (Walker & Osei, 2017).

3. The ethical challenge

Policymakers involved in formulating the Mental Health Act also desired to transform the existing, highly institutional mental health system into one that was more community-based, human-rights focused and prevention-oriented. In so doing, they faced with an ethical dilemma. How should the government, which is the main provider of formal mental health services nationwide, engage with or not engage with the prayer camps? Should the prayer camps even be formally recognised in the law? Should they be actively regulated? Should they become partners in promoting a culturally-appropriate, holistic approach to mental health care?

Policymakers reasonably felt faced with competing commitments. On one hand, government recognised an opportunity to engage the prayer camps and leverage both their location and reputation to develop a more robust system of mental health outposts than the government would be able to build in the near term. With such a system in place, more Ghanaians could be diagnosed with and receive some form of treatment for mental illness. On the other hand, the government had also made several public commitments to protect its citizenry from human rights abuses and may therefore have wanted to remain clear of any association with the camps. In so doing, the government could be confident that they were not inadvertently raising the profile of the prayer camps and putting additional Ghanaians at risk for being mistreated.

3.1 The argument for engaging

Many in government saw substantive opportunities for policy "wins" by engaging with the prayer camps in order to reform their practices. The pro-engagement argument had at least two parts to it, both of which featured arguments about the ability to intervene in the care of people with mental illness before serious harm or intensive treatment needs to occur.

First, some argued that engagement with the prayer camps as the only way to reduce or eliminate the torturous practices that Human Rights Watch and others had identified. According to this view, government must create a regulatory or accrediting body that first created a comprehensive index of prayer camps and then performed systematic site visits to all camps in order to ensure the safety of patients staying there. In order to be effective, the regulatory or accrediting body would likely need both carrots and sticks, where carrots could be small grants of money to improve facilities or public recognition and sticks would be fines for violations of human rights, criminal arrests of staff and ultimately closure of the facility. Policymakers recognised that this was not an inexpensive oversight system to create but mused aloud about a lack of realistic alternatives. While some high-ranking mental health advocates argued that the camps should simply be shut down immediately, policymakers worried that this would only result in the Prophets and their staff going underground and beginning new camps in new places with new names (Edwards, 2014). Thus, an elaborate game of cat and mouse may begin – costing the government still more valuable money and time.

Second, proponents of engagement argued that the prayer camps could be transformed into a network of mental health outposts where talk therapy, psychotropic drugs and other interventions could be delivered (Ofori-Atta et al., 2018). This vision of the future was particularly attractive because prayer camps existed largely in areas that were otherwise void of formal psychiatric care. To deliver mental health care in these regions would likely require some degree of task-shifting and the prayer camps were known to be a popular resource for families confronting mental illness. Thus, one might reasonably have argued that engaging the prayer camps to deliver some form of mental health care would have been a strategic use of existing resources.

Relatedly, proponents of turning prayer camps into mental health outposts further argued that the resulting outposts could be a testament to the principles of community-based mental health care. The prayer camps that would serve as outposts are often near to small towns and offer mentally ill people ample opportunity to be engaged with shopkeepers, church communities and civic life. Moreover, some stakeholders argued that the prayer camps operated with a more holistic view of mental illness than did psychiatric hospitals. They offered pastoral care and counseling and took seriously spiritual dimensions of illness, which are important to many Ghanaians. Both of these attributes stood in stark contrast to the system of institutionalised care available to people with serious mental illness.

Traditional public health texts often distinguish between primary, secondary and tertiary prevention (see Table 1; Institute for Work and Health, 2015). Using this framework, it is possible to describe the goals of the pro-engagement contingent as preventative in nature, although there are two prevention arguments operating simultaneously. Those who were most concerned about reducing or eliminating torturous practices motivated

Table 1 Levels and timing of primary, secondary and tertiary prevention.

Level of prevention	Timing of activity	Goal
Primary prevention	Prior to onset of disease or injury	Prevent the onset of illness or injury before the disease process begins
Secondary prevention	After the onset of disease or injury	Early diagnosis and prompt treatment of a disease, illness or injury to prevent more severe problems developing
Tertiary prevention	After the onset of disease or injury	Rehabilitation following significant illness

their argument for engagement based on the need for primary prevention of *abuse* – meaning they believed that engaging the prayer camps could enable preventive measures that would avoid the onset of abuse and the ensuing injury. Those who were more concerned about transforming the prayer camps into mental health outposts argued that the camps could become sites of secondary and tertiary prevention for people living with mental illness – meaning they believed that engaging the prayer camps could enable early detection and prompt treatment of existing disease in lowest-intensity setting possible.

Angela Ofori-Atta, a clinical psychologist and professor at the University of Ghana, summarised both aspects of the argument for engagement in a situational analysis of Ghanaian mental health published in 2010:

> *In order to reduce the abuse of people with mental illness within informal treatment facilities, it is recommended that a mental health authority and tribunal is established as stipulated in the new Mental Health Bill to regulate the practices of traditional and faith healers. Closer collaboration between faith and traditional healers and orthodox psychiatric care could help to protect the human rights of those with mental illness and ensure that those who would benefit from psychiatric treatment are referred to appropriate services*
>
> ***Ofori-Atta et al. (2010)***

3.2 The argument against engaging

The prayer camps were also have been recognised as national liabilities on account of their sizeable track record of human rights abuses. Many in the Ghanaian government may have not wanted to acknowledge or engage with the prayer camps out of a concern for political ill-will or backlash. (Silence, as previously mentioned, had historically been the chosen strategy.)

Particularly between 2010 and 2012, when this bill was being debated, Ghana was one of the fastest growing economies on earth (2011) and was on the brink of moving from a low-income to middle-income country according to the World Bank (World Bank, n.d.). To acknowledge the prayer camps and publicly grapple with their practices over the course of the policy development risked drawing unwanted national attention to practices that may appear primitive and regressive to a Western audience.

Like the argument for engagement, the argument against engaging the prayer camps had several facets. First, policymakers worried that in drafting a Mental Health Bill that appeared to make the camps a sanctioned sector would be a violation of the government's formal, legal commitment to international human rights treaties and conventions. By virtue of being a UN member since 1957, Ghana had pledged to uphold the UN's Universal Declaration of Human Rights which was first adopted by 52 member states in 1948. The Declaration consists of 30 articles affirming an individual's rights which, although not legally binding in themselves, have been elaborated in subsequent international treaties and have provided political leverage for advocates to shame governments that fail to protect citizens from violations. More recently, Ghana signed and ratified the Convention against Torture and Other Cruel, Inhuman or Degrading Treatment or Punishment in 2000 and signed the Optional Protocol to the Convention against Torture and Other Cruel, Inhuman or Degrading Treatment or Punishment in 2006 (UNOHCR, n.d.). In these documents, torture is defined as "any act by which severe pain or suffering, whether mental or physical, is intentionally inflicted upon a person" for reasons other than those lawfully sanctioned (OHCHR, n.d.). In 2007, Ghana also became a signatory to the UN Convention on the Rights of People with Disabilities, which was later ratified in 2012. Said Convention asserts that "State Parties shall ensure that all facilities and programmes designed to serve persons with disabilities are effectively monitored by independent authorities" and "State Parties shall take all appropriate measures to promote the physical, cognitive and psychological recovery, rehabilitation and reintegration of persons with disabilities who become victims of any form of exploitation, violence or abuse."

Second, policymakers who wished to avoid engaging worried loudly about lending undue legitimacy to the prayer camps. In their view, to recognise prayer camps as mental health providers under the law would be tantamount to putting them on the same legal and regulatory footing as psychiatric hospitals and sending a signal to the community that government officials took the camps seriously as sites of healthcare intervention. Desperate people living with mental illness and caring for family members with mental illness could not afford to be misled by their government into trusting these camps. As soon as the government recognised these entities, they would have done them an enormous favor. In the words of one member of the Mental Health Authority, "There is a fine line between collaborating and being used (Taylor, 2016b)."

Third, some policymakers conceded that transforming the prayer camps into mental health outposts may be an exciting opportunity, providing the prayer camps with government financing to improve their facilities would signal to the formal psychiatric sector that the efforts they had made to avoid abusive practices in spite of limited resources had been for naught. Health spending in Ghana throughout the 2000s remained between 5% and 6% of Gross Domestic Product, with mental health budgets receiving less than 3% of that figure under ordinary circumstances (Roberts et al., 2013; World Bank, n.d.). The WHO's Mental Health Atlas reported that 1.10% of the national health budget went to mental health in 2017. This under-funding created severe constraints on mental health care availability and manifested in the overcrowding in psychiatric hospitals, long lines to see clinicians in outpatient settings, meager public formularies of psychiatric medications and limited and virtually no capital investment in the system writ large (Ofori-Atta et al., 2010; Roberts et al., 2013). Thus, particularly among mental health policymakers and advocates who worked in the severely hampered public psychiatric system, the idea that government would offer resources to prayer camps to improve their facilities was offensive.

4. Bargaining with the devil

One way to conceptualise the questions inherent in this case is to consider the question in the language that many of the Ghanaian mental health policymakers did at the time. The prayer camps had developed such a reputation for abuse that many policymakers viewed the dilemma surrounding the mental health bill as one deeply wrapped up in the question of "Should we bargain with the devil?" Such a question harkens back to the medieval German legend of Faust in which a human agrees to sacrifice his soul to the Devil in exchange for worldly pleasures and unlimited knowledge.

Negotiations scholar Robert Mnookin has written substantively on this question in a book aptly titled *Bargaining with the Devil: When to Negotiation and When to Fight* (Mnookin, 2010). Recognising that the language of the "devil" can be inflammatory, he suggests that devil can be a placeholder in his analysis for an enemy who has "intentionally harmed you in the past or appears will harm you in the future." Usually, the devil also bears the characteristics of being untrustworthy and unwilling to behavior rationally. Mnookin uses the term bargain to refer to any effort to resolve conflict by mutually agreeing on a negotiated outcome.

For our purposes, it seems safe to assume that the prayer camps can reasonably be cast in the role of the devil in the eyes of at least some policymakers. In the Ghanaian case, the precise boundaries around what is a bargained solution versus a non-bargained solution are more nebulous, but I find the concept of bargaining to be roughly analogous to what I have been calling engaging. Working with the prayer camps to develop them into mental health outposts would surely be a form of bargaining. Even creating a regulatory scheme in which prayer camps have an opportunity to weigh in on the expectations and timelines would be a form of bargaining. If the government were to not acknowledge the prayer camps or attempt to forcibly shut them down, these outcomes would not be forms of bargaining.

Mnookin recognises the deep, ethical dilemmas inherent in the question of whether to bargain with the devil in both personal and professional life. Unlike other negotiations scholars who wish to believe that negotiation is *always* possible, Mnookin is sympathetic to the view that not all issues are negotiable. As a result, his summative answer to the question of whether one should bargain with the devil is "Not always, but more often than you feel like it."

Mnookin explicitly recognises the desire among potential bargainers to avoid legitimising their enemy, which was a core argument among the Ghanaian policymakers who wanted to avoid recognising the prayer camps in the legislation. He describes this concern for legitimation as a primal emotion that, while unavoidable, should be balanced wherever possible with a clearheaded assessment of the costs and benefits that negotiation (or, bargain) could create. Eschewing a full commitment to either teleological or deontological ethics, he suggests that would-be decision makers consider both the pragmatics and virtues of a situation. However, as his terse answer to the question of whether to bargain with the devil reveals, Mnookin preferences a forward-looking, instrumental and consequentialist view of negotiation's value. Having acknowledged an empathy for the desire to punish those who have wronged us, he writes:

> *When we are caught between the demands of principle and pragmatism, what we really need to ask ourselves is "To what extent should we focus on the future?" There's often an inescapable tension between achieving justice for past wrongs and the need for resolution … If you want to resolve the conflict and move forward, you may have to give the devil something you feel he doesn't deserve. This is a bitter pill to swallow (p262).*

What is unsatisfactory about Mnookin's logic in relation to the Ghanaian case is that policymakers who were concerned about legitimising the prayer camps were not only interested in righting past wrongs. They were also concerned about the potential for an enhanced status for the prayer camps would bring more people living with mental illness to the camp's doorsteps and thereby put more people at risk for being chained or starved. Hence, while the focus on the future can eliminate some extraneous concerns from the analysis, it does not take the legitimation concern entirely off the table.

In his framework for decision making, Mnookin also provides guidance for those making decisions on behalf of others, such as policymakers. He writes "When deciding on behalf of others, don't allow your own moral intuitions to override a pragmatic assessment (p265)." Here, he is distinguishing between an analysis of whether to bargain with the devil *oneself* versus an analysis of whether to bargain with the devil in cases in which case one would be bargaining *on behalf of* a third party. This distinction is important to Mnookin because even as he recognises people's moral intuitions and commitments to principle as relevant factors, he does not see those intuitions and commitments as necessarily representative of the third party who could stand to gain from a negotiated agreement. Applying this logic to the case of the Ghanaian policymakers, Mnookin's insight is to distinguish between the policymaker's intuitions and commitments and those of the Ghanaian people who they have been chosen to represent. Therefore, the policymakers' own intuitions and commitments should not be relevant in the analysis of whether or not to bargain.

This piece of Mnookin's advice appears to fly in the face of policy norms that suggest that experts are chosen to draft policies because they have relevant experiences and education that inform their decision-making such that it is superior the judgement of the average person. However, on closer inspection, a distinction can be made between Mnookin's urging to not allow individual's moral intuitions to stand in for third parties and conventional wisdom's desire to have individual's knowledge inform a policy on behalf of third parties. In the practice of policymaking, the line between the application of moral intuition and knowledge may be a difficult one to draw, but the concepts are distinguishable.

Returning the case of the Ghanaian prayer camps, it is reasonably clear what advice Mnookin may have offered a room full of policymakers as they considered the mental health legislation. He would have urged them to begin with a bias towards bargaining, or working with, the prayer camps in order to reach a negotiated agreement about their role in the mental health system. This bias would not need to be absolute, and Mnookin would have urged the group to remain open to rebuttals based on strong moral intuitions. The mainstay of the analysis he would have prescribed is a cost-benefit analysis of negotiating, which takes account not only what can be achieved through the bargaining but the costs of bargaining (and not bargaining itself) to each party. Given the details of the case, it may be that said cost benefit analysis features substantial costs to negotiating in the form of reputational benefits for the prayer camps. However, the cost to not bargaining may also be perceived as high among policymakers who, personally if not publicly, recognise that their unwillingness to bargain may be contributing to the ongoing abuses people are suffering in the camps.

5. Application to global mental health

Although the specifics of this case may be foreign to readers with limited exposure to mental health provision in low- and middle-income countries, I would argue that the underlying structure of this case is representative of global mental health challenges in a wide range of settings. That structure of the dilemma can be summarised as follows: First, policymakers recognised that sustainable change to the mental health sector required a confrontation of pervasive religious and cultural beliefs which, in the Ghanaian case, were fueling the popularity of the prayer camps and putting citizens in danger of abuse. Second, changing the culture around mental health in order to build out a more community-based delivery system required some form of engagement with the very actors whose behavior was creating harm to people living with mental illness — namely, prayer camps. This requirement is particularly common in cases where the resources are not available for a top-down prohibition or such a prohibition is thought to be ineffective, as it was in the Ghanaian case. And third, the engagement of said actors by the government posed substantial ethical risks. Once again, while the specifics of this case may be particular to the context in question, the structure of this ethical challenge is no doubt commonplace and therefore deserving of considered attention.

5.1 Sustainable change requires confronting religion and culture

Perceptions of mental illness and commitments to mental health care are built upon and deeply enmeshed in religious and cultural beliefs that can vary widely by geography and population. In Ghana, perceptions of mental illness have been heavily influenced by both traditional African cosmologies and Christian Pentecostal theologies emphasising the omnipotence of God, the dangers of evil spirits and witches, the potential for physical healing via deliverance and inerrancy of the Holy Bible. This has resulted in a large portion of the population conceptualising mental illness as spiritual in nature and therefore viewing the prayer camps appropriate providers of care.

In a particularly evocative study on the context-dependence of mental illness, Tanya Luhrmann et al. (one of whom is Ghanaian) analyze the quality of the voices that people with schizophrenia hear in the United States, India and Ghana. The research team found distinct variation in the emotional tone of voices, or auditory hallucinations, heard:

> *In a California sample, people were more likely to describe their voices as intrusive unreal thoughts; in the South Indian sample, they were more likely to describe them as providing useful guidance; and in our West African sample, they were more likely to describe them as morally good and causally powerful*
>
> ***Luhrmann, Padmavati, Tharoor, and Osei (2014), p. 648.***

While the sociological and anthropological literature on mental health and illness has long recognised these categories as culturally-constructed, Lurhmann et al. put forth a more precise hypothesis of how local context shapes symptoms.

> *What we think we may be observing is that people who fall ill with serious psychotic disorder pay selective attention to a constant stream of many different auditory and quasi-auditory events because of different 'cultural invitations'*
>
> ***Luhrmann et al. (2014), p. 648.***

The authors define invitations as variations in ways of thinking about persons, minds, spirits and so forth. To participate in some Ghanaian church services, for instance, is to be invited to think about spirits as omnipresent — which the authors suggest may impact how the population experiences mental illness.

In addition to influencing how the symptoms of mental illness manifest, culture and religion can also play a role in determining the kinds of care that people living with mental illness seek. In Ghana, spiritual afflictions are commonly perceived as contagious - either over the course of generations (similar to genetic mutations) or within a household (more like a germ) (Arias et al., 2016). This increases the stigma associated with disclosing mental illness among loved ones, as doing so may hinder the entire family's social standing and economic opportunities. Such stigma is thought to drive down the rates at which people seek care, particularly at the highly-visible psychiatric hospital in the center of Accra (Barke et al., 2011).

The question of whether these spiritual and cultural beliefs, which manifest in at least some cases as a preference for prayer camps over biomedical treatment for mental illness, would endure in the face of a robustly funded, formal psychiatric sector remains contested. Some global health experts assume beliefs in faith healing to be little more than a coping mechanism in the face of profound poverty. Those who hold this view are generally unworried about the impact that shutting down the camps would have on people's ability to exercise their religion and primarily focused on improving the resource availability in the formal psychiatric sector. Others argue that religious beliefs are enduring and cannot be explained by socioeconomics alone. In this case, preserving the prayer camps indefinitely as a mode of care available to people is important. Whatever the reader's view may be, the profound impact of spiritual and cultural beliefs remains a fact that those interested in improving global mental health must reckon with for the time being.

5.2 Culture change requires engagement of leaders

The second feature of the case that warrants comment flows from the first, and is that policymakers recognised that sustainable change in mental health required changing the attitudes and behavior of the population. But how can a government influence, never mind create, sustainable change in attitudes and behaviors? One approach is to first change the attitudes and behaviors of a relatively smaller group of influential people — be that a set of celebrities in Hollywood who then go on to tweet and post on Instagram about their newfound beliefs or a set of religious leaders who enjoy substantial cultural credibility.

There is an irony embedded in this realisation — which is that it is quite reasonable to expect that the people with the greatest cultural authority are also the ones more strongly steeped in the status quo attitudes and beliefs that an intervention seeks to change. Surely this was the case in Ghana, where leaders of prayer camps had substantial cultural authority to shape public perceptions of mental illness —they had been doing this work for decades - and may have also been among the most resistant to an effort attempting to change their own behavior. In other words, the highest leverage partners may well also be the most dangerous. The ethical question, which Mnookin's framework speaks to, then becomes whether a group's past participation in perpetuating unhealthy attitudes and behaviors makes them illegitimate allies or the most critical allies in promoting wider change.

5.3 Engagement of leaders poses ethical risks

The third challenge is that it's not at all obvious how to create sustainable attitudinal and behavior change without engaging those whose behavior you wish to change. Ignoring problems — in the sense of not giving them attention in either public or policy debates — has rarely proven effective. Some acknowledgement is necessary, at least for the purpose of problem identification, and with acknowledgement comes the most minimal form of relationship. If we accept that even this level of relationship to problematic attitudes and behaviors is too much, then our ability to bring about change is all but gone.

As mental health continues to garner more attention in global health circles, it can be assumed that the central problem raised by the case presented herein will recur in other settings. The Ghanaian government officials, and others who follow in their path, should be praised for their efforts to take a root-cause approach to solving problems like those presented by the prayer camps and for thinking practically about how to create more community-based, prevention-oriented mental health systems with limited resources. Their experience also underscores the need for policymakers to be clear with themselves and one another about how pragmatic versus principled they feel the need to be in their roles. Mnookin celebrates both historical figures that prioritised principle and pragmatism — including Churchill's refusal to negotiate with Hitler leading up to and during the Second World War and Mandela's willingness to negotiate with his captors — but he emphasises that the pragmatic instinct to negotiate should be the default. Particularly when decision makers are government officials, whose roles are to serve citizens, overriding the potential benefits of a mutually negotiated agreements with personal principles should be rare. Finally, the principled argument that government recognition of illicit actors such as the prayer camp will unduly legitimise them should be closely scrutinised and the question asked: are such actors not already legitimate in the eyes of the citizenry?

6. Epilogue

In 2012, after an 8-year lobbying effort led by Ghanaian mental health advocates and technical advisors from the World Health Organisation, the Ghanaian legislature passed what is now known as Mental Health Act 846 (Walker & Osei, 2017). Therein, the government committed itself to becoming a regulator of the prayer camps with regard only to potential human rights abuses, including involuntary restraint. No general oversight, regulation or accreditation of the camp's non-torturous activities was established. Ostensibly this choice amounted to something of a pragmatic middle ground among policymakers with differing views as to whether regulating was legitimising — allowing the government to limit the human rights abuses without investing itself deeply in the future of the prayer camps. It is unclear whether this will ultimately be the first step of a larger strategy of partnering and reforming the prayer camps in a more systematic way, the first step of a larger strategy to shut non-compliant prayer camps down, or neither.

Three aspects of the Mental Health Law are especially relevant in considering the relationship between government and the camps (Walker & Osei, 2017).

(1) *Establishment of clear human rights standards*: The Act has been lauded by international policymakers and stakeholders for taking a "human rights approach" to mental health care provision. It sets clear guidelines for what is meant by voluntary and involuntary treatment (Section 39—42), appeal of an involuntary admission (Section 44), the need for a documented treatment plan (Section 45), assertion of basic human rights for people with mental illness (Section 55), guidelines about the use of seclusion and restraint (Section 58), statements of rights to confidentiality and privacy (Sections 60—61) and access to one's own records (Section 62). These human rights standards are lofty goals for virtually all psychiatric care facilities in Ghana and particularly so for the prayer camps.

(2) *Creation of Mental Health Tribunal*: Within the law, Section 24–32 reads: "There is established by the Board a Mental Health Review Tribunal ... The Board shall appoint members of the Tribunal ....

  a. The Tribunal shall hear and investigate complaints in respect of persons detained under this Act.
  b. The Tribunal shall review and monitor cases of involuntary admissions and treatment processes and long-term stay voluntary admissions.
  c. The Tribunal may approve requests for intrusive or irreversible treatments ...
  d. The tribunal shall ensure that informed consent is obtained for intrusive or irreversible procedures ...
  e. The Tribunal may direct the discharge of a person detained under this Act despite a previous order of a court ..."

(3) *Creation of Visiting Committees*: Within the law, Section 34 reads: "The Board shall establish visiting committees for mental health service in each region and ensure annual allocation of recurrent funding for travel and subsistence ...

  f. A visiting committee shall collaborate with the relevant agency to (a) ensure that the rights of persons with mental disorder within the community are protected; (b) to conduct inspections within the region of its responsibility at least one a year with or without prior notice, (c) to receive and enquire into a complaint preferred by or against an officer, staff or patient of a mental health facility, and (d) to interview a patient without the presence of an officer of the facility in a public or private mental health facility or an unorthodox centre like a spiritual mental health facility ...
  g. A visiting committee shall report and make recommendations after each meeting to the Board with copies of the report to the relevant Regional and District Health Director, to the head of the mental health facility and collaborating relevant agency"

While the Act created a Mental Health Fund to support the expansion of formal care services, it did not provide for or guarantee any funding. Funding would need to be appropriated through a funding bill, known locally as a Legislative Instrument (Doku et al., 2012).

As of February 2019, the Legislative Instrument associated with the Mental Health Act still had not been approved, leaving the Mental Health Act with minimal regulatory and enforcement clout(CItiNewsRoom, 2018; Laary, 2014). In March 2018, six years after passage of the Act, Dr Akawasi

Osei, Chief Executive of the Mental Health Authority and Head Psychiatrist at Accra Psychiatric Hospital published an opinion piece stating:

> *We promised Ghanaians that in five years, there would be a change in our treatment of the mentally-ill; there would be no patients on the streets of Accra … prayer camps would have stopped chaining patients, psychiatric hospitals would always have drugs to attend to patients …. These promises were made on the assumption that following the passage of the bill, the enabling [*funding *mechanism] would shortly be passed and* funds *would be available but that is not happening.*
>
> ***Osei (n.d.).***

Thus, despite the enactment of a comprehensive law, the reality on the ground in Ghana even in 2019 is that formal mental health care services remain unable to meet the needs of the populace and the prayer camps persist in their practices effectively unregulated.

References

Ae-Ngibise, K., Cooper, S., Adiibokah, E., Akpalu, B., Lund, C., Doku, V., & MHaPP Research Programme Consortium. (2010). 'Whether you like it or not people with mental problems are going to go to them': A qualitative exploration into the widespread use of traditional and faith healers in the provision of mental health care in Ghana. *International Review of Psychiatry, 22*(6), 558–567. https://doi.org/10.3109/09540261.2010.536149.

Arias, D., Taylor, L., Ofori-Atta, A., & Bradley, E. H. (2016). Prayer camps and biomedical care in Ghana: Is collaboration in mental health care possible? *PLoS One, 11*(9), e0162305. https://doi.org/10.1371/journal.pone.0162305.

Awenva, A. D., Read, U. M., Ofori-Attah, A. L., Doku, V. C. K., Akpalu, B., Osei, A. O., … MHaPP Research Programme Consortium. (2010). From mental health policy development in Ghana to implementation: What are the barriers? *African Journal of Psychiatry, 13*(3), 184–191.

Barimah, K. B. (2013). Traditional healers as service providers in Ghana's national health insurance scheme: The wrong way forward? *Global Public Health, 8*(2), 202–208. https://doi.org/10.1080/17441692.2012.761262.

Barke, A., Nyarko, S., & Klecha, D. (2011). The stigma of mental illness in southern Ghana: Attitudes of the urban population and patients' views. *Social Psychiatry and Psychiatric Epidemiology, 46*(11), 1191–1202. https://doi.org/10.1007/s00127-010-0290-3.

CItiNewsRoom. (October 10, 2018). *CDD-Ghana to parliament: Pass LI on mental health act.* Retrieved from: https://www.modernghana.com/news/889234/cdd-ghana-to-parliament-pass-li-on-mental-health-act.html.

Daily Guide. (December 12, 2008). *Mills changes prayer camp.* Retrieved from: https://www.modernghana.com/news/194454/mills-changes-prayer-camp.html.

Doku, V. C. K., Wusu-Takyi, A., & Awakame, J. (2012). Implementing the mental health act in Ghana: Any challenges ahead? *Ghana Medical Journal, 46*(4), 241–250.

Edwards, J. (2014). Ghana: Mental health patients confined to prayer camps. *The Lancet, 383*(9911), 15–16. https://doi.org/10.1016/s0140-6736(13)62717-8.

Ghana Mental Health Situation Analysis. (n.d.). Retrieved from: https://www.who.int/mental_health/policy/country/ghana/en/.

GHLI Assists with mental health care reform in Ghana.(August 10, 2011). Retrieved from: https://news.yale.edu/2011/08/10/ghli-assists-mental-health-care-reform-ghana.

Institute for Work and Health. (Spring 2015). *Primary, secondary and tertiary prevention.* Institute for Work & Health. Retrieved from: https://www.iwh.on.ca/what-researchers-mean-by/primary-secondary-and-tertiary-prevention.

Kessler, R. C., Aguilar-Gaxiola, S., Alonso, J., Chatterji, S., Lee, S., Ormel, J., ... Wang, P. S. (2009). The global burden of mental disorders: An update from the WHO World Mental Health (WMH) Surveys. *Epidemiologia e Psichiatria Sociale, 18*(1), 23–33.

Laary, D. (November 12, 2014). *No legal backing for Ghana's mental health law | West Africa.* Retrieved from: http://www.theafricareport.com/West-Africa/no-legal-backing-for-ghanas-mental-health-law.html.

Luhrmann, T. M., Padmavati, R., Tharoor, H., & Osei, A. (2014). Differences in voice-hearing experiences of people with psychosis in the USA, India and Ghana: Interview-based study. *The British Journal of Psychiatry: The Journal of Mental Science.* https://doi.org/10.1192/bjp.bp.113.139048.

Mendez, J. (2014). *Report of the Special Rapporteur on torture and other cruel, inhuman or degrading treatment or punishment (No. GE.14-11818).* UN General Assembly, Human Rights Council. Retrieved from: https://www.ohchr.org/en/hrbodies/hrc/regularsessions/session25/documents/a-hrc-25-60-add-1_en.doc.

Mnookin, R. H. (2010). *Bargaining with the devil: When to negotiate, when to fight* (1st Simon & Schuster hardcover ed.). New York: Simon & Schuster.

Ofori-Atta, A., Read, U. M., & Lund, C. (2010). A situation analysis of mental health services and legislation in Ghana: Challenges for transformation. *African Journal of Psychiatry, 13*(2). https://doi.org/10.4314/ajpsy.v13i2.54353.

Ofori-Atta, A., Attafuah, J., Jack, H., Baning, F., Rosenheck, R., & Joining Forces Research Consortium. (2018). Joining psychiatric care and faith healing in a prayer camp in Ghana: Randomised trial. *The British Journal of Psychiatry, 212*(1), 34–41. https://doi.org/10.1192/bjp.2017.12.

OHCHR. (n.d.). *Convention against torture.* Retrieved from: https://www.ohchr.org/en/professionalinterest/pages/cat.aspx.

Osei, D. A. (n.d.). *The Mental Health Act 846, six years on.* Retrieved from: https://www.graphic.com.gh/news/health/the-mental-health-act-846-six-years-on.html.

Read, U. M., & Doku, V. C. (2012). Mental health research in Ghana: A literature review. *Ghana Medical Journal, 46*(2 Suppl. l), 29–38.

Read, U. M., Adiibokah, E., & Nyame, S. (2009). Local suffering and the global discourse of mental health and human rights: An ethnographic study of responses to mental illness in rural Ghana. *Globalization and Health, 5*(1), 13. https://doi.org/10.1186/1744-8603-5-13.

Read, U. (2012). "I want the one that will heal me completely so it won't come back again": The limits of antipsychotic medication in rural Ghana. *Transcultural Psychiatry, 49*(3–4), 438–460. https://doi.org/10.1177/1363461512447070.

Roberts, M., Asare, J., Morgan, C., Adjase, E., & Osei, A. (2013). *The mental health system in Ghana. Ghana Ministry of health and the Kintampo project.* Retrieved from: http://www.mhinnovation.net/sites/default/files/downloads/innovation/research/The-Mental-Health-System-in-Ghana-Report.pdf.

Ssengooba, M., Barriga, S., Dufka, C., Peligal, R., Amon, J., Mcrae, A., ... Olugboji, B. (2012). *"Like a Death sentence": Abuses against persons with mental disabilities in Ghana.*

Taylor, L., & Sayeed, S. (2019). *Research and responsibility in global health: An analysis of the joining forces study in Ghana.*

Taylor, L. (2016a). RECONSIDERING SAMUEL a mental health caretaker at a Ghanaian prayer camp. *Perspectives in Biology and Medicine, 59*(2), 263–275. https://doi.org/10.1353/pbm.2017.0011.

Taylor, L. (December 2016b). *Anonymous psychiatrist not involved with JFR in Accra, Ghana.*
The most religious country in the World is....(May 17, 2015). Retrieved from: https://www.africanheritagecollection.com/blogs/main/18153717-the-most-religious-country-in-the-world-is.
UNOHCR. (n.d.). *Ghana ratification status.* Retrieved from: https://tbinternet.ohchr.org/_layouts/TreatyBodyExternal/Treaty.aspx?CountryID=67&Lang=EN.
Walker, G. H., & Osei, A. (2017). Mental health law in Ghana. *BJPsych International, 14*(2), 38–39.
WHO. (2018). *WHO mental health atlas - Ghana country profile.* Retrieved from: http://www.who.int/mental_health/evidence/en/.
World Bank. (n.d.). *Ghana | Data.* Retrieved from: https://data.worldbank.org/country/ghana.

CHAPTER THIRTEEN

Public mental health without the health? Challenges and contributions from the Mad Pride and neurodiversity paradigms

Ginger A. Hoffman*
Department of Philosophy, Saint Joseph's University, Philadelphia, PA, United States
*Corresponding author: Emails: drgingerhoffman@gmail.com, ginger.hoffman@sju.edu

Contents

Developments in Neuroethics and Bioethics, Volume 2
ISSN 2589-2959
https://doi.org/10.1016/bs.dnb.2019.07.003

1. Introduction

Must the field of public mental health always act with the aim of advancing public mental health? In this chapter, I will defend the (admittedly) grossly counterintuitive answer: *no*. More specifically, the central question I pose in this chapter is: 'What is the proper ultimate aim of the field of public mental health?' And the response I offer is *not* the obvious one; no matter how much it may seem like *increasing public mental health* is the only remotely plausible answer, it is not. Moreover, *increasing public mental health* is an answer that brings with it considerable difficulties, including the fact that it (surprisingly) clashes with some of the field's most deeply held values.

Clearly, I am by no means the first to consider the question of public mental health's[1] proper ultimate aim.[2] What makes my approach distinct, however, is that it, by design, incorporates the perspectives and expertise of a vital subset of public mental health users[3]—users whose perspectives and expertise have thus far apparently been overlooked.[4] Namely, this is the subset of users who support the Mad Pride and neurodiversity paradigms.[5] Although I will describe these paradigms in more detail in Section 2, very briefly, these paradigms are characterized by the view that what many label as mental illnesses (e.g. autism, schizophrenia, bipolar disorder, etc.) are, rather, non-pathological forms of mental diversity that are worthy of acceptance. And, depending on the context, these forms of

[1] Examples of previous works exploring this question include: [Brown, Learmonth, & Mackereth, 2015; Jenkins, McCulloch, Friedli, & Parker, 2002, (especially p. 27); Levin, Hanson, & Hennessy, 2004, (especially p. 5)]. In addition, (Brülde, 2011) and (Schneider, 2014) are two examples of works which contain helpful discussions of the ultimate aim of the field of public health more broadly (as opposed to its subfield of public *mental* health specifically). Also, note that, throughout this chapter, I will sometimes use the phrase *public mental health* to stand for *the field of public mental health* for the sake of brevity.

[2] Examples of previous works that explore the question of public mental health's ultimate aim include the following: Brown, Learmonth & Mackereth, 2015; Jenkins, McCulloch, Friedli, & Parker, 2002, (especially) p. 27; Levin, Hanson & Hennessy, 2004, (especially) p. 5. In addition, Brülde, 2011 and Schneider, 2014 provide informative discussions of the ultimate aim of public health broadly (as opposed to the subfield of public *mental* health specifically). Finally, the following are some helpful introductions and analyses of public mental health more generally, many of which at least implicitly address the question of the field's ultimate aim: Mendelson et al., 2012; Diego, Forlini & Meurk, 2018; Friedli, 2009; Faden & Shebaya, 2016; Leaf et al., 2012; Radden, 2017, 2016; Saxena, Jané-Llopis & Hosman, 2006.

diversity can be worthy of more than just acceptance—they can be worthy of appreciation and esteem. As such, they have a stronger affinity with other forms of diversity like race and gender than with illnesses like gout and diabetes.[6] Notably, the Mad Pride and neurodiversity paradigms fully accept that such forms of mental diversity can, and often do, come with

[3] Who counts as a public mental health user? Technically, everyone is or could be a public mental health user, since the field of public mental health targets entire populations for many of its initiatives (e.g. prevention efforts). However, depending upon the context, the term *public mental health user*, along with its abbreviation *user*, also refer specifically to those persons who are considered by psychiatry to have (in psychiatry's words) one or more psychiatric conditions or mental illnesses (note that there is precedent for use of the abbreviation *user* as a neutral, respectful designation; for example, see Tait & Lester, 2005). To be specific: this is the class of persons who meet criteria for one or more diagnostic categories in the Diagnostic and Statistical Manual of Mental Disorders, Fifth Edition (hereafter, DSM-5) (American Psychiatric Association, 2013) or Chapter V of the International Statistical Classification of Diseases and Related Health Problems, 10th revision (hereafter, ICD-10) (World Health Organization, 2004). Such categories include recognizable entities like bipolar disorder, autism, major depressive disorder, obsessive compulsive disorder, and schizophrenia, to name just a few. Currently, there is significant disagreement about what counts as appropriate terminology for both the diagnostic categories, and the persons who fit them, largely due to the significant disagreement over whether or not such diagnostic categories actually represent illnesses (or other medical conditions). Thus, the terms *mental differences* and *mentally different persons/persons with mental differences* are advantageous insofar as they are neutral on that issue (since *difference* can signify either an illness or a non-pathological type of divergence). Unfortunately, however, these terms have yet to gain widespread recognition; the more commonly recognized terms are mental illness and mental disorder [and the corresponding *person with a mental illness* (*disorder*)]. Thus, in this chapter, I aspire to use the term preferred by the individuals it denotes in the particular context in which it is deployed. However, in contexts where such preferences are unclear, or where the preferred term might cause significant confusion for the reader, I will generally either use *mental difference*, *mental difference/illness*, or *Madness*. I will follow this same approach for referring to the individuals who are designated. Also, in deciding between terms like *mentally different/ill person* (e.g. identity-first language) and *person with a mental difference/illness* (i.e. person-first language), I will do my best to choose the option preferred by the individual(s) in question.

[4] In saying *I actively seek to incorporate perspectives from a vital subset of public mental health users*, the phrase *incorporat[ing] perspectives* denotes my inclusion of quotations and views from accounts authored by mentally different individuals within the Mad Pride and neurodiversity movements, as well as from perspectives I've encountered through my own involvement in the Mad Pride movement. Importantly, however, I in no way wish to imply that *this type of* relatively minimal incorporation is sufficient to count as the type of bona fide user inclusion and involvement referenced throughout this chapter. Indeed, bona fide user inclusion is significantly more robust, involving active, frequent, and continued participation of actual individuals in the ongoing work of principle, policy, and practice development that I will maintain is necessary to the field of public mental health. In other words, I am well aware that my inclusion here can serve at most as a supplement to, and not in any way as a *substitution* for, a significantly more robust user involvement.

tremendous suffering. However, these paradigms also emphasize that the existence of this suffering is fully compatible with their main tenets—for example, their tenet that mental differences need not be bad *overall*.[7]

[5] Although my approach is indeed distinct, I believe it is actually driven by some of the same values driving the field of public mental health in similar investigations. For example, the field of public mental health also explicitly values user input (see Section 2), and insists on the importance of factors well beyond biological functioning (e.g. subjective, social, political, etc.) in individuals' lives (see my discussion of the *conflation strategy* in Section 4.6). However, thus far, input from the subset of users belonging to the Mad Pride and neurodiversity movements, to my knowledge, has not been sought [The one article (of which I'm aware) that comes closest to doing so is Friedli (2000), which briefly mentions the importance of Mad Pride perspectives for "transform[ing] the terms of the debate about mental health" (p. 15). For the rest of the article, however, Friedli transitions into a discussion of the benefits of mental health promotion. Although it is clear that Friedli intends for the term *mental health* to mean much more than "the absence of mental illness," it is also *unclear* exactly what its meaning does include, and thus to what extent it is compatible with a Mad Pride perspective].

[6] However, not all subspecies of these paradigms maintain that mental differences must be non-pathological. See Section 3 for an elaboration of this possibility.

[7] A subset of classic works within the Mad Pride and neurodiversity movements (virtually all of which are authored by self-described Mad Pride and/or neurodiversity advocates), along with a few additional descriptions and analyses of the Mad Pride and neurodiversity paradigms include: Adbusters, 2002; Armstrong, 2010; Autistic Self Advocacy Network, 2019; Bascom, 2012; Beazar, 2017; Costa, 2015; Curtis et al., 2000; DuBrul, 2014; Cutler, 2019a, 2019b; Farber, 2012; Glaser, 2008; Hoffman, 2017; MindFreedom International, 2006; Montgomery, 2018; Newsweek Staff, 2009; Nudel et al, 2009; Prism, 2017; Rashed, 2019; Roberts, Major, & Brady, 2008; Sinclair, 1993; Singer, 2017; Waddingham, 2017; Walker, 2012; 2014. Although there are some more or less consistent differences between the Mad Pride and neurodiversity paradigms, I believe it's nevertheless possible to identify at least one core principle spanning both movements (see Section 2, and also Cutler 2019a). And it is this principle upon which I focus in this chapter. As such, referring to this principle with both the terms *Mad Pride* and *neurodiversity* is technically redundant. Thus, when brevity requires, I will refer to both movements with just the term *Mad Pride*. When viable, however, I will employ both *Mad Pride* and *neurodiversity*, since I in no way wish to undercut the importance of the neurodiversity movement. In contexts where I drop one of the terms, I chose to keep *Mad Pride* over *neurodiversity* for a few reasons: (1) Mad Pride is more commonly interpreted to apply to all mental differences/illnesses, whereas *neurodiversity* is sometimes meant to refer only to autism, or to autism and other mental differences considered to be *developmental disorders*, (2) *neurodiversity* is currently more likely to be misinterpreted as simply a synonym for *autism spectrum* ***disorder***, rather than correctly interpreted as a distinctive and arguably revolutionary movement holding that mental differences like autism are worthy of acceptance, and (3) technically, the term *neurodiversity* is something of a misnomer for the central commitments of the Mad Pride and neurodiversity paradigms. These commitments would be much more accurately captured by the term *mental diversity*, since neurodiversity (or, diversity among neurological features) can actually manifest as mental *similarities* rather than *diversities* (see Hoffman, 2012, 2019; Rashed, 2019). Finally, the terms *Mad Pride* and *neurodiversity* refer to both: (1) theoretical models describing the nature of mental differences, including their features that are relevant to what we should do about them/how we should act towards them, and (2) social movements promoting those paradigms (or certain central aspects of them). I will typically specify whether I am referring to (1) or (2). In several contexts, however, the distinction between the two isn't important. In those contexts, I will sometimes just use the term *Mad Pride* to refer to the Mad Pride paradigm and/or movement.

There are strong moral and prudential reasons to adopt this approach of incorporating Mad Pride and neurodiversity perspectives and expertise. And these reasons apply both to my own approach in the present chapter, but also, I maintain, to the approaches adopted by public mental health organizations to diverse aspects of their own work [more precisely, the moral reasons apply to both of our approaches, while some (but not all) of the prudential reasons are directed exclusively to the field of public mental health]. The diverse aspects of the field of public mental health's own work include, for example, multiple stages of the development, formation, analysis (including scholarly analyses), implementation, assessment, and revision of public mental health principles, values, and practices. Sections 1.1 and 1.2, directly below, describe the most prominent of these reasons.

1.1 A *moral* reason to incorporate Mad Pride and neurodiversity perspectives

Most importantly, this inclusive approach deliberately seeks to oppose the aforementioned widespread exclusion of Mad Pride perspectives—not only from the discourse of public mental health, but from the broad swath of professional and public discourse more generally. This is an exclusion that's both deeply entrenched and multilayered. To see this, first note that mentally different/ill individuals *in general*, regardless of whether or not they support the Mad Pride paradigm, are specially prone to epistemic injustice (for example, see Crichton, Carel, & Kidd, 2017; Radden, 2012; Santi, & Kyratsous, 2015). This is due in part to the pervasive and false assumption that having a mental illness/difference renders one globally irrational, and, as such, not to be trusted. But, second, note that this distrust and dismissal can be all the more extreme for the subset of mentally different/ill individuals who subscribe to the Mad Pride paradigm. Such individuals are discounted both for the reason above (i.e. due to having a mental difference/illness in the first place), and also (presumably) for their adherence to a minority view (which is what the Mad Pride paradigm currently is). As such, their perspectives are often unabashedly dismissed both by society generally, as well as by other mentally different/ill individuals specifically.

In opposing such exclusion, I strive for my inclusive approach to heed the exhortation of the well-known disability rights slogan *Nothing About Us Without Us* (see Charlton, 1998; Fleischer & Zames, 2011), along with its attendant moral demands. This slogan calls for the inclusion of individuals in the making of certain decisions—namely, in the making of those decisions that will directly affect them. As such, this slogan calls for basic fairness as a moral demand that is far from strenuous, requiring what is arguably the minimum modicum of respect.

1.2 A *prudential* reason to incorporate Mad Pride and neurodiversity perspectives

I believe the moral injunction above, alone, provides sufficient reason for the inclusion of Mad Pride proponents' perspectives. In addition, however, there are prudential reasons for such inclusion as well (where prudential reasons, unlike moral reasons, are reasons based on self-interest—for example, in the present case, they are reasons based on benefits gleaned specifically by the field of public mental health). For instance, one prudential reason is that mentally different/ill supporters of the Mad Pride paradigm can contribute to the advancement of the field of public mental health by serving as a unique and valuable source of expertise (given, of course, that they choose to do so). Indeed, such individuals possess a great deal of expertise—for example, expertise regarding multiple aspects of the experiences of mental differences themselves. And this expertise is very much needed to complement the expertise of mentally different/ill individuals who do *not* subscribe to the Mad Pride paradigm—that is, it is necessary to round out the full spectrum of user views.

Furthermore, as removed as the Mad Pride paradigm seems to be from the field of public mental health, it is precisely the magnitude of this separation that prompts the field to all the more vigorously reconsider its most fundamental values and assumptions. And, in addition to prompting such values' reconsideration, it can subsequently creatively aid in their reformation. In fact, one could argue that, the greater the separation, the greater the likelihood of introducing new conceptual possibilities. Indeed, in what follows, I will argue that the Mad Pride paradigm offers at least one such conceptual possibility—one that exposes a previously hidden wrinkle in the field of public mental health's ultimate aim, thereby highlighting the need for its revision.

Although the objective of identifying the proper ultimate aim of public mental health is, in many respects, the central objective of this chapter, it is equally important to consider the *way* in which this objective is pursued. In other words, it is equally important to acknowledge the above methodological insight that inclusion of Mad Pride and neurodiversity user perspectives is of the utmost importance. Thus, to accompany this chapter's first objective, I set forth a second objective about the nature of the means used to pursue Objective 1—namely, that they must adhere to *Nothing About Us Without Us*. More specifically, Objective 2 is the objective to provide evidence for the following claim: *The inclusion of Mad Pride and neurodiversity perspectives is instrumental in efforts to re-sculpt and improve the values, principles, and practices of the field of public mental health.* Note that I do not intend to fulfill this second objective only by way of explicit argument. Rather, since it concerns the *method* by which I pursue Objective 1, I also hope to automatically adduce support for it through the very act of working towards Objective 1. To see this, note that Objective 1 is an objective to *re-sculpt and improve one underlying value/principle of the field of public mental health* (that of the identity of its ultimate aim). And, my strategy for satisfying Objective 1 is to *include Mad Pride and neurodiversity perspectives*. Thus, my hope is that the successful attainment of Objective 1 will, in itself, fulfill Objective 2. To summarize, the two central objectives of this chapter are:

Objective 1:	Identify the proper ultimate aim of the field of public mental health
Objective 2:	Provide evidence for the claim: *The inclusion of Mad Pride and neurodiversity perspectives is instrumental in efforts to re-sculpt and improve the values, principles, and practices of the field of public mental health.*

To work towards these two objectives, I will proceed as follows. First, before being able to enlist Mad Pride and neurodiversity perspectives in the pursuit of Objective 1, more must be said about what such perspectives entail in the first place. Thus, in Section 2 I distill and describe what I take to be the main tenet shared across the Mad Pride and neurodiversity paradigms, as well as across the social movements associated with them. In Section 3, I identify the specific *subclass* of Mad Pride and neurodiversity paradigms upon which I will focus. This subclass makes apparent a conceptual possibility that, in turn, spurs a re-examination and revision of the current ultimate aim of public mental health. Section 4 houses this re-examination and revision. In Section 5, I briefly gesture towards a further implication of Section 3's subclass of the Mad Pride and neurodiversity paradigms. Specifically, this is an implication that challenges the field of public mental health's staple *practices* of prevention and cure (as opposed to the implication in Section 4 that challenges one of the field's central *values/principles*).

2. An introduction to the Mad Pride and neurodiversity paradigms

Given that the novelty of my approach to the long-standing question of public mental health's ultimate aim rests exclusively on the content of the Mad Pride and neurodiversity paradigms, having a solid grasp of that content is crucial before proceeding. As such, I devote the bulk of the present section to identifying and describing what I take to be the principal tenet spanning both the Mad Pride and neurodiversity paradigms.

First, however, it's important to note that there is every reason to think that the field of public mental health would fully support this chapter's overall efforts to incorporate the perspectives of Mad Pride and neurodiversity advocates. Not only are there the prudential reasons mentioned in Section 1.2, but the field quite clearly commits itself to the moral demands broached in Section 1.1, affirming the importance of including the voices and perspectives of those individuals its policies affects. As such, it also embraces the aforementioned slogan *Nothing About Us Without Us*. The following are two representative examples of the field's explicit expression of this commitment. The first comes from the World Health Organization (WHO); specifically, it is item #6 in WHO's *Mental Health Action Plan 2013—2020*:

> *Persons with mental disorders and psychosocial disabilities should be empowered and involved in mental health advocacy, policy, planning, legislation, service provision, monitoring, research and evaluation.*
>
> **(*World Health Organization*, 2013, p. 10)**

And the second is from the World Psychiatric Association (*WPA*); specifically, it is item #9 in the *WPA Position Statement on Social Justice for Persons with Mental Illness (Mental Disability)*:

> *Involve persons with mental illness (mental disability) and their families in decision making process by ensuring their effective participation in planning, delivery, implementation and monitoring of mental health, health and other social services.*
>
> ***(World Psychiatric Association, 2015, pp. 1—2)***

Presumably, this commitment to the 'involvement of … persons with mental disorders' (World Health Organization, 2013, p. 36) is a commitment to the involvement of *all* persons with mental disorders; that is, to consulting the *full set* of representative views emanating from this entire group of individuals.[8] As such, it seems apparent that the field of public mental health would back my current efforts to bring in vital user perspectives—in this case, of Mad Pride and neurodiversity supporters—that were previously overlooked.

In order to begin to better understand the Mad Pride and neurodiversity paradigms, an initial adumbration can be gleaned from their appreciable contrast with the currently prevailing paradigm of mental differences in Western societies: what is called *the Medical Model* or *the Medical Cure Model.* This model dominates the current understanding of mental differences to such an extent that many people fail to realize that it is in fact just one model among many, and, as such, is not the final word on mental differences. The main elements of the Medical Cure Model are neatly packaged into the following quotation: 'Mental health conditions are legitimate medical conditions that need early intervention and evidence-based care. Just like cancer, just like diabetes' (Giliberti, 2017). Directly contrary to this idea that mental differences '***need*** early intervention and … care' (Giliberti, 2017, emphasis mine), a fundamental tenet of the Mad Pride and neurodiversity paradigms is that mental differences can be perfectly acceptable as they are; they do *not necessarily need* any sort of intervention, prevention, care, cure, treatment, or other alterations. I will call this tenet the *normative claim of acceptance.*[9]

This tenet's centrality to the Mad Pride and neurodiversity paradigms can be clearly seen in Mad Pride and neurodiversity advocates' expressed attitudes towards their own mental differences, as well as in their own understandings of these paradigms. Here, I present only the briefest of samples.

[8] At least one public mental health document specifies the necessity of involving the full range of users; in its Position Statement on Social Justice for Persons with Mental Illness (Mental Disability). WPA states: The UN Convention on Rights of Persons with Disabilities recognizes … the need 'to promote and protect the human rights of **all** persons with disabilities, including those who require more intensive support.' (World Psychiatric Association, 2015, p. 1 (*emphasis mine*)). And indeed one such human right is the right for all individuals to be involved in deliberations and decisions about those policies and practices that will affect them. This is most clearly expressed in CRPD, Article 4: In the development and implementation of legislation and policies to implement the present Convention, and in other decision-making processes concerning issues relating to persons with disabilities, States Parties shall closely consult with and actively involve persons with disabilities, including children with disabilities, through their representative organizations. (U.N., 2006).

[9] Note that the medical cure model and the Mad Pride and neurodiversity models are most certainly not the only two models of mental differences/illnesses; in addition to them, there are not only camps representing a range of views spanning the conceptual space roughly between the two upon which I focus, but also camps representing almost any combination of views on specific issues (e.g. issues including: the nature of mental differences; what (if anything) should be done about mental differences; the causes of mental differences, etc.).

First, an individual named E. C. articulates core elements of the Mad Pride paradigm as follows:

> *Mad Pride … question[s] the need for any kind of treatment or recovery at all. The Mad Pride movement is a paradigm that finds value in and even celebrates the traits, characteristics, states, thoughts, feelings, and experiences that are usually categorized as mental illness or madness by psychiatry, the DSM, and society. Instead of framing distress and difference as a problem that needs to be eliminated through medical or behavioral treatments, the Mad Pride movement argues that these experiences exist along the spectrum of human diversity.*
>
> **(Cutler, 2019*b*, p. *226*)**

E. C. also adopts the Mad Pride model in interpreting her own experiences of Madness:

> *As a Mad person … I am very sensitive and I do become easily overwhelmed by social situations, for better or for worse. That is not something I need to be treated for or overcome; that is something about myself that I accept.*
>
> **(Cutler, 2017)**

Next, Nick Walker—a central figure within the neurodiversity movement—proposes the following as a 'fundamental principle[s] … of the neurodiversity paradigm':

> *… Neurodiversity – the diversity of brains and minds – is a natural, healthy, and valuable form of human diversity. There is no 'normal' style of human brain or human mind, any more than there is one 'normal' race, ethnicity, gender, or culture.*
>
> ***(Walker, 2012, p. 228)***

The normative claim of acceptance threads through what is actually a rich and wide-ranging variety of interpretations and instantiations of Mad Pride and neurodiversity perspectives. This diversity in many ways stems from the fact that Mad Pride and neurodiversity are not just paradigms—they are also social movements. Thus, each individual activist group, and potentially each member within each individual group, may have fostered a different notion of what Mad Pride or neurodiversity means to them. Thus, there is currently no one universal definition of *Mad Pride*, nor of *neurodiversity*, etched in stone.[10] Furthermore, the Mad Pride and neurodiversity

[10] And this lack of an *official* definition is arguably also partially due to the fact that both Mad Pride and neurodiversity are relatively young movements. For example, Mad Pride has arguably been around since at least the 1990s (Schrader, Jones, & Shattell, 2013), although it certainly existed before within the corners and some of the more central conceptual threads of the older and broader c/s/x (consumer/survivor/ex-patient) movement (Morrison, 2005).

paradigms[11] each possess certain views that tend to more or less consistently differ across the two paradigms. However, there is nevertheless some overlap between them. For the purposes of the present chapter, I will focus upon the commonalities found within the entire range of Mad Pride and neurodiversity perspectives.[12] Finally, it's critical to note that there is most likely a sizable contingent of mentally different individuals who possess *Mad Pride-esque* perspectives (i.e. perspectives identical to or otherwise closely aligned with those of the Mad Pride paradigm) even though they may have never heard of Mad Pride or neurodiversity by name. That is, many individuals express anywhere from a vague inkling to a strong conviction that the way their mind works is not necessarily *bad* overall, and thus would almost certainly support at least some version of the Mad Pride and/or neurodiversity paradigms if given the opportunity. Thus, I consider these individuals as members of the group I refer to as *Mad Pride and neurodiversity supporters*.[13]

It's crucial to reiterate that, as underscored above, the Mad Pride and neurodiversity paradigms are perfectly compatible with the fact that mental differences/illnesses are very often associated with tremendous suffering. In no way does either movement as a whole wish to discount this suffering: 'It is not the intention of the Mad Pride movement or neurodiversity field to suggest that the experiences constructed as madness, mental illness, or neurodivergence do not come with suffering or limitations' (Cutler, 2019). If this seems counterintuitive, consider the following two ways in which one might understand this compatibility. First, for some mentally different individuals, their suffering may be counterbalanced by the presence of what they regard as deeply valuable features of their mental differences (e.g. enhanced creativity, empathy, access to novel and/or profound and/or liberating perspectives on the world, spiritual experiences, etc.) (for example, see Armstrong, 2010; Waddingham, 2017). Second, this suffering may not be (primarily) caused by the mental differences themselves, but instead (largely or completely) by the relentlessly vicious stigma, sanism,

[11] See Walker, 2014 for more information about neurodiversity terminology.

[12] In doing so, I am admittedly papering over what are arguably critical differences well worth discussing in a different context. However, my strategy of putting these differences aside for future work is not without precedent, and indeed is the strategy adopted by Cutler, 2019.

[13] Or Mad Pride and neurodiversity *adherents*, or *followers*, etc. I think this inclusion makes sense given that such individuals can be considered to be *potential* supporters (or adherents, followers, etc.).

discrimination, and social shunning that mentally different individuals experience from the surrounding environment (Chapman, 2019; Rashed, 2019).

As such, it's important to note that the word *Pride* in Mad Pride often assumes a non-standard meaning. Specifically, the *Pride* in *Mad Pride* need not imply an attitude that invokes feelings of celebration, nor need it imply a sense of honor, achievement, or a belief that its object (in this case, Madness/mental difference/illness) is more valuable than its lack.[14] And finally, the *Pride* in *Mad Pride* (as well as all other aspects of the Mad Pride and neurodiversity movements and paradigms) absolutely in no way implies that all individuals with mental differences/illnesses are committed to possessing pride in their mental differences/illnesses. Even though most Mad Pride adherents perceive their mental differences as positive or neutral overall, this does not preclude the possibility that, depending upon the presence or absence of myriad contextual factors, mental differences can indeed serve as overall negative contributions to certain individuals' lives. In fact, again depending upon the presence or absence of various factors, mental differences can theoretically contribute to the overall *badness* or *goodness* of a given individual's life in copious different ways (see Hoffman, in preparation).

Thus, keeping in mind the above clarifications, I propose the following as one possible Mad Pride model:

Minimal Mad Pride Model:

Mental differences can be worthy of acceptance; they need not be cured, treated, otherwise changed, or prevented in the first place. Moreover, depending upon certain factors, mental differences can also be worthy of esteem, appreciation, and other types of positive valuation beyond mere acceptance.[15] (This is equivalent to the normative claim of acceptance.)

On this Minimal Mad Pride Model (as well as on the other two Mad Pride models I introduce in Section 3), mental differences are generally regarded by individuals who support the Mad Pride paradigm as overall neutral or positive contributions to their lives, such that most (if not all) Mad Pride adherents would prefer a life with their mental difference over a life without it.

[14] However, the word *Pride* in Mad Pride does indeed sometimes imply these things for certain individuals.

[15] For example, they could be positively valued as fundamental parts of one's identity, or for a multitude of benefits they produce (at least for select people).

3. Differentiating between two subclasses of the Mad Pride paradigm: The Traditional and Hybrid Mad Pride Models

Importantly, note that there is something non-standard in my above rendering of the Minimal Mad Pride Model that requires explanation. Namely, I have omitted any claim about the illness status of mental differences—that is, about whether mental differences are, or are not, illnesses (or other medical pathologies[16]). Such a maneuver renders the Minimal Mad Pride Model especially wide-ranging (here, *Minimal* refers to a minimal set of requirements), but also unusual, and perhaps even controversial. The latter two qualities arise from the fact that the vast majority of written accounts of Mad Pride and neurodiversity do, indeed, include an additional tenet stating that mental differences are neither illnesses nor pathologies of any sort. Indeed, many such advocates vehemently embrace this tenet (sometimes maintaining that it is necessary for a view to count as Mad Pride), especially (and understandably) if they also maintain that something's identity as a *non-illness* is necessary for it to be acceptable just as it is. In addition, many such advocates also implicitly maintain that if mental differences *did* turn out to be legitimate medical illnesses, then they would *not* be worthy of acceptance. Leaving aside the question of whether such advocates actually do possess these additional beliefs, however, I will identify the above strand of Mad Pride that emphasizes a condition's non-illness status as the *Traditional Mad Pride Model.* The Traditional Mad Pride Model is simply the Minimal Mad Pride Model plus the additional metaphysical claim that mental differences are not illnesses:

Traditional Mad Pride Model:

1. **The metaphysical claim about illness-status:** Mental differences are neither illnesses nor medical problems/pathologies of any kind. Rather, they are non-pathological forms of mental diversity.
2. **The normative claim of acceptance:** Mental differences can be worthy of acceptance; they need not be cured, treated, otherwise changed, or prevented in the first place. Moreover, depending upon

[16] *Other medical pathologies* include diseases, disorders, injuries, and all other types of medical conditions. For the purposes of this chapter, these terms, along with the term *illnesses*, will all be used interchangeably. This is because the relevant contrast is between that which is *medical* and that which is *non-medical*, rather than between different types of medical conditions.

certain factors, mental differences can also be worthy of esteem, appreciation, and other types of positive valuation beyond mere acceptance.

Even though the Traditional Mad Pride Model is considered the canonical Mad Pride model (as its name indicates), I offered the Minimal Model in order to introduce a more inclusive rendition of Mad Pride. Specifically, I think that the tenet that mental differences/illnesses can be worthy of acceptance (i.e. the normative claim of acceptance) is the minimum necessary commitment for maintaining a Mad Pride perspective, but that it is also sufficient for such a perspective (at least for the Minimal Model of Mad Pride). It is important to note that there are individuals who embrace the normative claim of acceptance coupled with (depending on the individual) varying views about the metaphysical tenet. Thus, although not always widely recognized, not all individuals who embrace the normative claim of acceptance additionally insist that mental differences cannot be illnesses. While some such individuals are presumably agnostic about the metaphysical nature of mental differences, others unmistakably veer in the opposite metaphysical direction from Traditional Mad Pride advocates, expressly regarding mental differences as bona fide illnesses. This latter subclass of individuals, therefore, believes that something can be an illness and can nevertheless still be worthy of acceptance.[17] I will refer to this particular amalgam of views as the *Hybrid Mad Pride Model.*

Hybrid Mad Pride Model:

1. **The metaphysical claim about illness-status:** Mental differences are, indeed, bona fide illnesses.
2. **The normative claim of acceptance:** Mental differences can be worthy of acceptance; they need not be cured, treated, otherwise changed, or prevented in the first place. Moreover, depending upon certain factors, mental differences can also be worthy of esteem, appreciation, and other types of positive valuation beyond mere acceptance.

So there are at least these two major variants (or *subclasses*) of Mad Pride beyond the Minimal Model—the Traditional Model and the Hybrid Model[18] in addition to a potentially large number of others that vary with

[17] I acknowledge that many traditional Mad Pride advocates may challenge my characterization of the hybrid Mad Pride perspective as properly counting as *Mad Pride* in the first place. Although I do have several reasons for my characterization, if they aren't compelling, then I am very open to revising my terminology to best cohere with the values of the Mad Pride movement. Thus, for the time being I consider this classification as merely provisional.

[18] Some users espouse this hybrid model unequivocally, while others embrace something very close to it [Craft, 2018; Cutler (personal communication, 30 June 2019) and Mrs TeePot, 2017].

respect to their position on one or more additional details. Given that each of these could easily have its own vast set of consequences for public mental health, it's only feasible to explore one variant within the space of a single chapter (more precisely, it's only feasible to explore one *consequence* of one variant in the space of a single chapter). Thus, in this chapter, I choose to restrict my exploration to select consequences of the Hybrid Mad Pride Model for public mental health, and to leave exploration of the Traditional Mad Pride Model (and other potential models) to future work.

At this point, one may understandably wonder why I have chosen the Hybrid Model, especially since it is arguably a minority, fringe view [and may not even be considered to count as Mad Pride proper in the first place (see footnote 16)]. In response, one reason for my choice is that the Hybrid Model, compared to the Traditional Model, is likely to be much more palatable to the field of public mental health, since both it and the field (by and large) share the foundational assumption that mental differences are illnesses.

Second, if I had chosen the Traditional Model, it would have actually required the examination of the consequences of *two* different tenets on public mental health: the normative claim of acceptance (that mental differences/illnesses are worthy of acceptance), *and* the metaphysical tenet that mental differences/illnesses are not pathological in any way. As mentioned above, space considerations preclude the exploration of the consequences of more than one tenet.

However, one may understandably still be reluctant to embrace my choice. Indeed, what about the fact that the Hybrid Model seems, to put it plainly, untenable? After all, how can something that's a bona fide illness actually be worthy of acceptance? To respond, first, I want to reemphasize that this Hybrid Model, seemingly untenable or not, is indeed embraced by a number of individuals[19] and these individuals are ostensibly no less deserving of representation than are any others within the full range of user perspectives. Second, there are several reasons to think that the Hybrid Model is actually not that untenable in the first place. Consider what it is that characterizes a bona fide *illness*. Granted, there is no one agreed-upon

[19] For example, one might possess a malfunctioning of their *emotion dampening function* (if that is indeed a function), such that they have difficulty lowering the magnitude of intensity of their emotions. Depending on the particular individual, they may well prefer to experience intense emotions, and thus may not mind that their *emotion dampening function* is *broken*. They may, in fact, prefer that it be broken.

definition of *illness* (that is, there is no one agreed-upon set of criteria that distinguishes medical conditions from non-medical conditions). However, of numerous contenders, the ones that are arguably considered the most promising (at least within the field of psychiatry, as represented in the DSM-5) all prominently feature the idea of (biological) *malfunctioning* (see for example, Boorse, 1975). For example, such a definition might dictate that a mental difference counts as an illness if and only if that mental difference is an instantiation of some type of *malfunctioning*. Assuming these types of definitions, it's not terribly difficult to see how one could, indeed, accept that they possessed a mental malfunctioning (and therefore an illness), while also accepting that the type of functioning disrupted simply wasn't a type of functioning that they considered to be terribly important in their lives.[20] Furthermore, the entire Disability Pride movement (focusing mainly on physical disabilities) provides ample precedent and additional considerations in favor of the combination of a condition's status as a medical condition with its status of being worthy of acceptance and perhaps esteem and appreciation (see Barnes, 2016; Fleischer & Zames, 2011 for analyses related to Disability Pride. For an expanded defense of the Hybrid Mad Pride Model more generally, see Hoffman, in preparation).

Having now specified the subclass of Mad Pride upon which we will focus—namely, the Hybrid Mad Pride Model—we are now in a position to begin to identify how it indicates the need for a revision of public mental health's currently-accepted ultimate aim. The details underlying this connection are the focus of the next section: Section 4.

4. Using the Hybrid Mad Pride Model to identify public mental health's proper ultimate aim

Before proposing a new candidate ultimate aim of public mental health, and the way in which the Hybrid Mad Pride Model signals the need for it, I first endeavor to undermine the plausibility of the field's currently-accepted aim—specifically, the aim of increasing the public's mental health. Prima facie, this task may seem patently unachievable, since the claim that *public mental health's ultimate aim is to increase public mental health*

[20] Moreover, this definition is cited almost verbatim on the page titled "Mental Health" on the American Public Health Association website (American Public Health Association, 2019).

seems simply true by definition. Indeed, what would the field of public mental health's ultimate aim be, if not to increase mental health?

Yet, even though it's certainly true that a very *important* aim of the field of public mental health is to increase the public's mental health, there is still room to ask whether it is really the field's *ultimate* aim. Obviously, before posing this question, we must get clear on the distinction between ultimate and other aims. Roughly, I take a field's (proper) ultimate aim to be the final and most important end goal towards which that field should strive. Indeed, the ultimate aim's status as *the most important* goal clearly implies that its importance supersedes that of all other aims (including what are known as *intermediate aims*). Intermediate aims, in contrast, are those aims that lie between various causal factors (e.g. particular public mental health interventions and programs) and an ultimate aim.

Intermediate aims can in fact be important aims, but, by definition, they are important only insofar as they help bring about an ultimate aim. Thus, hypothetically, if a particular action (for example, a particular public mental health intervention) happened to increase the likelihood of the realization of the field's *intermediate* (albeit seemingly important) aim, but also happened to decrease the likelihood of the realization of the field's *ultimate* aim, then, presumably, according to the above definitions, that action should not be performed (or, that public mental health intervention should not be implemented), since the ultimate aim's importance trumps that of the intermediate aim.

In the case of the field of public mental health, I will argue, as I intimated above, that its proper ultimate aim is, surprisingly, *not* to increase public mental health (as highly counterintuitive as that sounds). However, this does not mean that increasing public mental health isn't an important aim of the field—it is. It is important insofar as it, as an intermediate aim, facilitates the achievement of the field's *ultimate aim* (and it's the intermediate aim, of course, that the field of public mental health is best equipped to achieve). So, what is the ultimate aim of the field of public mental health? I will argue that the field's ultimate aim is to increase *well-being*—[and *not* (mental) *health*].

At this point, one pressing question is what the difference between health and well-being is in the first place. Indeed, both the fields of public health and medicine, along with common parlance, tend to rampantly use these two terms interchangeably, often glossing over certain distinctions between them, or conflating them entirely. One of the most spectacular examples of such a conflation comes from WHO, in its famous [and infamous (e.g., see

Bickenbach, 2015)] main definition of *health* from 1948 (which WHO still uses to this day):

> *Health is a state of complete physical, mental and social well-being and not merely the absence of disease or infirmity.*
>
> **(World Health Organization, 1948)**

And WHO carries over this definition into mental health as well:

> *Mental health is defined as a state of well-being in which every individual realizes his or her own potential, can cope with the normal stresses of life, can work productively and fruitfully, and is able to make a contribution to her or his community.*
>
> **(World Health Organization, 2018)**[21]

However, the terms *health* and *well-being* can also mean quite distinct things, and I believe that this distinction is critical in many ways, including its ability to prompt a more precise assessment of the ultimate and intermediate aims of public mental health. Thus, before laying out the considerations in favor of the claim that well-being, as opposed to health, is the proper ultimate aim of the field of public mental health, more must be said about exactly what *health* and *well-being* can mean in the first place.

4.1 The precise meaning of *health*

Despite its aforementioned common conflation with *well-being*, *health* has a more precise, and arguably primary, definition: 'the absence of illness or any other type of pathology,' [indeed, the primary definition of "health" from the Lexico Online Dictionary (formerly the Oxford Dictionary Online) is nearly identical. It is: "The state of being free from illness or injury." (Health, 2019)], or [what I take is mostly equivalent for our present purposes (depending upon the appropriate definition of *illness* or *pathology*)] 'the state of being free from chemical, biological, and psychological malfunctioning.' This more precise definition captures what I believe is the intended meaning of *health* used within the context of the field of medicine, and will be (and

[21] This can be spelled out more precisely as follows (here, using the already familiar examples of mental illness in cases of individuals who support the Hybrid Model of Mad Pride). Regarding the former—that it's possible to have well-being without health—it is possible to possess substantial levels of well-being even while largely lacking health (specifically, in the sense of having a mental illness that one accepts and appreciates—of course, importantly, this is assuming that the Hybrid Mad Pride Model is indeed correct in its metaphysical claim that mental differences are bona fide illnesses).

has been) the definition I adopt in this chapter (unless the context clearly indicates otherwise).

4.2 The precise meaning of *well-being*

The more specific (and arguably primary) definition of *well-being* is 'the state of one's life going well' or 'the state of one's life being good.' Of course, it is often the case that health (as defined directly above in Section 4.1 as 'the state of being free from chemical, biological, and psychological malfunctioning') causally contributes to one's life going well, and thus to one's well-being. Nevertheless, *health* and *well-being*—as defined in the present (4.2) and the previous (4.1) sections—are assuredly not the same thing. An initial way to grasp the important difference between them is to note that well-being—that is, how well one's life is going—arguably consists in much, much more than how well one's body (and mind) are functioning.

There are numerous accounts of what constitutes well-being, primarily offered by the field of philosophy, but also by psychology, sociology, public health, and other disciplines. When summarizing the conceptual terrain of well-being, contemporary philosophers typically identify at least three central accounts: hedonism, desire-fulfillment theory, and objective list theory. Although these tend to be considered the main accounts within contemporary philosophy, of course there are many others, as well as a multitude of variations and combinations of these three (see Bradley, 2015; Fletcher, 2016a, 2016b; Parfit, 1984; Crisp, 2017). In contrast, contemporary psychologists (often echoed by public health scholars) tend to define *well-being* with one or more quality of life scales, yielding what is known as the quality of life theory as their primary account (see Bradley, 2015, especially pp. 79–82). The following descriptions provide snapshots of each of these four accounts:

Hedonism: The life that goes best for a given person X is the life where X is happiest overall. (Adapted from Parfit, 1984, pp. 493–494)

Desire-fulfillment theory: The life that goes best for a given person X is the life where X's overarching desires (specifically, X's desires concerning how their life should go overall) are fulfilled to the greatest extent. (Adapted from Parfit, 1984, pp. 494–499)

Objective list theory: The life that goes best for a given person X is the life where X possesses as many items on an 'objective list' as possible (where the list in question is *objective* insofar as it applies to everyone). An example of one of many possible objective lists is a list including the following items (taken directly from Parfit, 1984, p. 499): 'moral goodness,' 'the development of one's abilities,' 'knowledge,' and 'the awareness of true beauty.' (Adapted from Parfit, 1984, pp. 499–501)

Quality of life theory: The life that goes best for a given person X is the life where X would give the highest ratings to various items on a 'quality of life scale'—specifically, items meant to assess one's quality of life. As an example, the simplest version of such a scale would contain only the following item: 'How would you rate your overall quality of life?' (see Bradley, 2015, especially pp. 79–82).

4.3 The possibility of well-being without health

Notably, it is possible on all of the above four accounts to have well-being without health, as well as health without well-being.[22] It is also possible, on three of the above four accounts (that is, on all but the objective list theory), to have situations where an increase in health causes a decrease in well-being, or where one common cause simultaneously produces both an increase in health and a decrease in well-being. All of these possibilities could easily be illustrated with ample examples from individuals who embrace the Mad Pride paradigm. However, I offer instead the following preliminary (fictional) examples from outside the realm of mental differences/illnesses, in order to demonstrate that the divorce between health and well-being extends beyond the realm of mental health and mental illness.

First, consider hedonism: it may be that what makes Caliph happiest is eating whatever he wants, even if this results in high cholesterol levels—levels that he knows are high enough to very likely shorten his life. Caliph is simply a person who would be much happier with a shorter life filled with exotic cheeses, French fries, and freshly fried doughnuts, the luxury of eating them whenever he wants, and the freedom from the stress, time, and energy needed to diligently monitor his diet. Here, hedonism would dictate that the best life for Caliph is the life that includes his frequent consumption of an ample supply of high cholesterol foods. Moreover, a health intervention that would drastically limit Caliph's intake of his favorite fried foods would be one that simultaneously: (a) (by stipulation) caused an increase in Caliph's

[22] Although they are not entirely immune from controversy; it's important to note that *all* of the above theories face significant challenges (see Bradley, 2015; Fletcher, 2016a; Heathwood, 2016).

health, and (b) caused a decrease in Caliph's overall happiness, and thus a decrease in his well-being.

Second, the severance of well-being from health can also be demonstrated in terms of desire-fulfillment theory. Specifically, Crystal may desire for her life overall to be as spiritually rich as possible. If Crystal happens to experience (what she considers as) deeply spiritual phenomenon via her epileptic auras, and indeed desires a life with epilepsy and these deep spiritual experiences over a life with neither, then, according to desire-fulfillment theory, Crystal's well-being would be higher with epilepsy. Here, Crystal's illness (i.e. her lack of health) is actually causally responsible for a portion of her well-being. Third, moving to objective list theory, we can see that it, too, allows for the combination of (relatively) high well-being and (relatively) low health. Indeed, for most objective lists, one can see that health is, at most, one of *many* goods in life (i.e. one of many items on the objective list). Finally, on quality of life theory, Connie has cerebral palsy, and rates their quality of life just as highly as they would have if they had not had cerebral palsy. For Connie, then, the presence of a disability [which could be construed as a state of lower health (although this conclusion is not without its detractors)] has no effect on their well-being.

Importantly, the types of scenarios in the above examples are not merely speculative; on the contrary, they are borne out in real life. For example, cases like Crystal's are discussed in Barnes, 2016 and Sacks, 2012 (to name just a few). And the notion that Connie's disability would not lower their quality of life rating is supported by several studies showing that people with physical disabilities rate their life satisfaction (which is more or less equivalent to quality of life, depending on the life satisfaction and quality of life scales in question) just as highly as do people without physical disabilities (see Bagenstos & Schlanger, 2007 for a review of many such studies). In the same general vein, other studies provide evidence that a significant number of disabled individuals would refuse a cure for their disabilities. Such refusals may suggest that, not unlike Connie, these individuals hold that their disabilities do *not* negatively affect their well-being. On the contrary, they may even maintain that their disabilities *enhance* their well-being. One such study of disability activists found that, of those individuals surveyed, 47% strongly to weakly agreed with the statement 'even if I could take a magic pill, I would not want my disability to be cured' (Hahn & Belt, 2004, p. 455).

Although I won't officially adopt any one of the above four accounts of well-being in this chapter, I will often focus upon desire-fulfillment theory, and sometimes quality of life theory. This is because these two accounts tend to be especially favored in the philosophy and psychology communities, and are thus arguably the least controversial choices.[23,24] Furthermore, these two theories (along with hedonism) are plausibly the most consistent with the spirit of most public mental health documents, which tend to emphasize that well-being can only be determined with respect to the desires, emotions, beliefs, and other internal states of the individual herself, rather than by some universal objective standard.

4.4 Why think that increasing well-being, rather than mental health, is the proper ultimate aim of the field of public mental health?

There are several strong reasons in favor of the claim that increasing well-being, rather than mental health, is the proper ultimate aim of public mental health. First, quite simply, of the two options (well-being and health), well-being is the greater, more all-encompassing good. According to the more precise definitions I presented and adopted in Section 4.2, a high level of well-being can include a high level of health, but a high level of health, by and large, does not automatically include all of the goods that constitute well-being. And because well-being is defined as the state of one's life going best, well-being is almost certainly more important to most people than is mental health. Of course, many people take mental health to be necessary for their well-being, but (barring highly atypical realizations of certain accounts of well-being), they also recognize that mental health contributes only one part of that well-being. Beyond this, however, other people—namely, individuals who support the Hybrid Mad Pride Model identified in Section 3— *do not* take mental health to be necessary for their well-being. Rather, most such supporters consider mental health (as realized by the reduction of their own mental illness/difference) to be *antithetical to* their well-being (for example, insofar as their own mental illness/difference is something they desire, and/or that raises their happiness and/or quality of life). That is, most Hybrid Mad Pride supporters would consider an increase in mental health to cause a *decrease* in their well-being.

[23] Additionally, many of the points I make will hold true on all of these four accounts of well-being.

[24] Well-being is arguably more important to individuals *by definition* (that is, depending in part on the account of *well-being* assumed).

Finally, then, this brings us to the issue of how the Hybrid Mad Pride Model signals the need for the revision of public mental health's currently-accepted ultimate aim. Specifically, the recognition that *increasing such supporters' mental health may actually decrease well-being* reveals that an ultimate aim of *increasing mental health* will not always lead to a felicitous outcome (in terms of well-being).

Here's another way to think about the need for a revised ultimate aim. In light of the above considerations, why would citizens agree to support government-sponsored public health initiatives that aimed to increase the public's mental health simply for its own sake (and that, consequently, simultaneously decreased some individuals' well-being)? Or, to put it another way, why would citizens agree to support governmental initiatives that would likely increase something important to *some* of them (e.g. mental health) at the cost of possibly (selectively) reducing something *more* important to *all* of them (e.g. well-being[25])?

4.5 Why think that a portion of the field of public mental health already agrees that increasing well-being is its correct ultimate aim?

Thus, there are good independent reasons backing the claim that well-being should be the ultimate aim of the field of public mental health. Perhaps surprisingly, however, I believe that this claim is *already* backed by much of the field of public mental health itself. That is, to the extent that the field assumes WHO's canonical definition of *health* [as 'a state of well-being ...' (World Health Organization, 1948)], it presumably regards *health* and *well-being* as more or less equivalent. And thus, to that extent, it also presumably regards the (widely currently-accepted) ultimate aim of *increasing (mental) health* as more or less *equivalent to* (my proposed) ultimate aim of *increasing well-being*.

Thus, I am positing well-being as the proper ultimate aim of public mental health not just as a conclusion of my own considerations, but also as what I believe is a reflection of the stance of much of the field itself. Thus, in this (and many other) respects, much of the field and I seem to agree that well-being is the proper ultimate aim of public mental health. Where we differ, however, is on the matter of whether health and well-being are

[25] Or*the widely accepted aim*, *the current popular aim*, or similar expressions.

fundamentally equivalent states: much of the field of public mental health appears to treat them as such, while I maintain that they are not.[26]

4.6 A possible justification for the conflation of *health* and *well-being*

In fact, the idea that the field of public mental health is, like me, trying to emphasize the importance of well-being, and is, I believe, trying to posit well-being as its ultimate aim, might *theoretically* explain why WHO defined *health* as *well-being* (in other words, why they *conflated health* and *well-being*). (In other words, it might serve as one *possible* explanation—I am not attempting to relate any historical facts here.) Thus, when I say that public (mental) health tends to conflate the terms *health* and *well-being*, I do not intend to imply that this necessarily happens due to some sort of accidental sloppiness. I suspect, rather, that the field's definition of *health* as *well-being* is likely an intentional rhetorical strategy that was actually deployed in order to emphasize the very same central point I'm trying to emphasize: that well-being, rather than *just* health, is the proper ultimate aim of public (mental) health. Consider the following passage from Hutchinson et al.:

> *In general, public health definitions of wellness are extremely comprehensive and broad, indicating an underlying philosophy that suggests wellness implies living a balanced, fulfilling and optimal life.*
>
> *… In contrast to the field of mental health, the field of public health has long embraced a holistic, multidimensional, future-oriented view of health …. Even*

[26] It's important to emphasize that not *all* of the field of public mental health regards health and well-being as equivalent. Certainly some constituents in the field regard the two as fundamentally different (as I do). And, similar to my own conclusion that, of the two, *well-being* should be the ultimate target of the field's work, at least one pair of scholars who maintains the separateness of health and well-being *also* concludes that public (mental) health must aim beyond simply increasing health. Specifically, Faden and Shebaya, in their analysis of the ethics of public health, claim: '… public health has an obligation to evaluate the impact of its policies and practices on human well-being broadly and not only on health' (Faden & Shebaya, 2016). Additionally, Roache and Savulescu advance a tightly analogous claim for the field of medicine in general (rather than that of public health specifically): One perhaps surprising point highlighted by this consideration of medicine's purposes is that, whilst the primary purpose of medicine may seem prima facie to be to restore health, this purpose … guides only part of current medical practice. It is the more general aim of improving the lives of patients and others via welfare and autonomy enhancement that unites medical practice. Of course, at this level of generality, medical practice resembles many other fields: education, politics, and law are all motivated (at least in theory) by the general aim of improving people's lives. What distinguishes medical practice from such other fields is, then, not its aims, but its employment of medical knowledge, techniques, and tools to achieve its aims. (Roache & Savulescu, 2018, p. 257).

> *definitions of the word 'health' in research articles or position papers in public health are oriented towards a wellness perspective.*
>
> **(Hutchison et al., 2006, p. 244)**

Thus, the decision to define *health* as *well-being* (which ends up being a decision to conflate *health* and *well-being*) actually *may* be carefully calculated to maximize rhetorical power—in particular, the power to incite a greater acceptance of 'a holistic, multidimensional, future-oriented view of health' throughout the general population. Arguably, such rhetorical power is greatly needed at this time—a time when much of the population is infatuated with biomedical and reductionist approaches to helping people live better lives.

I will refer to this possibility that conflation was (and is) an intentional strategy as *the conflation strategy*. Specifically, the conflation strategy is what I take to be a deliberate maneuver to conflate the terms *health* and *well-being*, presumably in order to underscore the point that activities carried out in the pursuit of increased health are, and should be, ultimately carried out in the pursuit of something even greater (e.g. well-being).

However, even assuming the existence of the conflation strategy, I believe that any positive rhetorical power it generates is likely ultimately insufficient to justify its attendant harms. I will discuss these harms in the conclusion (Section 6), as well as in Section 5—a brief detour that explores a possible additional implication of the inclusion of Mad Pride perspectives.

5. A brief exploration of how the Mad Pride paradigm could reveal aspects of public mental health practices in need of revision

In this section, I introduce an additional potential ramification of bringing Mad Pride perspectives to bear on the field of public mental health (that is, a ramification that extends beyond the domain of the revision of one of public mental health's central *principles/values* (i.e. the identity of its proper ultimate aim), and into that of its actual *practices*. My discussion here will indeed be preliminary. However, I include it in hopes of: (1) making manifest the pressing need for more thorough analysis of public mental health practices, and, (2) to garner additional support for Objective 2—that is, to more forcefully convey a sense of just how important Mad Pride perspectives are to public mental health. Not only do such perspectives call for and subsequently inform the revision of one of the central conceptual underpinnings of public mental health—namely, its ultimate aim (as discussed above in Section 4)—but they also unmask the possibility of grave consequences of certain staple public mental health practices.

The particular public mental health practices to which I'm referring are those designed to prevent, cure, or otherwise mitigate mental illnesses; I will refer to these practices collectively as *prevention and curative treatments*, or, for brevity, *P&C* (or *P&C interventions, initiatives, practices,* etc.). Of course, current P&C interventions have not matured nearly enough to ensure anything close to the complete prevention of mental illnesses, nor of their total elimination or cure. However, efforts to develop more efficacious interventions are widespread; in fact, there have already been around 400 randomized controlled trials testing the efficacy of mental health-related prevention practices, many of which have been deemed successful (Mendelson et al, 2011, p. 494). Furthermore, several of these prevention practices and initiatives have already been put into practice both across the U.S. and internationally (see Embry, 2011; Mendelson et al., 2012; Radden, 2016).

The P&C interventions in which I am particularly interested include school and workplace behavioral interventions (e.g. stress-reduction exercises, mindfulness practices, and cognitive-behavioral therapy-like strategies to monitor and challenge negative thoughts)—see Mendelson et al., 2012 for a particularly helpful overview. This interest stems largely from the fact that these types of practices tend to be implemented *throughout* one's lifetime (consider especially workplace interventions). Indeed, the concept of *lifelong prevention* is frequently invoked and championed by the field of public mental health (for example, see Mendelson et al., 2012, pp. 467–469). The reason that this is especially relevant here is that it means that individuals who have *already* come to support the Mad Pride paradigm are among those affected.

Of course, one might understandably question whether such individuals really would be among those affected; after all, as (official or unofficial) Mad Pride supporters—that is, as individuals who presumably have no interest in curing or preventing the continued existence of their mental illnesses/differences—wouldn't they simply refuse such interventions to begin with? This question brings me to the second noteworthy feature driving my interest in these types of P&C practices: their coercive or semi-coercive nature (including the fact that they are often administered in the absence of informed consent). I think it is this feature that provides an answer to the preceding question: indeed, Mad Pride supporters would indeed refuse such interventions, *unless* they were coerced or partially coerced into them, and/or they didn't realize that they were availing themselves of ways to cure their mental differences in the first place.

Why think that such P&C practices are coercive or semi-coercive? Consider the following examples. First, many classroom interventions are presumably simply included as just another normal activity during class time; as such, they are considered to be required (in the same way that the rest of the school day's lessons and activities are). And workplace interventions are often incentivized, at least implicitly (e.g. I imagine via pressure to be considered *a team player*).

Furthermore, in the cases of both school and workplace interventions, it is highly unlikely that their identities *as* mental illness prevention and cure initiatives are made known. Again, school place interventions are often simply fit into the overall class schedule, appearing as one of many types of normal class lessons. And workplace programs, similarly, often fail to be explicitly presented as mental illness prevention or cure initiatives. Instead, they tend to be introduced as, for example, general stress reduction techniques. As such, these types of P&C interventions are arguably administered in the absence of informed consent, since the relevant *informing* about their nature—that is, about their very existence *as* prevention and/or cure practices—fails to occur (informed consent is likely compromised in several other ways; for instance, Radden, 2017, Part 2(i) provides additional such examples).

It is not difficult to see how such P&C efforts could lower the well-being of Mad Pride supporters. In fact, I believe that this can be demonstrated with any of the four aforementioned accounts of well-being; here, however, I will focus on desire-fulfillment theory. Briefly, individuals who subscribe to Mad Pride (virtually always) desire to live a life *with* their mental difference/illness; specifically, preferring (and desiring) such a life over a life without their mental difference/illness. Several quotations from individuals belonging to the Mad Pride and neurodiversity movements strongly evidence the existence of these desires. Here are two:

- 'We don't want to be normal.' (Hall, quoted in Robinson & Rodrigues, 2009)
- 'If there was a magic pill that would make me neurologically typical, normal, I wouldn't take it.' (Ne'eman in Roberts et al., 2008; see also Newsweek Staff, 2009)

Additionally, many advocates take a different, perhaps darker, tack in voicing their opposition to P&C efforts, especially to anything like a universal cure. These advocates regard their mental differences as essential to who they are as a person, and thus, by extension, as essential to their existence. So their opposition to P&C takes a particularly strong form. Specifically, they claim that any completely effective cure would erase their existence. Arguably one of the most well-known examples of this type of expression is from neurodiversity activist Jim Sinclair, in his essay titled 'Don't Mourn for Us':

> I invite you to look at our autism … from our perspective: …
>
> Autism isn't something a person *has* … There's no normal child hidden behind the autism. Autism is a way of being. It is *pervasive*; it colors every experience, …and encounter, every aspect of existence. It is not possible to separate the autism from the person—and if it were possible, the person you'd have left would not be the same person you started with. …
>
> Therefore, when parents say,
>
> > I wish my child did not have autism,
>
> What they're really saying is,
>
> > I wish the autistic child I have did not exist, and I had a different (non-autistic) child instead.
>
> Read that again. … This is what we hear when you pray for a cure. (Sinclair, 1993, p. 16)

Thus, consider those P&C efforts implemented during stages of life where one will have already formed a commitment to Mad Pride. And consider especially those efforts that are implemented non-discriminately without informed consent. It is not difficult to see that such efforts (assuming they were efficacious) would directly defeat any desires individuals might possess to live a life with their mental difference/illness intact [or, indeed, to live a life (to exist) at all! (see Sinclair, 1993)]. As such, desire-fulfillment theory would straightforwardly conclude that these P&C efforts would lower the well-being of those individuals desiring to retain their mental differences/illnesses.

Of course, public mental health's ultimate goal of increasing well-being concerns *everyone's* well-being (or, at least, everyone who is potentially impacted by the intervention in question). Even though there are no clear-cut equations detailing how to correctly balance different effects on the well-being of different individuals, ideally, any full consideration of P&C's impact on overall well-being should take the well-being of all affected individuals into account, including those who do not adhere to Mad Pride.[27]

In sum, consideration of the perspectives of users who support the Mad Pride paradigm reveals that the staple public mental health initiatives of prevention and curative treatment will almost certainly decrease the well-being of many of these users. Assuming that the ultimate aim of public mental health is indeed to increase well-being, then P&C, arguably the main tools of public health, may, ironically, undermine the field's own ultimate aim, at least in these cases. Given this possibility, P&C initiatives demand serious scrutiny and re-evaluation, perhaps even leading to their discontinuation.

6. Conclusion

6.1 The status of the two main objectives

In this chapter, I hope to have satisfied the two main objectives I advanced in the introduction (Section 1). The first objective was to evaluate and subsequently offer a refinement of the field of public mental health's ultimate aim. In executing this first objective, I argued that the proper ultimate aim of the field of public mental health is to increase well-being (and *not*, perhaps surprisingly, to increase mental health per se). The second objective was to demonstrate that the field of public mental health has excellent reasons to incorporate the perspectives of Mad Pride and neurodiversity

[27] However, depending upon which method for properly comparing different effects on different individuals' well-being is used, it may well be that the balance is tipped in favor of discontinuing P&C efforts (again, focusing on the types of widely-propagated (semi-) coercive P&C efforts detailed above). For example, note that two likely available options would be: (a) to continue to implement P&C efforts and to make them more effective, and (b) to discontinue P&C efforts but to offer (free, easily obtainable) treatment for any individuals who desired it. In this scenario, presumably most Mad Pride advocates would deeply desire to retain their mental differences/illnesses; in contrast, most individuals with mental differences/illnesses who do not adhere to Mad Pride (for brevity, *Mad Pride detractors*) would deeply desire to rid themselves of their mental differences/illnesses. While continuing P&C efforts (option a) would satisfy these detractors' desires, it would also *permanently* defeat Mad Pride advocates' desires to retain their mental differences. On the other hand, discontinuing P&C efforts (option b) would allow both groups of individuals' desires to *ultimately* be fulfilled, although Mad Pride detractors would presumably have to have their desires defeated temporarily, for a relatively short period of time (e.g. between the point of their realization of their mental illness and the point when their desired treatment commenced).

proponents into its work. I approached this second objective by offering three such reasons: (1) the presence of a basic moral requirement for such incorporation expressed by the slogan *Nothing About Us Without Us*, (2) the imperative for the field of public mental health to adhere to its own explicitly championed value of user involvement, and (3) the fact that Mad Pride perspectives can provide a unique set of tools and methods that the field can employ in scrutinizing and ultimately advancing its foundational values, principles, and practices.

In fact, I devoted the bulk of this chapter to developing one example of this third type of reason—that is, to proposing one way in which the Mad Pride and neurodiversity movements (I believe) could and should assist the field of public mental health in refining one of its foundational values—specifically, the value of the identity of its own ultimate aim. To elaborate, I drew from the perspectives of a subset of Mad Pride supporters (supporters of the Hybrid Model) to raise the possibility that one can have well-being without full health; in fact, one's well-being could even *require* a certain lack of health (in this case, realized by the presence of a mental illness). And this raised the further possibility that public mental health's widely-accepted ultimate aim (i.e. the aim to increase mental health) needed revision, given that health and well-being don't always accompany one another, and yet both are obviously goods towards which public mental health intends to aim. In order for the ultimate aim of public mental health to accommodate both of these values, I proposed replacing it with a more nuanced version. Namely, I argued that the widely-accepted ultimate aim of 'increasing public mental health' should be replaced with the new (or, newly phrased) ultimate aim of 'increasing well-being.' In doing so, I endeavored to support Objective 2 by way of using Mad Pride perspectives to achieve Objective 1.

6.2 The vicious circle: a hypothetical explanation for the exclusion of Mad Pride and neurodiversity perspectives

At this point, there is at least one vexing question still lingering—the question of how it came to pass that the field of public mental health (apparently) overlooked a contingent of its users (i.e. the contingent supporting the Mad Pride paradigm). In fact, this lacuna is all the more perplexing given the field's explicitly expressed dedication to the inclusion of user perspectives (see Section 2).[28] I am in no position to know the actual historical, political, sociological, psychological, and/or other types of antecedents that may have lead to this (apparent) exclusion. I can, however, offer a purely hypothetical explanation of the field's (presumably unintentional) temporary departure from its adherence to the slogan *Nothing About Us Without Us*. This explanation rests on the presumption that the apparent oversight in question is in no way the result of sheer negligence (or any other type of blameworthy behavior) on the part of the field of public mental health.

Rather, one possible way to justify this apparent oversight has the explanatory structure of a vicious circle that contorts public mental health's initial good intention (admittedly, this preliminary description is quite abstract; however, it will become less so as I fill in its details.) To elaborate, first, the vicious circle I'm envisioning revolves around two mutually reinforcing loci:

1. The conflation of the meanings of health and well-being
2. The reduced likelihood of noticing Mad Pride and neurodiversity perspectives and their supporters.[29]

Second, the initial good intention is simply public mental health's conflation strategy, introduced in Section 4.5—to review, the conflation strategy refers to the field's intention behind its merging (conflation) of the meanings of *health* and *well-being*. It also, quite noticeably, serves as a possible intention behind the first of the vicious circle's loci.

Now that I've labeled the main elements of this explanatory structure, I can now offer my hypothetical explanation for the apparent neglect of Mad Pride perspectives in more detail. Again, I am simply offering this as a

[28] Something else that adds to the mysteriousness of the situation is the fact that WHO has authored a document explicitly naming two Mad Pride-affiliated organizations as exemplary defenders of human rights. WHO's document is titled *Understanding Human Rights* and it is self-described as a training to "act, unite and empower for mental health." (World Health Organization, 2017, title page). The first Mad Pride-affiliated organization highlighted by WHO is *MindFreedom International*, which is arguably one of the main forces of the Mad Pride movement in the U.S. (see MindFreedom International, 2006; https://mindfreedom.org/). The second organization that WHO publicizes is the World Network of Users and Survivors of Psychiatry (*WNUSP* -see http://www.wnusp.net/). Although not as visibly connected to Mad Pride, it commits to "celebrat[ing] the positives of madness" (World Network of Users and Survivors of Psychiatry, 2009, p. 3) and "continu[ing] to build the user / survivor movement world wide" (World Network of Users and Survivors of Psychiatry, 2009, p. 3).

[29] As an aside, note that each of these two loci is essentially a diagnosis of a root cause underlying the current lack of fulfillment of Objectives 1 and 2 of this present chapter (see Section 1). As such, one way to conceptualize the overall aim of this chapter is for it to serve as an antidote to any future vulnerabilities to becoming trapped in this circle.

hypothetical explanation, and in no way to intend to imply any sort of conveyance of historical fact. My explanation goes as follows. First, as described in Section 4.5, certain agents within the field of public (mental) health formulated and acted upon a good intention—the intention to rhetorically emphasize the fact that the notion of health transcends a constricted set of biological facts, and, instead, exists within a vast social, political, economic, and interpersonal network. As discussed, the action chosen to realize this good intention was to define health as (a state of) well-being (see World Health Organization, 1946). This definition arguably serves as an example of the conflation of the meanings of *health* and *well-being*, and, as such, marks the point at which public mental health metaphorically slipped into the vicious circle.

Now, how is this vicious circle perpetuated? The first part of the generation of the circle is the arc from (1) to (2). When the conflation of health as well-being is taken up, individuals (including public mental health professionals) consequently become more likely to consider health and well-being as identical, or otherwise very tightly coupled, and thereby presumably become less likely to notice instances where health and well-being diverge from one another (I will offer more detail on how this decrease in likelihood could happen after the current initial sketch). Of course, one instance where health and well-being diverge is (presumably) in the experiences of individuals who subscribe to the Hybrid Mad Pride Model. In other words, this is the experience where one's mental illness (assuming that mental differences are illnesses—see footnote 21) or lack of mental health, is entirely consistent with, and perhaps even conducive to, one's well-being.

In turn, (2) can reinforce (1), since an increased invisibility of examples where health and well-being diverge will result in an absence of noticeable counterexamples, which will almost certainly serve to impel the conflation further. One reason for this would be that, presumably, the longer one goes without encountering a counterexample, the stronger the inductive evidence (in this case, for conflation) will appear to be.

At this point, it's worth further describing the causal path from the vicious circle's first locus [(1) the conflation of the terms *health* and *well-being*] to its second locus [(2) the neglect of the perspectives of individuals supporting Mad Pride]. One of the reasons I think this particular part of the circle is especially important is because it serves as an example of a way in which language and conceptual distinctions could potentially end up having concrete effects on our behavior—in this case, on how we, as a society, treat certain marginalized groups (and whether we even notice them in the first place). As such, it offers one way to counter the common accusation that language and conceptual distinctions have little or no effect on real-world social justice problems.

By equating the terms *health* and *well-being*, public mental health is allowing the word *health* to signify at least two different things: (1) proper biological functioning, and (2) living a good life (that is, possessing well-being). This type of equivocation can (potentially) lead to the reshaping of certain beliefs and implicit biases (for example, in this case, it's possible that it could lead to a belief, or an implicit bias, that one cannot live a good life without proper biological functioning). More specifically, here's one possible way in which this could occur. If one very often sees, hears, uses, and/or asserts the terms *health* and *well-being* interchangeably, it would make sense that the technically distinct conceptual categories of health and well-being could, as a result, start to meld or merge together in one's mind. And this melding, presumably whether conscious and/or subconscious, would almost certainly affect one's interpretation, usage, and overall understanding of the meanings of health and well-being. For example, this melded mental concept could prime one to pay less attention to instances where health and well-being were disjoined from one another, and hence failed to mesh with one's mental concept of health and well-being as merged together. Accordingly, individuals with this fused category of health and well-being would presumably be less likely to think about, seek out, and notice scenarios where these two states were distinct from one another. Again, notably, instances of Hybrid Mad Pride are very often precisely such scenarios.

This phenomenon of the ability of pre-established mental categories, schemata, implicit biases, beliefs, and other conceptual frameworks to influence and prime our behavior—often resulting in its decreased accuracy and increased irrationality—has been repeatedly supported by a vast body of psychological research. For just a small sample of some overviews of this research, see Ariely, 2008; Wilson, 2002; and Chabris & Simons, 2009.

One relevant observation here is that a strict insistence upon the distinction between health and well-being (and thus, a resistance to the conflation of health and well-being locus of the vicious circle) would presumably effectively halt the generation of this vicious circle. And of course, halting the circle would ostensibly reduce the neglect of Mad Pride perspectives, and, as such, would enable a more complete realization of Nothing About Us Without Us. Thus, in addition to satisfying the two main objectives of this chapter, I hope to have also provided an example of how the adherence to philosophical distinctions and nuances (in other words, to what some call *nitpicking*) can have, despite being the target of such name calling, deep and positive implications for social justice.

Acknowledgments

I owe thanks to the following individuals for their diverse forms of unyielding support during my writing of this chapter: Kathleen Lowenstein, Emily Cutler, Bhob Rainey, Juliana Sasken, Serife Tekin, Marie Nicolini, Marisa Egan, and S. Ghouli. Additionally, I offer special gratitude to Jennifer Radden and Kelso Cratsley for: their initial idea for this chapter, their dedication to including Mad Pride and neurodiversity perspectives in the current volume, their confidence in my expertise on these perspectives, their multiple incisive suggestions about my arguments and my presentations of them, their ongoing encouragement and kindness, and their generous devotion of copious amounts of time and energy to the furtherance of this chapter and the larger project in which it resides.

References

Adbusters: Journal of the mental environment. (2002, May/June). *Mad world/Mad Pride Issue, No. 41, 10*(3).

American Psychiatric Association. (2013). *Diagnostic and statistical manual of mental disorders* (5th ed.). Washington, DC: Author.

Armstrong, T. (2010). *The power of neurodiversity: Unleashing the advantages of your differently wired brain*. Philadelphia: Da Capo Press.

American Public Health Association. (2019). *Mental health*. https://www.apha.org/topics-and-issues/mental-health.

Ariely, D. (2008). In Rev (Ed.), *Predictably irrational: The hidden forces that shape our decisions*. New York: HarperCollins Publishers.

Autistic Self-Advocacy Network. (2019). *Position statements*. https://autisticadvocacy.org/about-asan/position-statements/.

Bagenstos, S., & Schlanger, M. (2007). Hedonic damages, hedonic adaptation, and disability. *Vanderbilt Law Review, 60*(3), 745–797.

Barnes, E. (2016). *The minority body: A theory of disability*. New York: Oxford University Press.

Bascom, J. (Ed.). (2012). *Loud hands: Autistic people, speaking*. Washington, D.C.: The Autistic Press.

Beazer, J. (2017). Mad Pride and depression. *Asylum, 24*(2), 8–9.

Bickenbach, J. (2015). WHO's definition of health: Philosophical analysis. In T. Schramme, & S. Edwards (Eds.), *Handbook of the philosophy of medicine* (pp. 961–974). Dordrecht, Netherlands: Springer.

Boorse, C. (1975). On the distinction between disease and illness. *Philosophy and Public Affairs, 5*(1), 49–68.

Bradley, B. (2015). *Well-being*. Cambridge: Polity Press.

Brown, J. S., Learmonth, A. M., & Mackereth, C. J. (2015). *Promoting public mental health and well-being: Principles into practice*. London & Philadelphia: Jessica Kingsley Publishers.

Brülde, B. (2011). Health, disease, and the goal of public health. In A. Dawson (Ed.), *Public health ethics: Key concepts and issues in policy and practice* (pp. 20–47). New York: Cambridge University Press.

Chabris, C. F., & Simons, D. J. (2009). *The invisible gorilla: How our intuitions deceive us*. New York: Broadway Paperbacks.

Chapman, R. (2019). Neurodiversity theory and its discontents: Autism, schizophrenia, and the social model of disability. In Tekin Ş, & R. Bluhm (Eds.), *The Bloomsbury companion to philosophy of psychiatry* (pp. 371–390). London: Bloomsbury Press.

Charlton, J. I. (1998). *Nothing about us without us: Disability oppression and empowerment*. Berkeley and Los Angeles: University of California Press.

Costa, L. (2015). *Mad Pride in our Mad culture,* Consumer/Survivor Information Resource Centre Bulletin *No. 535. Mad Pride Issue, Mad Pride Toronto Bulletin*. http://www.csinfo.ca/bulletin/Bulletin_535.pdf.

Craft, T. (2018, August 22). *I celebrate disability pride, but am I really proud of my mental illness? [Blog post]. Rooted in rights*. https://rootedinrights.org/i-celebrate-disability-pride-but-am-i-really-proud-of-my-mental-illness/.

Crichton, P., Carel, H., & Kidd, I. J. (2017). Epistemic injustice in psychiatry. *BJPsych Bulletin, 41*, 65–70. https://doi.org/10.1192/pb.bp.115.050682.

Crisp, R. (2017). Well-being. In E. N. Zalta (Ed.), *The Stanford encyclopedia of philosophy (Fall 2017 edition)*. https://plato.stanford.edu/archives/fall2017/entries/well-being/.

Curtis, T., Dellar, R., Leslie, E., & Watson, B. (Eds.). (2000). *Mad Pride: A celebration of Mad culture*. Truro: Chipmunkapublishing.

Cutler, E. (2017, June 3). *Breaking free from the stigma paradox. [Blog post]*. www.madinamerica.com/2017/06/breaking-free-stigma-paradox/. (Accessed 5 June 2017).

Cutler, E. S. (2019a). Listening to those with lived experience. In S. Steingard (Ed.), *Critical psychiatry: Controversies and clinical implications* (pp. 179–206). Cham, Switzerland: Springer.

Cutler, E. (2019b, February). Mad pride. In MindFreedom International (Ed.), Organizing guide for psychiatric survivors (pp. 226–233). Eugene, OR. Retrieved from mindfreedom.org/organizing-handbook.

Diego, S. S., Forlini, C., & Meurk, C. (2018). Public mental health ethics: Helping improve mental health for individuals and communities. *Public Health Ethics, 11*(2), 121–125.

DuBrul, S. A. (2014). The Icarus Project: A counter narrative for psychic diversity. *Journal of Medical Humanities, 35*(3), 257–271. http://doi.org/10.1007/s10912-014-9293-5.

Embry, D. D. (2011). Behavioral vaccines and evidence-based kernels: Nonpharmaceutical approaches for the prevention of mental, emotional, and behavioral disorders. *Psychiatric Clinics of North America, 34*(1), 1–34.

Faden, R., & Shebaya, S. (2016). Public health ethics. In E. N. Zalta (Ed.), *The Stanford encyclopedia of philosophy*. Fall 2017 edition https://plato.stanford.edu/archives/win2016/entries/publichealth-ethics/.

Farber, S. (2012). *The spiritual gift of madness*. New York: Inner Traditions.

Fleischer, D. Z., & Zames, F. (2011). *The disability rights movement: From charity to confrontation*. Philadelphia: Temple University Press.

Fletcher, G. (2016a). *The philosophy of well-being: An introduction*. New York: Routledge.

Fletcher, G. (Ed.). (2016b). *The Routledge handbook of philosophy of well-being*. New York: Routledge.

Friedli, L. (2000). Mental health promotion – re-thinking the evidence base. *The Mental Health Review, 5*(3), 15–18.

Friedli, L. (2009). Future directions in mental health promotion and public mental health. In I. Norman, & I. Ryrie (Eds.), *The art and science of mental health nursing: A textbook of principles and practice* (2nd ed., pp. 43–61). U.K.: McGraw-Hill Education.

Giliberti, M. (2017, October 6). *The importance of getting the right treatment for you. [Blog post]*. https://www.nami.org/Blogs/From-the-CEO/October-2017/The-Importance-of-Getting-the-Right-Treatment-for. (Accessed 3 August 2019).

Glaser, G. (2008, May 11). *'Mad Pride' fights a stigma*. The New York Times.

Hahn, H. D., & Belt, T. L. (2004). Disability identity and attitudes toward cure in a sample of disabled activists. *Journal of Health and Social Behavior, 45*, 453–464.

Health. (2019). *Lexico online dictionary (formerly Oxford dictionary online)*. https://www.lexico.com/en/definition/health.

Heathwood, C. (2016). Desire-fulfillment theory. In G. Fletcher (Ed.), *The Routledge handbook of philosophy of well-being* (pp. 135–147). New York: Routledge.

Hoffman, G. A. (2012). What, if anything, can neuroscience tell us about gender differences? In R. Bluhm, A. Jacobson, & H. Maibom (Eds.), *Neurofeminism: Issues at the intersection of feminist theory and cognitive science* (pp. 30–55). Basingstoke, U.K.: Palgrave-MacMillan.

Hoffman, G. A. (2017). Proud to be crazy, proud to be ill?. *Exploring the conceptual justifications of Mad Pride, Disability Pride, and neurodiversity. Invited keynote lecture at the 19th Annual Meeting of the.* Madrid, Spain: International Network for Philosophy and Psychiatry.

Hoffman, G. A. (2019). "Aren't mental disorders just chemical imbalances?," "Aren't mental disorders just brain dysfunctions?," and other frequently asked questions about mental disorders. In Ş. Tekin, & R. Bluhm (Eds.), *The Bloomsbury companion to philosophy of psychiatry* (pp. 59–92). London: Bloomsbury.

Hutchison, D. S., Gagne, C., Bowers, A., Russinova, Z., Skrinar, G. S., & Anthony, W. A. (2006). A framework for health promotion services for people with psychiatric disabilities. *Psychiatric Rehabilitation Journal, 29*(4), 241–250.

Jenkins, R., McCulloch, A., Friedli, L., & Parker, C. (2002). *Developing a national mental health policy*. East Sussex, U.K.: Psychology Press, Ltd.

Leaf, P. J., Ostrow, L., Manderscheid, R. W., Shern, D. L., & Eaton, W. W. (2012). Progress made, but much more to be done. In W. W. Eaton (Ed.), *Public mental health* (pp. 510–530). Oxford: Oxford University Press.

Levin, B. L., Hanson, A., & Hennessy, K. D. (2004). Overview of prevention, integration, and parity. In B. L. Levin, J. Petrila, & K. D. Hennessy (Eds.), *Mental health services: A public health perspective* (2nd ed., pp. 3–14). New York: Oxford University Press.

Mendelson, T., Pas, E. T., Leis, J. A., Bradshaw, C. P., Rebok, G. W., & Mandell, W. (2012). The logic and practice of the prevention of mental disorders. In W. W. Eaton (Ed.), *Public mental health* (pp. 459–509). Oxford: Oxford University Press.

MindFreedom International. (2006). *Mad Pride FAQs*. http://www.mindfreedom.org/campaign/madpride/other-info/frequently-asked-questions/mp-faq.

Montgomery, C. (2018, July 1). *Pride. Together. [Blog post]. Radical Abolitionist: A Cognitive Liberty Blogspace.* https://www.radicalabolitionist.org/radical-abolitionist/2018/7/1/pride-together.

Morrison, L. J. (2005). *Talking back to psychiatry*. New York: Routledge.

Nudel, C., Benner, A., Poore, M., Ball, B., Knighton, D., Long, Y., & Ware, M. (Eds.). (2009). *Firewalkers: Madness, beauty, and mystery*. Charlottesville, VA: VOCAL, Inc.

Parfit, D. (1984). *Reasons and persons*. Oxford: Oxford University Press.

Prism, Imogen (2017, August 27). Kind of like being a witch or wizard': How I learned to value my neurodivergence." [Blog post]. *The Body is Not an Apology*. https://thebodyisnotanapology.com/magazine/kind-of-like-being-a-witch-or-wizard-how-i-learned-to-value-my-neurodivergence/.

Radden, J. (2016). Mental health, public health and depression: A bioethical perspective. *Ethics, Medicine and Public Health, 2*(2), 197–204.

Radden, J. (2017). Public mental health and prevention. *Public Health Ethics, 11*(2), 126–138.

Radden, J. (2012). Recognition rights, mental health consumers and reconstructive cultural semantics. *Philosophy, Ethics, and Humanities in Medicine,* 7(6). http://www.peh-med.com/content/7/1/6.

Rashed, M. (2019). *Madness and the demand for recognition*. Oxford: Oxford University Press.

Roache, R., & Savulescu, J. (2018). Psychological disadvantage and a welfarist approach to psychiatry. *Philosophy, Psychiatry, & Psychology, 25*(4), 245–259.

Roberts, D., Major, M., & Brady, J. (2008). *June 10). Controversial new movement: Autistic and proud.* abcNEWS. https://abcnews.go.com/GMA/OnCall/story?id=5033594&page=1.

Robinson, I., & Rodrigues, A. (2009, August 24). *'Mad Pride' activists say they're unique, not sick*. ABCNews. http://abcnews.go.com/Health/print?id=8382903.

Sacks, O. (2012). *Hallucinations*. New York: Vintage Books.

Santi, A., & Kyratsous, M. (2015). June). Epistemic injustice in assessment of delusions. *Journal of Evaluation in Clinical Practice, 21*(3), 479–485. https://doi.org/10.1111/jep.12347.

Saxena, S., Jané-Llopis, E., & Hosman, C. (2006). Prevention of mental and behavioural disorders: Implications for policy and practice. *World Psychiatry, 5*(1), 5–14.

Schneider, M. (2014). *Introduction to public health* (4th Ed.). Burlington, MA: Jones & Bartlett Learning.

Schrader, S., Jones, N., & Shattell, M. (2013). Mad Pride: Reflections on sociopolitical identity and mental diversity in the context of culturally competent psychiatric care. *Issues in Mental Health Nursing, 34*, 62–64.

Sinclair, J. (1993). Don't mourn for us. In Julia Bascom, & The Loud Hands Project (Eds.), *Loud hands: Autistic people, speaking* (pp. 15–21). Washington, D. C.: The Autistic Press.

Singer, J. (2017). *NeuroDiveristy: The birth of an idea*. www.neuroredux.blogspot.com.

Staff, Newsweek (2009). *The growing push for 'Mad Pride.'*. https://www.newsweek.com/growing-push-mad-pride-79919.

Tait, L., & Lester, H. (2005). Encouraging user involvement in mental health services. *Advances in psychiatric treatment, 11*(3), 168–175.

TeePot, Mrs (2017). *Mental illness: Pride & anti-recovery. [Blog post]. Medium*. https://medium.com/@MrsTeePot/mental-illness-pride-anti-recovery-45a76794f6a1.

United Nations (UN) General Assembly. (2006). *Convention on the Rights of Persons with Disabilities (CRPD)*. https://www.un.org/development/desa/disabilities/convention-on-the-rights-of-persons-with-disabilities/convention-on-the-rights-of-persons-with-disabilities-2.html.

Waddingham, Rai. (2017). *Don't pity me: Psychosis gave me Mad skills. [Blog post]. Medium*. https://medium.com/@RaiWaddingham/dont-pity-me-psychosis-gave-me-mad-skills-60c0d5084a17.

Walker, N., & The Loud Hands Project. (2012). Throw away the master's tools: Liberating ourselves from the pathology paradigm. In J. Bascom (Ed.), *Loud hands: Autistic people, speaking* (pp. 225–237). Washington, D.C.: The Autistic Press.

Walker, N. (2014). *Neurodiversity: Some basic terms & definitions. [Blog post]. Neurocosmopolitanism blog*. http://neurocosmopolitanism.com/neurodiversity-some-basic-terms-definitions/.

Wilson, T. D. (2002). *Strangers to ourselves: Discovering the adaptive unconscious*. Cambridge, MA: The Belknap Press of Harvard University Press.

World Health Organization. (1948). Preamble to the constitution of WHO as adopted by the International Health Conference, New York, 19 June - 22 July 1946; signed on 22 July 1946 by the representatives of 61 States (Official Records of WHO, no. 2, p. 100) and entered into force on 7 April 1948. Retrieved from https://www.who.int/about/who-we-are/frequently-asked-questions.

World Health Organization. (2004). *ICD-10: International statistical classification of diseases and related health problems* (2nd ed.). Geneva: World Health Organization https://apps.who.int/iris/handle/10665/42980.

World Health Organization. (2013). *Mental Health Action Plan 2013-2020*. Geneva: World Health Organization. https://www.who.int/mental_health/publications/action_plan/en/.

World Health Organization. (2017). *Understanding human rights – WHO QualityRights training to act, unite and empower for mental health (pilot version)*. Geneva: World Health Organization.

World Health Organization. (2018). Mental health: Strengthening our response. https://www.who.int/news-room/fact-sheets/detail/mental-health-strengthening-our-response.

World Network of Users and Survivors of Psychiatry. (2009). *WNUSP Kampala Declaration (Extended version).* http://www.wnusp.net/documents/WNUSP_KampalaDeclaration2009.pdf.

World Psychiatric Association. (2015). WPA position statement on social justice for persons with mental illness (mental disability). http://www.wpanet.org/uploads/Position_Statement/WPA%20Social%20Justice_FINAL%20Sep%202015.pdf. (Accessed 28 November 2018).

Made in the USA
Monee, IL
12 September 2026

62977635R00205